Dictionary of Waste
and Water Treatment

Dictionary of Waste and Water Treatment

John S. Scott, FIL, BSc, ARSM, MIStructE
AMINinE, AMISWM

Consultant; formerly Technical Press Officer,
National Coal Board

Paul G. Smith, BSc, MSc, MIPHE

Lecturer in Public Health Engineering,
University of Strathclyde, Glasgow

Butterworths

London   Boston   Sydney   Wellington   Durban   Toronto

First published 1981

© John S. Scott and Paul G. Smith, 1980

British Library Cataloguing in Publication Data

Scott, John Somerville
   Dictionary of waste and water treatment
   1. Sanitary engineering
   I. Title   II. Smith, Paul
   628'.03   TD9   80-40515

ISBN 0-408-00495-9

Typeset by Reproduction Drawings Limited, Sutton, Surrey

Printed in England by Hartnoll Print Ltd., Bodmin, Cornwall

Preface

Waste and water treatment, better known in Britain as public health engineering, covers the design, building and operation of plants for water treatment and supply, sewerage, sewage treatment and disposal, and solid waste treatment and disposal. It aims to control pollution of air, water and land, and to improve amenity. This book, with many cross-references, explains its terminology and should help laymen as well as scientists or engineers qualified in other subjects to understand its literature. It touches on a variety of many branches of engineering—chemical, civil, electrical, environmental, mechanical, structural and water—as well as the sciences of chemistry, biology, hydrology, microbiology, and virology, to name only a few.

Reclamation and re-cycling of solid waste, although not truly of concern to public health, is nevertheless important because of the increasing scarcity of natural resources.

In this dictionary we have followed the convention for biological latin where the genus and species is written in italics. All other terms concerned with biological classification are printed in roman type. All cross-references are printed in bold italic type.

The authors are grateful to many colleagues in the National Coal Board and the University of Strathclyde, especially Professor D. I. H. Barr and Dr P. Coackley, of the Civil Engineering Department.

John S. Scott
Paul G. Smith

A

abattoir wastes   *See meat-processing wastes.*
abiotic   Description of the non-living part of an *ecosystem.*
ABS   *Alkyl benzene sulphonate.*
absolute viscosity   US term for dynamic *viscosity.*
absorption   Entry of a fluid into a solid, of a gas into a liquid, etc. Whatever enters is held, unlike *adsorption.*
absorption pit, seepage p., absorption well, dumb well, soakaway   A hole in the ground for disposal of rainwater, *sullage* or treated sewage *effluent.* It is dug in country districts where there is porous soil, a *water table* not less than 2 m below ground and no other danger of contaminating water supplies. To increase the volume and make the digging easier, a trench may be dug instead of a pit. The trench is 0.3 to 0.9 m wide, 0.75 m deep, and filled with rubble or gravel, in which at 0.45 m from the bottom are *land drains* that remove the water. The trench is covered with topsoil separated by a thick plastics sheet from the rubble so that the roots do not block it. A *grease trap* upstream prevents grease from sullage overloading the soakaway. *Percolation tests* are needed to ensure that the ground is porous enough. *Evapo-transpiration* can be encouraged by planting trees, shrubs or crops near the trench. The slope of the bottom of the trench is important and should be steeper in more porous soils. In clay 1 in 130 may be enough but in sands or gravels 1 in 40 may be needed to ensure that the far end of the trench receives some effluent and that the near end is not overloaded. *See subsurface irrigation* and page 291.
absorption tower   A type of *wet scrubber.*
absorption trench   *See absorption pit.*
abstraction (verb **abstract**)   Removal of water from a river, lake, well, borehole or other source to a waterworks or *impounding reservoir.* It thus becomes *raw water.*
accelerated clarifier   A *solids contact clarifier.*
accelerated composting   *Mechanical composting.*
Accentrifloc clarifier   A circular *solids recirculation clarifier.*
acceptable daily intake, ADI   Of a chemical, that daily intake which during a complete lifetime seems without risk (World

1

Health Organization).

access eye An access hole at a bend in a drain pipe, ordinarily covered by a metal plate wedged or bolted over it, which enables the pipe to be rodded. *See rodding.*

acclimatisation Adaptation to an altered environment, a term often applied to *bacteria* and other *microbes.*

ACGIH The *American Conference of Governmental Industrial Hygienists.*

Achari Mites which may be found in *trickling filters.*

Achorutes subviaticus See Hypogastrura viatica.

Achromobacterium A *genus* of bacteria commonly found in *activated sludge, trickling filters, de-nitrification* processes and *anaerobic sludge digestion.*

acid *See acidity.*

acid dewpoint *Sulphur dioxide,* SO_2, and sulphur trioxide, SO_3, in flue gas can raise its *dewpoint* considerably, even above $200°C$. The exit temperature of the gas then must also be held above $200°C$, lowering the efficiency that is possible for any boiler in the gas circuit because the gases remove useful heat up the chimney. Each fuel has its own acid dewpoint in given furnace conditions, below which corrosive *condensation* forms in flues, chimneys, boilers, etc. *See acid soot.*

acid fermentation One stage of *anaerobic* decomposition of sewage sludge, resulting in the biodegradation of complex *organic compounds* to simpler organic acids, as in *anaerobic sludge digestion. See* below and *biodegradable.*

acid formers A group of facultative and *obligate anaerobes*, capable of *hydrolysis*, which ferment the complex *organic compounds* in *sewage* to organic acids, including acetic acid, CH_3COOH, and propionic acid, CH_3CH_2COOH. For effective *anaerobic sludge digestion* they should be in equilibrium with the *methanogenic bacteria.* The main acid formers are the *obligate anaerobes*, which are 10 to 100 times more numerous in digesting sludge than the facultative anaerobes.

acid hydrolysis A method of converting the *cellulose* in municipal refuse into a non-polluting fuel—ethanol (ethyl alcohol). After pulping in water, the cellulose can be extracted fairly easily from the refuse and converted into fermentable sugars in only about 70 s in a *stainless steel* or Monel metal reactor, with 0.4% sulphuric acid at $230°C$. After other processes the sugars are fermented and concentrated to a strong alcohol.

acidising *Stimulation of a well* in dolomite or limestone with acid, to remove incrustation.

acidity Acidity of a water is measured by hydrogen ion concentration, usually called *pH.* Waters with pH below 7.0 are

2

acid. In unpolluted water, acidity comes from dissolved *carbon dioxide* or *organic* acids leached from the soil. Atmospheric pollution also may cause acidity. Acid waters corrode metal or concrete. *See alkalinity.*

acid mine drainage Polluted mine-drainage water is often acidic. The pyrite (iron sulphide) present in mines (especially coal or sulphide mines) is oxidised by air, water and *bacteria (Thiobacillus)* to sulphuric acid and ferrous sulphate. Consequently the water leaving the mine may contain from 100 to 6000 mg/litre of sulphuric acid, from 10 to 1500 mg/litre of ferrous sulphate, $FeSO_4$, from 0 to 350 mg/litre of aluminium sulphate, $Al_2(SO_4)_3$, and from 0 to 250 mg/litre of manganese sulphate, $MnSO_4$. Treatment of the water may include *sedimentation* after *neutralisation* by lime or limestone. *See discharge prevention, oxidation prevention, recharge prevention.*

acid soot, a. smut Large soot particles that contain sulphurous or sulphuric acids formed in a furnace where sulphur in the fuel burns to SO_x. Emissions of acid soot can be reduced or eliminated either by reducing the amount of excess air, by using low-sulphur fuel, by eliminating air leaks into *flues*, by raising *flue gas* temperature, by insulating flues or chimney or by any of these in combination. Acid soot will be emitted from furnaces burning residual oil if any flue surface is cooler than 145°C. Soot will catch on surfaces wetted with acid and will later flake away.

actinomycetes Unicellular, *filamentous organisms* that are ordinarily listed as *bacteria*, although they reproduce like *fungi*. They are common in earth and may be responsible for unpleasant taste or smell in drinking water, and may also be found in *activated sludge*. Some species cause illnesses such as diphtheria, leprosy and *tuberculosis*.

activated alumina Alumina (Al_2O_3) that has been treated to increase its surface area and thus to enhance its adsorptive properties. It can remove phosphate and *arsenic* compounds from liquid wastes and can dry gases.

activated carbon, active c. Carbon that has been derived from anthracite coal, wood, charcoal, etc., by treatment with live steam, or in some other way has had its *porosity* and thus its surface area increased. Its very large surface area per unit of mass enables it to adsorb smells, trace *organics* and other undesirable components of a water intended for drinking. It can be used as a *filter* in granular form in a *carbon adsorption bed* or merely thrown as a fine powder into the water, after which it is removed by *filtration*. In granular (pellet) form the spent carbon can be regenerated by steam treatment at 1000°C, with about 5% loss of carbon per cycle. *See iodine number, phenol adsorption*

3

test, regeneration.

activated silica A *coagulant aid* that, added to water, forms a long-chain inorganic polymer. It is prepared by adding a small quantity of acid to sodium silicate.

activated sludge A sewage *sludge* made by continuous re-circulation of the sludge from the *secondary sedimentation tank* to the *aeration tank*, thus acquiring many useful active aerobic *bacteria*, from 100 to 1000 million per millilitre of mixed liquor. The bacteria are embedded in or on a slime that forms 90% of the solids content of the sludge. The slime, mainly *poly-saccharides*, is produced by the bacteria as they consume the sewage. *Protozoa*, especially *ciliate protozoa*, may also be present in the sludge, feeding on the bacteria. *See* below.

activated sludge process A continuous, aerobic biological treatment for *sewage*, dating from 1913, that uses a culture of *bacteria* suspended in *settled sewage* with some *protozoa* in an *aeration tank* to adsorb, absorb and transform the organic pollutants. They thus form *flocs* that may reach 0.1 mm diameter; these are both mixed and kept in suspension either by air blown in at the bottom of the tank (*diffused-air system*) or by *mechanical aeration* (*see Figure A.1*). The solids in the aeration tank are normally at a concentration (*MLSS*) of 1500 to 5000 mg/litre. These cultures are then settled in a *sedimentation tank* from which most of the sludge returns to the aeration tank. The *dissolved oxygen* in the aeration tank should be at least 0.5 mg/litre, preferably 1 to 2 mg/litre (see *oxygen-activated sludge*). The aeration tank may be designed on *aeration period*, preferably on *sludge loading rate* or *mean cell residence time*. If the *ammonia* present is to be oxidised to *nitrate*, the plant must be designed for *nitrification*. In the start-up of an activated sludge plant the time needed for establishing the appropriate bacteria and protozoa can be greatly reduced by *seeding* the new

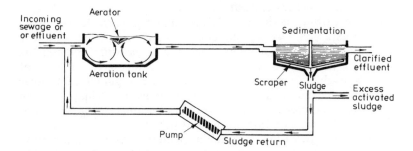

Figure A.1 Activated sludge process (flow diagram)

4

aeration tank with sludge from one that is working well. Activated sludge treatment demands only about one-seventh of the land occupied by *trickling filters*, so is used in city treatment plants. It also does not suffer from *filter flies* and has little smell. The loss in *head of water* is smaller than with a trickling filter, as is the construction cost, but the operating cost is higher and skilled attention is essential. Many varieties of activated sludge treatment exist, including *contact stabilisation, extended aeration, modified aeration, oxygen-activated sludge. See also oxygenation efficiency, return activated sludge, sludge production.*

activated sludge settling tank   *See secondary sedimentation tank.*

active carbon   *See activated carbon.*

acute toxicity   A condition of polluted air or water that kills quickly. For fish, *ASTM 31* states a limit of 96 h. Acute toxicity can be measured by a *median tolerance limit. Compare chronic toxicity.*

adenosine triphosphate, ATP   The energy-storing substance required to carry out the biochemical reactions in a cell. When its energy is released, it breaks down to adenosine diphosphate (ADP), ready again to capture energy that will raise it to the level of ATP.

ADF   *Alternating double filtration.*

ADI   *Acceptable daily intake.*

adiabatic   A description of something that occurs without heat transfer to or from the surroundings.

adiabatic lapse rate   A *lapse rate* under *adiabatic* conditions, which is calculated as about 0.5°C temperature drop per 100 m of climb for air saturated with water vapour and 1°C per 100 m climb for dry air. A sub-adiabatic lapse rate (involving less cooling of the air with height) makes for *stability* of the air. An atmosphere with a superadiabatic lapse rate (when the temperature drops more than adiabatically) is unstable and so disperses pollutants quickly. *See inversion.*

adit, sough   A tunnel driven into a hillside or from a well into water-bearing ground, at a slight upward slope so that *groundwater* drains naturally out of the soil. Adits of disused mines can be used as suppliers of *raw water*, as can *ghanats.*

adsorption (verb *adsorb)*   Adhesion of gases, vapours, liquids or solids on to a solid surface, the adsorbent (e.g. *activated alumina, activated carbon*). The substance adsorbed, the adsorbate, can thus be extracted from a fluid stream. Adsorbents can be used once and then discarded, or used and reactivated (*see regeneration*). In *biological treatment* it is an important mechanism for *sewage* purification. For example, within about

5

30 min of entering an *aeration tank* the solid and liquid *organics* are adsorbed on to the *flocs* of *activated sludge*. After this adsorption, extracellular *enzymes* can start to break down the organics in such a way that they are absorbed and transformed by the bacteria. Chemisorption is adsorption by chemical bonding on to the surface of the adsorbent. *See iodine number.*

adulticide A *pesticide* that kills adult insects.

advanced wastewater treatment, AWT, advanced treatment Further action on sewage effluent that has undergone *biological treatment*. *De-nitrification* is one part of AWT, although it may have already begun in the *secondary treatment*. AWT may include *physico-chemical treatment* such as *ammonia stripping, activated carbon* or *deep-bed filters, ion exchange* or *membrane processes*.

Aëdes aegypti A *mosquito* that breeds in water storage jars in which water is fetched from a stand pipe. It carries *virus* diseases including *dengue* and *yellow fever*.

aerated channel An *aerated launder*.

aerated contact bed A *trickling filter*.

aerated grit chamber Several makes of *grit channel* are aerated with coarse air bubbles so as to improve the separation of the grit

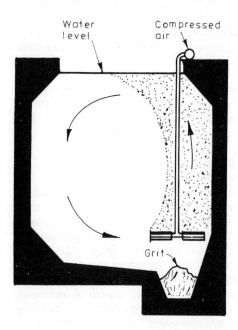

Figure A.2 Aerated grit channel (cross-section) (from 'Disposal of Sewage and other Waterborne Diseases' by Imhoff, Muller & Thistlethwaite (Butterworths) 1956)

Figure A.2 Aerated grit channel (cross-section)

6

from the organic solids (*see Figure A.2*). The air flow imparts a spiral motion to the sewage. The spiralling flow governs the size of particle of a given specific gravity that will be removed, and adjustment of the air flow can produce almost 100% clean grit. *Detention periods* of 3 min at maximum flow are common. *Aeration* that takes place after grit removal is called *pre-aerating* because it immediately precedes the main *sewage treatment*.

aerated lagoon A *waste stabilisation pond* through which sewage flows, supplied with oxygen by floating *surface aerators* and sometimes also by diffusers or submerged air pipes. It is an approach to a *complete mixing system* of activated sludge treatment but with no *return activated sludge*. Settlement follows when the sewage leaves the lagoon. If aeration is not widespread enough, it is possible for parts of the lagoon to become *anaerobic*, especially in the bottom mud. *See high-rate aerobic lagoon.*

aerated launder, a. channel A channel for distributing sewage at a treatment works, with air injected throughout its length along the bottom to hold the solids in suspension. The velocity in the channel is then not critical and the sewage is well oxygenated. Aerated launders also may be used for distributing *mixed liquor* to settling tanks.

aeration Addition of air to sewage or water so as to raise its *dissolved oxygen* level. Many methods exist—the use of aerators or *spray irrigation*, or (in an *aeration tank*) *diffusers* or *mechanical aeration*.

aeration period The *detention period* in an *aeration tank*. From 6 to 8 h at *dry-weather flow* is typical for domestic *settled sewage*.

aeration tank A tank for mixing and *aeration* of sewage and *activated sludge*—normally 3 to 5 m deep. Tanks may be divided into square pockets three times as deep as their width. Alternatively they may be continuous rectangular channels or *oxidation ditches*.

aerator A machine or other device to dissolve oxygen in sewage or water. It may also purge undesirable gases (hydrogen sulphide or carbon dioxide). Four types exist—*cascade* or free-fall *aerators, spray aerators, injection aerators* and *surface aerators*. All types create useful turbulence at the air—water interface.

Aerobacter *Facultative* bacteria that flourish both in *activated sludge* and in *anaerobic sludge digestion tanks*.

aerobe (adjective *aerobic*) A micro-organism that needs free or dissolved oxygen to develop (in *aerobiosis*).

aerobic—anaerobic lagoon A *facultative lagoon*.

aerobic digester An *aeration tank* that treats waste activated *humus* or *primary sludges* or a mixture of them, usually in a

7

small plant with *extended aeration* or *contact stabilisation* treatment. The high cost of aeration and the absence of fuel gas are disadvantages compared with *anaerobic sludge digestion*, but the reduction of *volatile solids* is about equal and the *supernatant liquid* has a lower *BOD*. Also, the sludge has no smell, is stable and easily de-watered, there are fewer operational problems and the *detention period* is shorter than with anaerobic digestion. *See* below.

aerobic digestion (1) *See aerobic digester.* (2) One of the five steps in every refuse *composting* process. It follows 'preparation' (removal of iron, steel, other metals, glass, etc., and the addition, for wet pulverisation, of water or sewage *sludge*). During digestion the refuse heats up and is converted or half-converted to *compost* by *aerobic* microbes. The later steps are *curing*, finishing and storage.

aerobic lagoon (1) A *maturation pond.* (2) An *aerated lagoon.* (3) A *high-rate aerobic lagoon.*

aerobiosis Any life process that must have free or dissolved oxygen, usually with the production of carbon dioxide, CO_2; nitrate, NO_3^-; sulphate, SO_4^{2-}; phosphate, PO_4^{3-}; etc.

aerosol *Fog, mist, smoke* or any other *dispersion* of solid or liquid particles so small that they remain suspended in the air. *See respirable dust.*

aerosol propellant A liquid of low boiling point, used in an aerosol canister to expel its contents as an *aerosol* spray. Aerosol propellants include fluorocarbons, *chlorofluorocarbons*, chloro-fluoromethanes, etc., which have been banned for this use in some American states because they may damage the ozone layer of the atmosphere at 15 to 30 km altitude. *See trichloro-fluormethane, Freons.*

afterburner A device fitted in the *flue gas* circuit of some incinerators between the furnace and the chimney, so as to burn tars and other volatiles without *smoke* or smell. It may need to be gas fired. *See also two-stage incinerator.*

agar-agar A substance made from seaweed and used in solid culture *medium* for bacteria.

ageing of lakes Lakes may become marsh and eventually dry up because of the stimulation of their plant life by natural *eutrophication*. The addition of human wastes or farm fertilisers may accelerate the ageing.

agglomeration Coalescence or *flocculation.*

aggregate-bed filter A *gravel-bed filter.*

aggressiveness of water *Corrosiveness of water.*

agricultural drain A *land drain.*

agricultural runoff The water that flows from cultivated land. In the USA it is the greatest single source of *pesticides* in water, although their content is usually below 1 μg/litre. The effects of pesticide accumulation (*biomagnification*) in fish or water birds may be serious. In the UK runoff is the source of the troublesome *nitrates* in the water of East Anglia.

agricultural water   *See boron, electrical conductivity, irrigation.*

airborne dust   *Dust* larger than 10 microns settles quickly in air. At 1 micron size, however, dust settles so slowly (about 0.03 mm/s, varying with density) that it can be considered airborne at and below this size. *See fume.*

air chamber, a. vessel   An air pocket near the outlet from a reciprocating pump, or in a water main, etc., which smooths out *pressure surges in pipe flow.*

air changes   Living rooms as well as workshops are ventilated by replacing their stale air with fresh air. Where the air is severely contaminated or in a hot underground kitchen, there may need to be sixty changes per hour. A library or typing office may need only two changes per hour. One air change is a volume of fresh air equal to the volume of the room in question.

air classification, a. separation   Separating materials by an air blast on falling particles, usually in *countercurrent.* Although

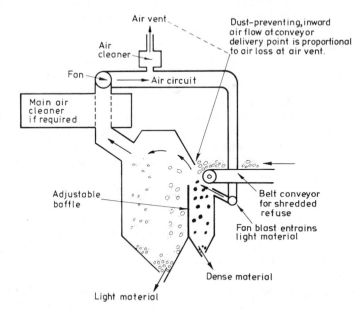

Figure A.3   Air classifier showing one method of circulating air to reduce pollution by dust

9

called *classification*, it is a true separation and, as winnowing, has been known since biblical times. Air classifiers (*see Figure A.3*) work on shredded refuse; they may be preceded by a *dryer* and are usually early in any *mechanical sorting plant*. Before the air separation the dust or ash is removed in a *trommel* and then the magnetic material. Although cyclone *dust arrestors* can themselves be air classifiers, many air separators are followed by a *cyclone. See air column separator*.

air cleaning plant *See gas cleaning plant*.

air/cloth ratio, gas/cloth r. A variable in the design of fabric filters, expressed in m^3/min of gas filtered per m^2 of filter surface. Having thus the dimension metre/minute, it is in fact the average speed of the gas through the fabric. One value used at a municipal incinerator is 0.6 m/min but multiples of this value have been used, depending on type of cloth, type of dust, etc.

air column separator A vertical pipe in which an upward air current separates paper, plastics film, etc., by blowing them up away from the 'heavies', which fall to the bottom of the pipe. The speed of the air current may be adjustable in two ways: (a) by the fan speed; (b) by adjusting the width of the pipe. This unit is well suited to separating all the lights from all the heavies. The *zig-zag separator* is more useful for separating plastics from paper. *See floating velocity*.

air diffuser A *diffuser*.

air drying The use of *lagoons, sludge drying beds*, etc., for drying sludge.

air filter A device that catches dust from air passing through it, on a mesh of textile fabric, felt, wire, paper, etc., rather than by *dust arrestors*. Filters are used either for measuring the amount of dust in the air or for reducing it. Their use in *activated sludge* plants reduces blocking of the *diffusers. See fabric filter*.

air flotation Usually *dispersed-air flotation* but sometimes *dissolved-air flotation*.

air-gap protection A way of preventing contamination of water by *back siphonage* in a main. It usually means the protection given by a *cistern* controlled by a ball valve. The bottom of the mains water inlet to the cistern is higher than the highest level that can be reached by the cistern water—i.e. the overflow. Similar protection is provided at baths, sinks and washbasins.

air injection, oxygen i. (1) Air is injected into long sewers to keep the sewage *aerobic*, especially in pumped rising mains and in the rising limbs of siphons, both of which run full and may become *anaerobic*. Pure oxygen has at times been used because the volume to be blown in to provide a given *dissolved oxygen* level is only one-fifth of the corresponding volume of air. *See sulphide corrosion*. (2) *Diffused-air system*.

air-lift pump A pump with no moving parts, consequently most useful for lifting abrasive material such as gritty water in sewage works. Two pipes of different diameter hang in the well. The smaller is sometimes within the larger. Air is blown down the smaller pipe and from it passes up the larger one. Air bubbles and water rise together in the larger pipe. Large volumes of very dirty fluid can be pumped though at high expense in compressed air and at low mechanical efficiency. For pumping sewage the air-lift pump has the advantage of *aeration* of the water, reducing smells.

air pollutants The bulk of air pollutants produced by human activity come from furnaces, the internal combustion engine and the smelting of metals, mainly *flyash; sulphur dioxide,* SO_2; *carbon monoxide,* CO; and hydrocarbons. CO_2, *carbon dioxide,* although produced in very large quantities, is not regarded as a pollutant. In industrial areas of Britain that produce *metal-plating wastes*, toxic metals, especially cadmium, found in lettuce leaves, are believed to have been brought by polluted rain. See also *acid smut, aerosol propellant, dust, fume, hydrogen sulphide, particulate, smog, smoke.*

air pollution control *See gas cleaning plant*.

air-pulse cleaning *Pulsed-air cleaning* of fabric filters.

air quality standards, ambient a.q.s. Maximum levels of pollutants in air. They can be achieved by stating and enforcing maximum *emission standards*—e.g. for *particulates, carbon monoxide,* hydrocarbons, *nitrogen dioxide, sulphur dioxide.*

air release valve, air v. A valve provided at a high point in a main to allow air to be vented from it automatically as the pipe fills with water. It is fitted with a floating ball that closes the vent when the water rises to it. It also admits air as the pipe empties.

air sampling *See Ringelmann chart, sampling train, tape sampler.*

air saturation value The solubility of *dissolved oxygen* in water, in equilibrium with air. It decreases as the temperature and *salinity* increase. The solubility of oxygen in water decreases also with decreasing atmospheric pressure (and increasing altitude). At 15°C and sea level the value is 10.1 mg/litre but at 0°C it reaches 14.6 mg/litre.

air scouring *See backwashing*.

air separation *See air classification*.

air stripping (1) *Ammonia stripping*. (2) Blowing air through effluent, sewage, etc., with an *aerator* to remove unwanted gas such as *carbon dioxide* or *hydrogen sulphide,* volatile *organic compounds* or (by *foam fractionation*) *synthetic detergents.*

air valve An *air release valve*.

AISI sampler The *tape sampler* devised by the American Iron and Steel Institute.

Aitken nuclei Tiny particles a few nanons across, produced when anything solid burns. They can amount to a few hundred thousand per ml in the air near a large boilerhouse.

albedo The fraction of the radiation that is reflected from a surface.

albuminoid nitrogen analysis An obsolescent but quick analysis of sewage—determining the fraction of the combined organic nitrogen that is easily decomposed by alkaline potassium permanganate, $KMnO_4$, after the release of any *ammonia* by distillation.

Alcaligenes **Facultative anaerobic bacteria** that grow well in *activated sludge, trickling filters* and *anaerobic sludge digesters*.

aldehydes *Organic compounds* that can be formed when hydrocarbons are incompletely burnt, as often happens in petrol engines. Consequently they occur in photochemical smog. Background formaldehyde should be limited to 0.015 p.p.m. measured over 24 h and acetaldehyde to 24 p.p.m. Some aldehydes are evil-smelling.

alga (plural **algae**) A large group of *protists*. They are primitive plants, being *photosynthetic*—i.e. in sunlight they release oxygen and use carbon dioxide from the air or water as the source of carbon for making their cells. They are single-celled or multicellular *autotrophs*. At night some algae react by *chemosynthesis*, consuming oxygen. Thus a water containing algae suffers diurnal variations in dissolved oxygen, and in sunlight its CO_2 content falls. The CO_2 originates from the bacteria (*see symbiosis*) or from bicarbonates releasing hydroxyl ions (OH^-) which tend to raise the pH of water:

$$HCO_3^- \rightleftharpoons CO_2 + OH^-$$

For most *waste stabilisation ponds* to function well, algae such as *Chlamydomonas, Chlorella, Euglena* are needed to supply oxygen to aerobic *heterotrophic* bacteria that consume and oxidise the sewage. The four most important classes of freshwater algae are the *green algae* (Chlorophyta) including *Chlorella*, the motile, bright green flagellate unicellular *Euglenophyta* including *Euglena*; the yellow-green or golden brown, usually unicellular *Chrysophyta*; and the *blue-green algae* (Cyanophyta). Sea-water algae (seaweed) may grow to a great size (the *brown algae*).

Freshwater algae are *microscopic*, although they mat together to cover a large water surface. In drinking water, algae are troublesome because they clog filters and may leave a taste when they die, or produce toxins that poison cattle. Algae in *reservoirs* can be reduced by oxygenating the water and reducing its CO_2 content or by adding a little copper sulphate, $CuSO_4$, or potassium permanganate, $KMnO_4$, or by *de-stratification of lakes and reservoirs.*

algal bloom An explosive summer growth of *algae,* water weed or other primitive plant, which may be undesirable because it is ugly or because on dying the algae decay, cause *de-oxygenation* of the water and thus may kill the fish. Before dying they prevent light from reaching the bottom plants and thus hinder their growth.

algal harvesting The growing of *algae* (e.g. in a *lagoon* of sewage effluent) so as to reduce its *nitrogen* and *phosphorus* concentrations. The algae are periodically removed. *See single-cell protein.*

algal pond A *waste stabilisation pond* treating raw sewage or effluent, to which an oxygen-rich effluent containing *algae* is added.

algicide A chemical to kill *algae. Copper sulphate,* the commonest, has to be used with care, as does *potassium permanganate.*

alkali, alkaline *See alkalinity.*

Alkali Inspectorate, Alkali and Clean Air I. UK qualified chemists or chemical engineers, first collected together as a group under the Alkali Act of 1863. Alkali inspectors supervise the more complicated industrial processes, mainly from the viewpoint of *air pollutants* but also for general amenity. The Inspectorate are now controlled by the Health and Safety Executive. In Scotland these duties are performed by the *Industrial Pollution Inspectorate. See best practicable means.*

alkaline fermentation In *anaerobic sludge digestion, methane fermentation.*

alkalinity Alkalinity in water has at least two meanings, as follows. (1) Any water with a *pH* value above 7.0, but *see also caustic alkalinity* and *total alkalinity.* (2) More often it means the total of the alkaline substances present (i.e. the bicarbonates, HCO_3^-; carbonates, CO_3^{2-}; and hydroxides, OH^-, all of which react with acids. $Ca(HCO_3)_2$ is the commonest of these but analyses are normally quoted as 'alkalinity as $CaCO_3$', meaning the total of substances determined by titration to pH 4.5 and expressed as the equivalent of $CaCO_3$. The commonest causes of alkalinity, $Ca(HCO_3)_2$ and $Mg(HCO_3)_2$, are the main cause of *carbonate hardness.* Consequently, when the figure of alkalinity equals the figure of hardness, all the hardness is carbonate. *See acidity.*

alkyl benzene sulphonate, ABS   A 'hard' *surfactant* that has been less used in *synthetic detergents* since 1964, when it began to be replaced by the more *biodegradable* (soft) surfactant *linear alkyl sulphonate,* LAS. ABS consists of a benzene ring with a sulphonate (SO_3) group substituted into it and a branch saturated carbon chain (normally 12 carbon atoms) attached to it. *Bacteria* have difficulty in attacking the branched carbon chain.

alligator shears   A unit of scrapyard plant with jaws like an alligator's or crocodile's that hold and chop heavy steel pieces to a length (1.5 m maximum) that can be fed into a steel furnace. It is also used for cutting thick timber or other bulky wastes before incineration or *shredding*. It may have one or several pairs of jaws.

allyl thiourea, ATU, thiourea   A chemical that has a strongly inhibiting effect on *nitrifying* organisms.

alpha activity   The *EPA* maxima in 1975 for drinking water were 15 picocuries per litre for alpha radiation and 5 pCi/litre for combined radium-226 and radium-228. Radium-226 is an alpha emitter.

alpha factor   For an *aerator* in an activated sludge plant, the rate of oxygen transfer in clean water divided by the rate of transfer in mixed liquor. *See oxygenation capacity.*

alpha particle   A particle emitted in *radioactivity*, which consists of two protons with two neutrons—i.e. a helium nucleus.

alternating double filtration, ADF   To increase the organic load on a *trickling filter*, and to avoid *ponding*, two filters can be placed in series, with a *humus tank* after each. When the primary filter shows signs of ponding, or after a set time (e.g. 1 week), the two filters are reversed, with the second filter operating before the first one. (*See Figure A.4.*) There is an additional pumping cost, but in a plant with several filters and humus tanks there

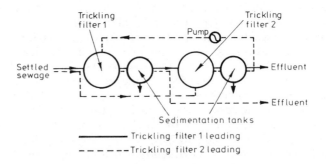
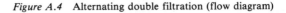

Figure A.4   Alternating double filtration (flow diagram)

14

may be no need to build extra humus tanks. To obtain an effluent of BOD_5:SS quality 20:30, organic loads up to 0.25 kg of BOD_5 per m^3 can be used. *Compare two-stage filtration.*

alum, aluminium sulphate, $Al_2(SO_4)_3$ A common *coagulant* for *raw water*. It has any number between 14 and 20 molecules of water in combination. Dissolved in water it hydrolyses into $Al(OH)_3$, the hydroxide, and sulphuric acid, but to precipitate the hydroxide, as required for *coagulation*, the water must be alkaline. The *alkalinity* may be naturally present in the water as *carbonate hardness*, thus:

$$Al_2(SO_4)_3 + 3Ca(HCO_3)_2 = 2Al(OH)_3\downarrow + 3CaSO_4 + 6CO_2$$

In a soft water with not enough carbonate hardness more can be added in the form of calcium bicarbonate or sodium carbonate (*soda ash*) or *lime*. If there is too much bicarbonate, which is possible in a hard water, it may be necessary to add a little sulphuric acid. *See corrosiveness of water.*

alumina, aluminium oxide, Al_2O_3 *See activated alumina.*

alumina fibre A *refractory* made of alumina, Al_2O_3, that can withstand furnace temperatures up to 1600°C.

aluminium chlorohydrate, $Al_2(OH)_2Cl_2$ A *conditioning* chemical added to sludge before *de-watering of sludge.*

aluminium recovery The contents of aluminium and steel in solid refuse amount, respectively, to some 20% and 10% of the manufactured output. The aluminium in refuse, although less than 1%, is probably its most valuable component. Nearly 25% of the aluminium cans in the USA are collected every year, amounting to 100 000 tonnes or 5500 million cans. Aluminium can be separated from the non-magnetic metals in a *mechanical sorting plant* by *float and sink treatment*. The sinks are then the copper, lead, tin, zinc and their alloys and any non-magnetic stainless steel. *See also eddy-current separator.*

alumino-ferric A chemical containing 92% hydrated aluminium sulphate, the remainder being mostly ferric sulphate, used as a *coagulant* for clarifying raw water or sewage effluent.

aluminum US spelling of aluminium.

alum sludge A *waterworks sludge* formed when *alum* is the *coagulant*.

alveole One of millions of tiny air-filled pores in the lung, where the blood gives up carbon dioxide to the air and takes in oxygen.

ambient air standards *Air quality standards.*

American Conference of Governmental Industrial Hygienists, ACGIH A US body that publishes annually *threshold limit values* for some 500 *air pollutants*.

amino acids Organic compounds that are the building blocks of proteins. Free amino acids contain an amino group ($-NH_2$) and a carboxyl group ($-COOH$) and are therefore *amphoteric*.

ammonia, NH_3 *Bacteria* readily decompose *urea* and *proteins* in sewage to form ammonia which may later be oxidised to *nitrites* and *nitrates*, as in the *nitrogen cycle*. Typical sewage contains 30 to 50 mg/litre of NH_3. It exists in water either as the ammonium ion, NH_4^+, or as dissolved ammonia, NH_3, thus:

$$NH_3 + H_2O = OH^- + NH_4^+ \text{ (ammonium)}$$

The equilibrium moves to the left as the pH rises. The reactions with chlorine are important in the *chlorination* of the water supply. Ammonia can be removed from water by *ammonia stripping*, chlorination, *ion exchange, nitrification–denitrification*, etc.

ammoniacal liquor Part of the liquid wastes from coke-making, containing many pollutants, including ammonia, *phenols, cyanide compounds*. The ammonia is distilled off, leaving the waste product, *spent liquor. See coke plant wastes*.

ammoniacal nitrogen Nitrogen as NH_3, ammonia, or NH_4^+, ammonium ion.

ammonia oxidisers *Nitrifying bacteria*.

ammonia stripping, air s. Removal of ammonia from sewage effluent by allowing it to flow down a *stripping tower* some 7 m high, up which 1500 to 3500 times its own volume of air is blown. The effluent is first made alkaline with lime to a pH of 11.0 so that the ammonia then exists all as NH_3. The method is very inefficient in cold weather and cannot be used if the water is likely to freeze in the tower. *See clinoptilolite*.

ammoxidation A reaction between the vapours of ammonia and hydrocarbons in the presence of oxygen.

amoeba A group of *protozoa* of the class of *rhizopods* including the genus *Amoeba*. Some of them produce diseases such as dysentery in humans. Others occur in *trickling filters* and possibly in *activated sludge*.

amoebiasis, amoebic dysentery Disease caused by *Entamoeba histolytica*. It can be caught from water that is devoid of *enteric bacilli*.

amoebic cyst A *cyst* of an *amoeba*.

amoebic meningo-encephalitis A usually fatal illness caused by the parasite *Naegleria gruberi*, an *amoeba* that enters the body by the nose and soon reaches the blood. Swimming and diving in polluted water have caused it. *See also waterborne disease*.

Amphipoda *Freshwater shrimps*.

ampholytic Description of a substance that carries positive and

16

negative charges.

amphoteric Description of substances that act as both acid and alkali.

anabatic wind An uphill wind, caused in the daytime by the sun warming a hillside.

anabolic metabolism, anabolism Metabolic activity—eating, to make *protoplasm*. *Compare* catabolic metabolism.

anaerobe (adjective **anaerobic**) A micro-organism that needs no free oxygen to develop (in *anaerobiosis*)—e.g. *Desulphovibrio desulphuricans*.

anaerobic contact process Industrial wastes with high *BOD*, such as those from meat processing, can sometimes be treated by *anaerobic sludge digestion*, after which the mixture may be separated by *sedimentation* or *flotation* and the sludge returned to the digester.

anaerobic fermentation, a. digestion *Anaerobic sludge digestion*.

anaerobic filter A tank, filled with packing media, like a *trickling filter* which is flooded, therefore without air and containing

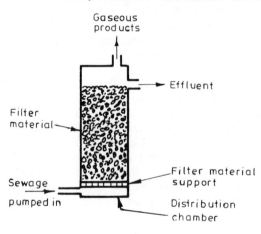

Figure A.5 Anaerobic filter

anaerobes (*see Figure A.5*). It may be used to de-nitrify a *nitrified effluent* that flows upwards through it. Sometimes a *deep coarse filter bed* is used.

anaerobic lagoon, a. pond A *waste stabilisation pond* with a high *organic load*, e.g. 100 g of *BOD$_5$* per m^3 of pond per day. The pond is devoid of *dissolved oxygen* and has few *algae*. Some 50% of the BOD is removed in 24 h. If the pond is in stable *alkaline* condition, there is little smell because the sulphide is present as the ion S^{2-} and not as H$_2$S. For full treatment the anaerobic pond should be followed by a *facultative lagoon* and two **maturation ponds**. Anaerobic ponds are de-sludged every 3 to 5 years.

anaerobic oxidation The reactions in an *anaerobic* process that produces methane and carbon dioxide can be described as anaerobic oxidation, since, in the absence of air, oxygen is removed from *organic compounds* to produce CO_2. Whether in *acid fermentation* or in **methane fermentation**, any oxygen involved in biochemical reactions must come either from water or from organic matter—nitrates or nitrites being decomposed. *See anoxic.*

anaerobic sludge digestion *Anaerobic* decomposition of sewage sludge. Normally the process occupies two stages, the first being *acid fermentation*, in which *facultative* and *obligate anaerobes* break down complex organics in the sludge to form simpler, low-molecular-weight substances such as acetic acid. In the second stage *methanogenic bacteria* form methane and carbon dioxide from the organic acids. The *acid fermentation* tends to reduce the pH of the sludge but methane formers can exist only at pH above 6.5. Therefore the pH should be kept above 7.0 to obtain methane. The process is slow and temperature-dependent. It is common to maintain a temperature of about 35°C (*mesophilic digestion*) in the first (primary) digestion, which is therefore usually heated. The tank is closed and stirred (mixed) by some means. The second stage (secondary digestion) is an unheated open tank in which the sludge settles out, so there is no mixing. Most of the digestion takes place in the primary digester. Complete digestion reduces the *volatile solids* in the raw sludge by up to 70%. This is equivalent to a 50% reduction in the total sludge solids, assuming that 70% of the raw sludge solids are volatile. For medium-rate digestion the primary digester is designed on *detention period* (20 to 30 days) or preferably on solids loading—e.g. 1 kg volatile solids per m³ of tank per day

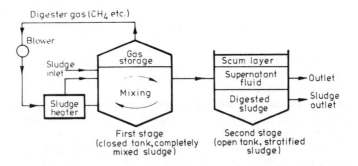

Figure A.6 Anaerobic sludge digestion (two-stage process)

(*compare* **high-rate anaerobic digestion**).

The secondary digestion tank has from 50 to 150% of the capacity of the primary tank. If a single-stage process is used (primary digestion only), the digestion tank is not mixed but stratified into layers. This process is much less efficient; the solids loading must be much lower than with two stages and the detention period longer. Many of the problems of operating anaerobic digestion arise from the difficulty of maintaining an adequate population of methanogenic bacteria. They are seriously affected by small quantities of toxic materials in the sludge. If the methanogenic bacteria are killed, the sludge is left in an obnoxious half-digested state. Starting an anaerobic digester can be difficult and pH control is critical to enable the methanogenic bacteria to flourish. The ratio of the concentration of organic acids to *total alkalinity* (both expressed as equivalents per litre) should be under 0.5, preferably about 0.2.

See also **acid formers, anaerobic contact process, cold digestion, digested sludge, methane fermentation, sludge gas, thermophilic digestion.**

anaerobic sludge digestion tank, a.s. digester An *anaerobic* tank that usually is heated to maintain the *anaerobic sludge digestion* process at 30 to 35°C, occasionally at about 50°C. In this primary stage the methane can be collected and used as a fuel. Primary tanks usually have scum breakers and a roof that may be floating or fixed. If the roof is floating, it forms a gasometer to hold the methane.

The primary digester normally is mixed by recirculating either the sludge or the gas or both. Sludge is heated by circulating it through an external heat exchanger. Tanks are usually circular, not less than 7.5 m deep at the centre or less than 6 m diameter, but up to 35 m diameter and 14 m deep. If a floating roof is provided, it has to be gastight round the rim where it meets the tank wall. *Cold digestion* tanks are unsuitable for large works.

anaerobiosis Life in the absence of free or dissolved oxygen, typically resulting in the production of methane, *ammonia, nitrogen, hydrogen sulphide,* and, like *aerobiosis,* also of *carbon dioxide,* due to *anaerobic oxidation.*

Ancylostoma duodenale   See Ankylostoma duodenale.

ångström A unit of length, 10^{-10} m, 10^{-4} micron, 0.1 *nanon.*

anion A negative *ion,* one that moves to the *anode* in *electrolysis.*

anion exchange Replacement of acid radical ions such as NO_3^- or SO_4^{2-} in water by using *ion exchange.*

anion exchange resin   *See ion-exchange resins.*

anionic detergent   *See synthetic detergent.*

Anisopus fenestralis   Sylvicola fenestralis.

Ankylostoma duodenale The hookworm responsible for ankylostomiasis. *See hookworm disease.*

annelid, Annelida Small segmental worms in *trickling filters*, which feed on the *biological film*. *Polychaeta* annelids are important in marine *bio-assays*.

anode (adjective **anodic**) The positive *electrode* in *electrolysis*, attracting *anions*.

anopheline mosquito, *Anopheles* **genus of mosquito** *Vectors* of *malaria* and sometimes *filariasis*, which carry the parasitic *spore-forming protozoa* that live in the blood and cause the illness. *See mosquitoes.*

anoxic Description of conditions without free oxygen. Oxygen then is available for respiration only from dissolved inorganic substances such as nitrate ions, NO_3^-. It is not *anaerobiosis*. Anoxic sections in an activated sludge plant may be used for *denitrification*.

antagonism (1) Interaction of two or more substances such that their total effect is less than the sum of their effects separately. (2) Prevention of the growth of an organism by the products or activities of another. *Compare synergism.*

anthracite capping US term for removal of the top 150 mm or so of a *rapid gravity sand filter* and its replacement with 150 mm of 0.9 mm anthracite. Anthracite is a hard coal of low volatile content (less than 10%) and a specific gravity of about 1.4. Its great advantage is that its low density compared with sands (s.g. 2.3) allows it to be used in a coarse size in the top layer of a downflow filter, where it remains after *backwashing*. This made possible the practice of *coarse-to-fine filtration* in downflow filters. *See dual-media filter.*

anthrax A disease caused by the bacterium *Bacillus anthracis* that may possibly be waterborne.

anticyclone An area of the earth's atmosphere with the highest atmospheric pressure at the centre. The winds blow in the opposite direction from a *cyclone*—i.e. clockwise north of the equator, anticlockwise south of it. Anticyclones usually bring calm weather and often accompany a temperature *inversion*.

anti-foaming agent A chemical added to an *aeration tank* to stop the production of foam. It usually consists of a mineral oil and a spreading agent.

application factor A factor applied to the concentration of toxicant in, e.g., a *median tolerance limit* (TL_m) of a *bio-assay*, that is considered to make it safe. The factor can vary from a tenth to a hundredth of the 96-h or 48-h TL_m figure. It is sometimes found by dividing the 96-h TL_m by the *maximum acceptable toxicant concentration*.

APWA American Public Works Association.

aquaculture Fish farming.

aqua privy A WC that discharges to a cesspool near the house—a
system used in some African countries. The overflow passes
usually to an *absorption pit* or trench (*Figure A.7*).

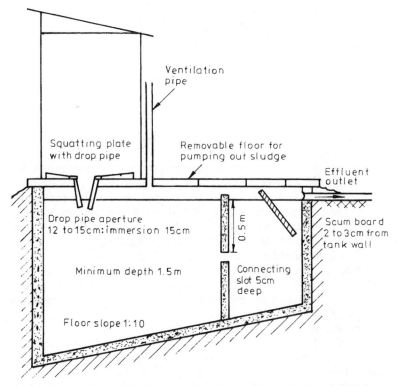

Figure A.7 Aqua privy (Reproduced from 'Water Treatment and Sanitation;
simple methods for rural areas', Intermediate Technology Publications Ltd.,
London)

aquatic Concerned with water.

aqueduct A conduit for water which may include lengths of
tunnel or bridge. *See ghanat.*

aquiclude *See aquitard.*

aquifer Rock or soil that not only contains quantities of water,
but also yields it easily to a pipe or well entering it (*see Figure
A.8*). It must therefore be porous—i.e. of high *permeability* like
uniformly sized gravel or fissured sandstone or limestone. For a

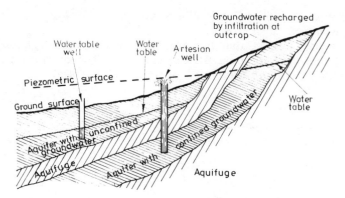

Figure A.8 Artesian well and water-table well

municipal water supply the aquifer should be at least 25 m thick over a large area.

aquifuge A rock or soil that neither contains nor transmits water in useful quantities.

aquitard, aquiclude A rock or soil of low *permeability* that delays the flow of water out of it. Even if it contains quantities of water, it releases water too slowly to be considered an *aquifer*. For clay this is shown by its high *wilting coefficient*.

arachnid, Arachnida A class of invertebrate animals that includes spiders, scorpions and mites, a subdivision of *arthropods*.

arbitrary flow Partial mixing of a fluid, resulting in a flow pattern that is intermediate between *plug flow* and a *complete mixing system*.

Archimedean screw A helix rotating in a close-fitting cylinder or semicircular channel. The lower end is immersed in the liquid or grains to be raised. The *screw pump* is one type.

area method of controlled tipping The main UK method of *controlled tipping*, also one of the two US methods. It differs from the *trench method* in not providing its own *cover*, but it economises land and can be used almost anywhere that cover is available. Cover may have to be imported if it is not available from a borrow pit on the site. The refuse is compacted on the ground in 50 cm layers as usual.

Arenicola See lugworm.

aromatic compounds *Organic compounds* formed from benzene rings. They may be not easily *biodegradable* and with attached chlorine atoms they often form the base of *pesticides*, antiseptics, etc.

arsenic, As Clams and shrimps may accumulate arsenic and have been found to contain 50 to 100 p.p.m. or more, because of *bio-*

magnification. *See also* Table of Allowable Contaminants in Drinking Water, page 359.

artesian well A well, nowadays usually a borehole to *confined groundwater* from which water flows without pumping because the *piezometric surface* is above ground level, implying that the water is under pressure (*see Figure A.8*). Often water is withdrawn too quickly from artesian wells and they become pumped wells.

arthropod, Arthropoda The largest *phylum* of invertebrate animals. Arthropods have segmented bodies and jointed limbs and include centipedes, *crustaceans*, *arachnids*, etc.

artificial recharge Replenishment of *groundwater* artificially through shafts, wells, pits, trenches, etc., often using water that has been treated chemically or by *sand filters*. An improvement to water quality may be achieved underground because the water may filter hundreds of metres through air-filled ground that provides biochemical and physical treatment. If sewage effluent is used for recharging, heavy bacterial growth may clog the *aquifer*. *See disposal well, spreading area.*

asbestos Many types of asbestos have been used for insulating steam pipes, etc. The most dangerous type to lungs is crocidolite, which is lavender blue. Amosite, chrysotile and tremolite are also a hazard. Thorough wetting of asbestos lagging reduces the amount of dust generated when it is stripped. A slurry of blue asbestos should not be allowed to dry out but should be disposed of immediately in double-sealed, thick polythene bags on a tip.

asbestos-cement pipe Pipe made of 10 to 15% asbestos fibre, bound with cement. It is stronger than concrete pipe but dearer. Cutting or drilling by hand ordinarily produces little or no *respirable dust* but power-operated carborundum disc saws can be very dusty. They should be covered by a guard plate and wetted by four water jets, two each side, passing through the plate.

ascariasis Intestinal illness, especially of children, caused by the nematode *Ascaris lumbricoides*. The illness can be waterborne and is acquired through the mouth. The worm's eggs are transmitted in human *faeces*. In 1972 ascariasis affected one-quarter of the world's population.

Ascaris lumbricoides A *nematode* round worm which is a *parasite* in humans, causing *ascariasis*. Its eggs can survive in the soil for months.

ASCE The American Society of Civil Engineers, the country's foremost civil engineering body. It has many members concerned with waste and water treatment.

Ascoidea rubescens A *fungus* that is common in the top 15 cm of *trickling filters.*

Asellus aquaticus, **common water louse, water bug, sow bug, hog louse** A *crustacean* that likes moderately polluted water and may be used as a *biological indicator of pollution.* It is not harmful but may harbour *viruses.*

Aspergillus niger A *fungus* that has been used as a tracer for estimating the growth of fungus in *trickling filters.* It has also been found that it can break down the pods of the carob bean to form *single-cell protein.*

Aspidisca A free-swimming *ciliate protozoon* that is often found in *trickling filters* and sometimes in *activated sludge,* especially if the effluent is of high quality.

assimilative capacity Of a receiving water, its ability to receive discharges of wastes without suffering pollution. *See self-purification.*

ASTM 31 One of the annual volumes published by the American Society for Testing and Materials. No. 31 is concerned with water and its treatment.

asymptomatic carrier A carrier of a disease, who shows none of its symptoms or signs.

atomic absorption spectrophotometer An instrument widely used for measuring trace concentrations of elements in water. A light source with a wavelength readily absorbed is directed through a flame in which the sample is being atomised. The intensity of light absorbed is related to the concentration of that element in the sample. The method is fast and generally free from interference between elements, but it requires a different light source for each element. The *EPA* recommends it for determining arsenic, barium, cadmium, chromium, lead, selenium and silver in water.

Atritor flash dryer A device for *flash drying* of sludge, with a fast-moving rotor carrying pegs that pass between other stationary pegs on a casing with hot gas flowing through it. The dried sludge is therefore pulverised. The dryer normally accepts only de-watered sludge.

attemperation Artificial cooling—e.g. of flue gases.

autecology The ecology of one species.

autoclave A laboratory autoclave is a 'pressure cooker' that reaches a pressure slightly above 1 atmosphere, at a steam temperature of 115 to 125°C, sufficient to kill *endospores.* It sterilises apparatus or chemicals. *See also tyndallisation.*

autolysis, (auto-lysis), self-digestion The disintegration of cells by the action of their own *enzymes,* part of the *endogenous phase of growth.*

automatic sampler A device that can sample either continuously or intermittently at set intervals. A *sampler* may be operated by electric battery or clockwork, or simply by immersing a hollow container in the fluid to be sampled and restricting the rate of displacement of the air from it.

automatic sorting of solid wastes   *See mechanical sorting plant.*

automobile emissions   The gases and solids mainly from automobile exhaust pipes but also from their petrol tanks, crankcase breathers, etc. Carbon dioxide, water and nitrogen are not pollutants and they form the vast bulk of the emissions. The pollutants mixed in with them are *carbon monoxide, lead* compounds, and oxides of nitrogen, NO_x, as well as unburnt and partly burnt hydrocarbons. Automobile emissions of all types are thought to contribute 60% of total *air pollutants* in western countries.

autotroph, autotrophic organism   An organism that uses carbon dioxide as a source of the carbon it needs for building new cells. Autotrophs do not, like *heterotrophs*, need organic carbon from dead vegetables, sewage, etc. They can consume dissolved nitrates or ammonium salts and they include nitrifying bacteria and *algae*.

autumn turnover   *See thermal stratification.*

available chlorine   *See chlorine residual.*

available dilution, dilution factor   The rate of flow of a *receiving water* divided by the rate of flow of an effluent entering it.

available residual chlorine   *See chlorine residual.*

AWT   *Advanced wastewater treatment.*

AWWA   The American Water Works Association, the foremost US professional association of water engineers.

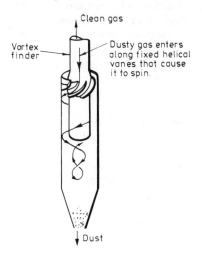

Figure A.9   Axial-inlet cyclone

axenic culture   A *culture* that includes only one kind of organism.

axial-flow pump, propeller p.   A rotodynamic pump which forces the water out in the same way as a ship's propeller. The passage between the blades is restricted, so this type is usually considered unsuitable for pumping water containing solids. It is most efficient for pumping large volumes of clean water or effluent against a low head.

axial-inlet cyclones, multi-cyclone, multi-cellular collector   A nest of small *cyclones* connected in parallel as *dust arrestors*, generally not larger than 60 cm diameter, usually much smaller, and occasionally as small as 7.5 cm diameter, grouped in one housing. The dirty gas enters axially but is given a tangential spin by vanes at the inlet (*Figure A.9*). More than 100 can be installed in one nest.

Azotobacter   The most important nitrogen-fixing *bacterium*. It extracts nitrogen from the air and may be present in *activated sludge. Compare nitrifying bacteria.*

B

Bacillariophyta   *Diatoms.*

bacillary dysentery   *Shigellosis.*

bacillus   A rod-shaped *bacterium*, but *see* below.

Bacillus   A genus of bacteria that are *facultative anaerobes*, common in *trickling filters, activated sludge* and sludge digestion. They form **endospores**. *See also* below.

Bacillus anthracis   The *bacterium* that causes *anthrax.*

Bacillus coli   An old name for *coliform group.*

back drop, drop connection, d. manhole, tumbling bay (in USA **chimney, sewer c.**)   A connection from a branch drain or sewer to a much deeper sewer. The sewage from the branch enters the *manhole* through a vertical pipe which may be inside or outside it but in any case has *access eyes* built in for *rodding* (*see Figure B.1*). Back drops prevent excessive scour and splashing in the manhole.

back end   The chimney end of a boiler or furnace.

backflow preventer   A *non-return valve* that prevents *back siphonage* in a water main.

background level   Of a pollutant, its concentration in air or water, excluding nearby sources.

back siphonage   Reversed flow in a water *main* caused by suction of water from a sink, bath, etc., into the main (*Figure B.2*). Back siphonage can be prevented by ensuring that the lowest outlet from any tap is above the highest point that the dirty water can

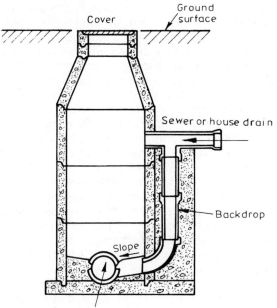

Figure B.1 Manhole with backdrop

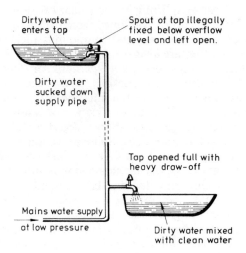

Figure B.2 Back siphonage

27

reach. Otherwise, if a tap submerged in dirty water is open while other taps below it are also open, the dirty water could be sucked into it. In the UK the chief defence against back siphonage is by *air-gap protection*; in each household one cold water tap for drinking is connected direct to the main. On the Continent and in the USA the main is often connected directly to the *WC* fitting as well as to the other cold water taps, an arrangement in which back siphonage is easier.

backwashing   (1) *Dual-media filters, multi-media filters* and *rapid gravity sand filters* are equipped with *underdrains* through which the flow can be reversed, at a multiple of the forward flow rate, usually expanding the filter bed by 20 to 50%. This reversal of the flow at the end of the filter run—backwashing—takes 3 to 15 min every 24 h or so, and the backwash water must run to waste. In modern plants the backwashing programme can be started automatically either by a clock or because of excessive head loss or excessive *turbidity*, as measured by a *turbidimeter*. Before backwashing, air is blown up to dislodge the dirt from the sand grains (air scouring). The *wash water*, which is *filtrate*, flows slowly at 0.4 m/min. It should not exceed 2% of the total flow, since it is lost. (2) Separating a mixed bed of *ion-exchange resins* into its components to prepare it for *regeneration*.

backwater curve   The curve of the water surface in a river or open channel, held up behind a dam, weir or other obstruction to the flow.

bacteria (singular **bacterium**)   Single-celled microscopic organisms that multiply fast by splitting in two (binary fission). In order to multiply they need carbon, obtained from carbon dioxide if they are *autotrophs*, or from organic compounds (dead vegetation, sewage, meat) if they are *heterotrophs*. Their energy comes either from sunlight if they use *photosynthesis* or from chemical reactions if they use *chemosynthesis*. They occur in air, water, earth, rotting vegetation and the intestines of animals, and are fundamental to the biological purification of sewage in *trickling filters*, activated sludge and sludge digestion, and the self-purification of rivers and lakes. Some cause illness in man.

Bacteria average about 1 μm long but vary typically from 0.5 to 10 μm and are classified by shape mainly into the following five groups: bacilli (singular bacillus), resembling rods; cocci (singular coccus), resembling spheres; spirilla (singular spirillum), resembling curved or spiral bacilli; spirochaetes, which are filament-shaped, spiral and flexible, some of which are responsible for *leptospirosis*; and *Vibrio*, which are curved. Some *Vibrio* cause *cholera*. Bacteria under ideal conditions may divide every 20 min, but they take up food so quickly that they

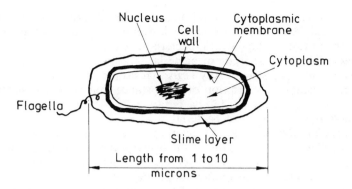

Figure B.3 Bacterial cell

are likely soon to be limited by shortages of food or oxygen or water. In raw domestic sewage there are about 10 million bacteria per ml, reducing to about 1 per ml in fully treated drinking water, 100 per ml in spring water or 10 000 per ml in a polluted river. Bacteria may be *aerobic* or *anaerobic*. They may be *obligate aerobes, obligate anaerobes, facultative anaerobes,* etc. (*See Figure B.3.*)

bacteria bed A *trickling filter.*

bacterial corrosion *See iron bacteria, sulphide corrosion, Thiobacillus, tuberculation.*

bacterial count *See bacteriological examination.*

bacterial leaching, bio-leaching The use of bacteria to attack dumps of mine *tailings* or other ore of low metal content. Bacteria dissolve the ore, enabling the metal to be extracted later from solution, and have succeeded with both copper and uranium ores.

bacterial purity British regulations require the maximum number of *Escherichia coli* or other coliform bacteria in 100 ml of drinking water to be zero in 95% of the tests, with never more than two *E.coli* or ten coliform bacteria per 100 ml.

bacterial tracer Any distinctive *bacterium* used to monitor water movements. *See tracer.*

bactericide Something that kills bacteria.

bacteriological examination Drinking and other waters are examined for their content of bacteria by the *membrane filtration* method, *most probable number* (*MPN*) and *plate counts.* The results of the methods are not comparable.

bacteriological pollution The presence of bacteria that cause disease is not easy to confirm on a routine basis, because there are so many different types. Consequently their presence

29

(bacteriological pollution) is usually inferred if *Escherichia coli* is found. *E.coli* can exist much longer outside the human body than most of the harmful bacteria. *See coliform group.*

bacteriology The study of bacteria.

bacteriophage, phage Any *virus* that attacks bacteria.

bacteriostatic Description of a substance that prevents bacterial growth but does not kill bacteria.

bacterium Singular of bacteria.

Bacterium dysenteriae A bacterium that causes *dysentery*.

Bacterium tularense One of the bacteria that cause *tularaemia*.

Bacteroides *Obligate anaerobe* bacteria that can form sulphides from protein in *anaerobic sludge digestion*.

Baetis rhodani A stone-dwelling *mayfly* which favours mildly polluted rivers. Most mayfly dislike pollution.

baffle A (usually) vertical board in a conduit for gas or liquid, which diverts or reduces flow, prevents eddying, etc.

baffled side weir A weir alongside a sewer, over which the sewage is encouraged to flow by a *baffle* across the main sewer. The baffle can be a vertical, square plate having its bottom edge level with the crest of the side weir, so that the flow over the weir increases as the sewage level rises in the main sewer.

baffle-type scrubber A *wet scrubber* that varies the direction of gas flow by baffles, often used as an *entrainment separator* and then called a baffle mist eliminator. The pressure drop is low, 2.5 to 7.5 cm water gauge, but many types do not collect particles smaller than 10 microns.

bag cleaning of fabric filters Filter bags in *fabric filters* must be cleaned regularly to reduce the gas pressure needed when they are dirty. The extra power consumed when bags are dirty is expensive; so is a bag burst because of excess pressure. Many methods exist, including mechanical shakers for shaking the dust off the bags, pulses of compressed air in the opposite direction from the general flow, etc. Bags through which the flow is normally outwards can be cleaned by collapsing them in momentary reversals of flow.

bag filter A *fabric filter*.

baghouse The steel gastight housing around a *fabric filter*. It contains tubular bags that filter *flue gas*, air or other dusty gas, either to remove pollutants or to recover valuable dust.

balanced draught In a furnace, a condition of the gas pressure within it such that the furnace, flues and chimney are under slight suction and any leakage is inwards. *Compare forced-draught fan, induced-draught fan.*

balancing reservoir A reservoir into which there is a steady delivery and from which the flow may be variable. It provides steady working conditions for the *filters* and other plant upstream of it.

balancing tank   A tank designed to reduce the variations in flow rate into a sewage treatment plant. It therefore stores sewage during times of high flow and releases it during low flows. *Stormwater tanks* may be used for flow balancing. Industries with batch discharges of effluent also may balance their flow before it enters the sewer.

bale, baling   *See high-density baling, low-density baling*.

bale breaker   A *hammer mill* that opens bales.

balefill   *Controlled tipping* by *high-density baling*. Because of the relative impermeability of bales compared with ordinary fill at controlled tips, they are likely to emit less *leachate*. The leachate squeezed from the bales in the press is, however, a liquid that could be polluting. With a density about double that of ordinary landfill, the *impermeability coefficient* of balefill was found by the American Public Works Association to be twenty times that of ordinary landfill. But the APWA conclude that the same environmental precautions are needed.

ballistic separator   (1) A type of *inertial separator* that makes use of the differences in fall velocity and fall path of materials (e.g. paper and metal) when they are thrown and fall freely. (2) A means of extracting from a *hammer mill* any objects that it cannot break. They are driven round by the hammers until they reach an opening and a chute intended for them.

band dryer   A *dryer* which dries sludge cake by hot air after pulverisation. It contains several horizontal conveyor belts of wire mesh above one another, with the sludge falling from the end of one to the beginning of the next, until it is removed by a screw conveyor at the foot.

band screen, belt s.   A sloping conveyor belt of wire mesh (e.g. 6-mm holes) through which sewage or water flows, leaving the larger solids on the screen.

Banks filter   A *pebble-bed clarifier*.

bankside storage   *Raw water storage* in a reservoir on the river bank. All river water should have at least 7 days' bankside storage before treatment. This allows solids to settle and some bactericidal action from *ultra-violet radiation*. The river intake can be closed to allow unusual river pollution to pass downstream.

bank storage   The ability of water to flow into the bank of a river at high water and to return to the river at low water, acting as groundwater in the interval. Gravelly or other porous soils have greater bank storage than clays, silts or impervious soils.

barium, Ba   *See* Table of Allowable Contaminants in Drinking Water, page 359.

Barminutor A type of *comminutor*.

bar screen, b. rack A screen for *raw water* or sewage, with parallel bars either horizontally or at a slope between 30° and 60° to the horizontal. It may be cleaned mechanically or by hand.

base exchange *Cation exchange*.

base flow In a stream, the part of the *runoff* that comes from the *groundwater*, long-term lake storage, etc.

basin lag *See time lag of a catchment*.

batch treatment Any treatment in which the process is completed and the products are discharged before more raw material can be taken in.

bathing water *See swimming water*.

bathythermograph An instrument that measures ocean temperature at depth.

Bdellovibrio bacteriovorus A widespread *predator*, a bacterium that can destroy other *bacteria*, including *Escherichia coli*.

beamhouse wastes In a tannery for cattle hides, the beamhouse work includes curing (drying the skin), fleshing (removal of fat), washing, de-hairing, bating (raising the pH by adding ammonium salts), pickling and de-greasing. The skins then are tanned and the wastes are combined with *tannery wastes*. *See fellmongering*.

Beccari system A *composting* system which is unusual in being partly *anaerobic*, although it is not foul-smelling.

bed load The weight or volume of silt, sand and gravel that rolls or slides or bounces (*saltation*) along a stream bed. *Compare sediment load*.

beet sugar wastes Some 70% of the water used in extracting sugar from beets is spent in washing them. Waste water is also produced in de-watering the pulped beets and in lime slurry residues and condensate from evaporators.

Typical mixed wastes have 400 to 500 mg/litre of *BOD$_5$*, 5000 mg/litre of *suspended solids*, 250 mg/litre of *alkalinity* and a pH of 8.0.

Beggiatoa A *filamentous organism*, a sulphur-oxidising *bacterium*. It can probably grow as *heterotroph* or *autotroph* but cannot be purely autotrophic, because it can be cultured only if *organics* are present. Hydrogen sulphide, H_2S, stimulates its growth and the H_2S is oxidised to sulphur, but it can be grown in the absence of sulphides.

bellmouth chamber A chamber at a junction between two or more large sewers, in which the side of the outgoing sewer forms a tangent to the incoming sewers.

belt filter press A *filter belt press*.

belt screen A *band screen*.

belt vacuum filter *See vacuum filter.*

benching The concrete or brick surround to a sewer channel in a *manhole*, etc. The benching slopes uniformly upwards away from the channel at a slope of about one horizontal in six vertical.

benthic deposits *Bottom sediments* that originate from dead or decaying organic material.

benthon, benthos The community of plants and animals living on the bed of a sea, stream, lake, pond, etc. Many insects or other benthic life are biological indicators, since the bed collects the sediments from the water. *Compare nekton, neuston, plankton.*

benzene hexachloride, BHC An *insecticide* that can be toxic to fish at concentrations as low as 35 μg/litre.

Berkefeld filter A domestic drinking water filter consisting of a metal cylinder containing compressed *diatomite* through which the water flows. It can be cleaned by *backwashing.*

best practicable means, b.p.m. The *Alkali Inspectorate* require works managements to use the best practicable means to avoid pollution. A new works, for instance, should pollute less than an old one. So far as *dust arrestors* are concerned, b.p.m. means that there should be efficient planned maintenance, with a well-organised stock of spare parts, prompt attention to breakdowns and duplicated plant where needed to maintain continuity of *pollution control. See infraction.*

beta activity In 1962–63 the total beta radiation in UK rivers and lakes was often around 20 picocuries per litre but, with the lower frequency of nuclear explosions, this has now diminished to about one-tenth (1 or 2 pCi/litre). *Raw water storage* and *slow sand filters* remove about half of the *radioactivity.*

beta factor The ratio of the *air saturation value* in an *activated sludge* to that in pure water at the same temperature and pressure.

beta particle A particle emitted during *radioactivity*, which may be positive (positron) or negative (electron).

bicarbonate hardness *Carbonate hardness.*

bi-flow filter A *sand filter* with upward flow in the lower half of the bed and downward flow in the upper half. The *effluent* pipes are in the middle of the sand bed and raw water is fed to the top and bottom of the filter. It is an attempt to obtain the advantages of an upflow filter while using the downflow to restrain the lower half of the bed. Hitherto the water quality has been less good than that from a *multi-media filter*, since the coarse sand settles to the lower part of the filter, many of the upflow advantages thus being lost. Bi-flow filters have been used in the USSR and the Netherlands. *See reverse current filter.*

33

big gun sprinkler   A *rain gun* used in the USA.

bilharzia, bilharziasis   *Schistosomiasis.*

billion   In the USA, one thousand million, 10^9. In Europe, formerly 10^{12}, one million million, but the US value is becoming accepted in the UK. In the USA one trillion = 10^{12}.

bimodal size distribution   A *particle size distribution* in which most of the grains are concentrated around two sizes.

binary fission   Dividing in two—the reproductive method of many *bacteria*. It exists also in *radioactivity*.

binary sorter   In a *mechanical sorting plant*, the usual device that divides the material into two parts, one with and one without the constituent being sorted.

bio-assay   A trial of a chemical or a pollutant, made by testing how much of it can be tolerated by a living organism. *See median tolerance limit, micro-nutrient.*

biocenose   *See biocoenosis.*

biochemical   Concerned with life, including growth and other activities.

biochemical oxidation   This usually refers to *biological treatment*.

biochemical oxygen demand   *See BOD.*

biochemistry   The chemistry of living beings—of physiology.

biocide   In the UK, a chemical that kills plants; in the USA, one that kills pests, a *pesticide*.

biocoenosis (USA **biocenose**)   The community present in a *biotope*. Thus, the water biocoenosis means all the water life.

bioconversion   The biological conversion of wastes. For solid wastes it can mean making *single-cell protein*, but it could also mean, for water, *biological treatment*.

biodegradable   Description of something that breaks down, decomposes, by biochemical changes normally involving *bacteria*. This is usually taken to be desirable in a waste product, but some decomposition products may be dangerous. Biodegradability is especially important for *synthetic detergents*. *Compare non-biodegradable.*

Biodisc   Trade name of a *rotating biological contactor* marketed in the UK by Ames Crosta Ltd. With primary and secondary settlement zones, it is a *packaged sewage treatment plant* suitable for communities of 50 to 1000 people. The discs are built of wire mesh with holes about 3 mm diameter.

biofiltration   The action of a *fixed-film reactor* (e.g. a *Biodisc*).

bio-flocculation   (1) The aeration of *settled sewage* with *activated sludge* for 1 h, followed by treatment of the *effluent* on *trickling filters*. The activated sludge is re-aerated. (2) *Flocculation* of a suspension containing *bacteria, algae,* etc.—e.g. flocculation of *activated sludge* in a *secondary sedimentation tank*.

biogas Gas from *anaerobic sludge digestion*, mainly *carbon dioxide* and *methane*.

bio-leaching *Bacterial leaching.*

biological amplification, b. concentration *See bio-magnification.*

biological film, b. slime, microbial film A film of slime that forms in a *micro-strainer, rotating biological contactor, slow sand filter, trickling filter,* etc. It is widely believed to be *zoogloeal* and contains metabolic products such as *polysaccharides* as well as a variety of microbes. *See Schmutzdecke.*

biological filter A *trickling filter.*

biological hydrolysis *See enzymatic hydrolysis.*

biological indicator (or **index**) **of pollution** The presence and frequency of a plant, animal, microbe or other form of life can give an idea of the level of pollution in its environment. But an organism that indicates pollution may provide no indication about poisons. Various classifications of organisms indicating levels of organic pollution have been produced—e.g. the *saprobic classification*, the *Trent biotic index.*

biological monitoring Repeatedly surveying plants and animals at a site to determine their abundance, at least roughly, and noting the physical, chemical and biological changes with time, as in a *transect.*

biological oxidation *See biological treatment.*

biological slime *Biological film.*

biological treatment, b. oxidation Any process using biochemical reactions, particularly *trickling filters, activated sludge,* or similar units for oxidising *sewage.*

bio-magnification, biological concentration Multiplication of the concentration of a chemical, usually toxic, as it moves up the *food chain* from the air or water eventually to the larger *predators.* Such *non-biodegradable* pesticides are likely to remain in natural waters for a long time and to be taken up by *bacteria, algae,* etc. They may then concentrate in those that are not affected by them and so rise up the chain. Thus concentrations of the *pesticides* aldrin, *Dieldrin* and endrin have been found in algae to be 120 or more times as great as they were in the water, and this within 30 min of the start of exposure. DDT may concentrate a hundredfold in birds. *Mercury* poisoning in *Minamata Bay* was caused by bio-magnification of *methyl mercury* in fish.

biomass The mass, measured by weight, of a population of animals or plants.

bio-oxidation *Biological treatment.*

bio-oxidisable *Biodegradable.*

Biosorption   A *contact stabilisation* method used in Texas. The *surplus sludge* is disposed of in *lagoons*, which function well in the sunny climate.

biosphere   Living animals, fishes, birds, trees, plants, *algae, bacteria* and any other life on earth.

biostimulants   *Nutrients.*

biota   The living parts of an *ecosystem.*

biotic index   A numerical classification of the life in a river, which indicates the quality of the water, as in the *Trent biotic index.*

biotope   A relatively small area where the environment is uniform—e.g. under large stones in river water of the Scottish Highlands; nearly always cold, well aerated and soft.

black fly (Simuliidae)   A pest that lives in rapid, warm rivers—e.g. *Simulium.*

black mud   Black mud, stinking of *hydrogen sulphide*, H_2S, is an indication of *anaerobic* conditions, with H_2S produced by *sulphate-reducing bacteria.* Even in a clean river, only the top 4 mm of mud can receive oxygen from the water over it.

black oil   Usually this means crude oil, occasionally heavy oil fractions, as opposed to *light oils.* A few hours after an oil spill, the light fractions evaporate and the petroleum acids dissolve in the water, leaving the heavy oil residue afloat.

black smoke   *Smoke* usually containing unburnt carbon, a fault in combustion which can be prevented by skilled operation of the furnace. It corresponds to a value of 5 on the *Ringelmann chart* but a much less severe type of pollution, *dark smoke*, can lead to prosecution in the UK.

black water   Water from the WC pan, so called, in modern systems that treat the two differently, to distinguish it from the 'grey' water of *sullage.*

bladder wrack, *Fucus vesiculosus*   A *brown alga*, or phaeophyte, a seaweed with blisters on its leaves. This *alga*, a *biological index of pollution*, flourishes about halfway between high and low water.

blanket   (1) A *fabric filter.* (2) *Cover* over a *controlled tip.*

blanket clarifier   A *sludge-blanket clarifier.*

bleaching powder   *Calcium hypochlorite.*

bleaching processes   *Hypochlorite*, a bleaching agent, is often found in *cotton textile wastes, synthetic textile wastes* and *woollen industry wastes.*

blinding   Blocking by fine material of a *slow sand filter* or of a wire or cloth mesh in a *filter press, screen, vacuum filter*, etc.

blood worms   Red aquatic *larvae* of the *chironomid* midge, which may indicate pollution by *sewage.*

bloom   A proliferation of *microscopic* life, usually of *algae* but

36

sometimes of other microbes.

blowdown (1) Removal of the dirtiest boiler water (from the bottom of the boiler), which often has a dissolved solids concentration of 10 000 mg/litre, including corrosion inhibitors such as chromates and polyphosphates. This practice, common with low-pressure and medium-pressure boilers, is unnecessary with high-pressure boilers, since they use no corrosion inhibitors and have to use *de-mineralised water*. (2) Hydrocarbons released from refinery units during shutdown or start-up.

blue-green algae, Cyanophyta A simple form of *algae* that can form large dense mats on the water surface in summer and spoil the look of the water. They resemble *bacteria* is some ways, being unicellular. They can use atmospheric *nitrogen* as a *nutrient* in building their cells, so the removal of nitrogen from *effluents* will not affect them. Other nutrients, particularly *phosphorus*, must be abundant in the water if a *bloom* is to appear. They may be found in *waste stabilisation ponds* and can give water an unpleasant taste and smell.

BOD, biochemical oxygen demand A measure of the amount of pollution by organic substances (*sewage*) in water. It is expressed as the number of milligrams of oxygen required by the micro-organisms to oxidise the *organics* in a litre of the water. In the standard test a sample of the water or a dilution of it is incubated at 20°C for 5 days (BOD_5). A blank sample shows how much the *dissolved oxygen* in the diluting water decreases with time. The dilution water is seeded with *bacteria*, typically by adding a few ml of sewage works effluent and some inorganic nutrients.

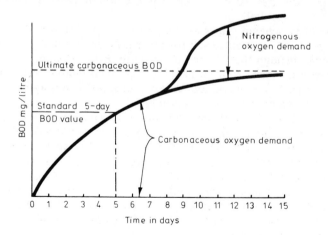

Figure B.4 Biochemical oxygen demand, hypothetical curve

37

An ideal shape of BOD uptake with time is seen in *Figure B.4*. For the first 6 to 10 days only carbonaceous matter is oxidised (*carbonaceous BOD*) but thereafter an increasing *oxygen demand* is exerted by the oxidation of ammonia compounds to nitrite and nitrate (*nitrogenous BOD*). BOD_5 is the standard way of measuring the oxygen demand of a water or for indicating its degree of sewage pollution. Disadvantages of the BOD test include: 5 days' wait for the result; no measure of the non-biodegradable organic substances. BOD results can be seriously inhibited by substances that poison bacteria.

body burden   The total load of a contaminant likely to be carried by people in a particular environment. Thus, if there should be fluoride in a local food or in the air, the body burden from those sources would be high and the allowable concentration of fluoride in water would be proportionately lowered for the locality.

Bohna filter   A *horizontal-flow sand filter*.

boiler feed water   *Feed water*.

borehole casing   *See well casing*.

borehole yield   The firm yield of a borehole is its lowest output of *raw water* in a given dry period or the average output which could be continually withdrawn.

borism   Illness caused by excess of *boron*, sometimes from water.

boron, B   Boron in water to the extent of only 1 or 2 mg/litre has been found to harm crops irrigated with it, even though as a *micro-nutrient* it is essential to them. Consequently, in Israel washing powders are not allowed to contain more than 5% perborate. *See* Table of Allowable Contaminants in Drinking Water, page 359.

bottle bank scheme   Special skips of 6 m³, partitioned into three locked compartments each for brown, green or white glass, with a hole through the lid for the bottle to drop through.

bottom ash   Ash from an ashpit. It is much denser and coarser than *flyash*; it can contain *clinker* and some of it may be coarse and strong enough for roadbuilding.

bottom sediments   Bottom muds in lakes, rivers or the sea tend to concentrate metal salts from the tiny percentages dissolved in the water above them. The reason is that many such substances form insoluble hydroxides or are absorbed by suspended particles that eventually settle on the bottom. For instance, the mud at *Minamata Bay* had 30 to 40 mg/litre of *mercury*, although the water over it had only from 1 to 10 µg/litre. The mud thus had a mercury concentration 3000 to 40 000 times as high as the water.

bound water   Water that is held at or near the surface of a solid.

The hydrophilic surfaces of proteins and polysaccharides have bound water on them, increasing their apparent surface area.

bourne   A stream along a hillside spring line, which extends up the valley as the underground water level rises.

b.p.m.   *See best practicable means.*

brackish water   Water that has a *salinity* about half that of sea-water or less, but is not drinkable by conventional standards. *ASTM 31* defines it as water with dissolved matter in the range 1000 to 30 000 mg/litre. *See brine.*

branch main   A water supply pipe that serves a small area and finishes in a dead end. Branch mains are undesirable because of *stagnation* in the pipe end. *Ring mains* are preferable also because the customers suffer less when there are bursts or repairs.

brass corrosion   *See de-zincification.*

breakbone fever   *Dengue.*

breakpoint chlorination   When *chlorine* is added to water or *sewage* to disinfect them, readily oxidisable substances (e.g. Fe^{2+} or Mn^{2+}) and organic matter react first with the chlorine. Thereafter the chlorine reacts with *ammonia* to form chloramines (monochloramine, NH_2Cl, and dichloramine, $NHCl_2$), which may be further oxidised by more chlorine to trichloramine, NCl_3 (nitrogen trichloride), or to *nitrogen* and its oxides. The breakpoint is reached when these reactions are complete so that continued addition of chlorine then produces *free residual chlorine*—that is, unreacted *hypochlorous acid* (HOCl) or *hypochlorite* ion (OCl⁻). (*See Figure B.5.*) The chlorine in the chloramines is the *combined residual chlorine.* Effective *disinfection* can be achieved with 0.1 mg/litre of free residual chlorine if it is present as HOCl. Hypochlorite ion,

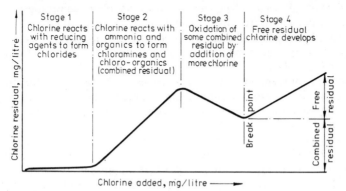

Figure B.5   Breakpoint chlorination

39

OCl⁻, and chloramines are less effective as disinfectants. The ammonia–chlorine reaction needs 10 parts by weight of chlorine to neutralise one part of ammonia expressed as nitrogen. *See chlorination.*

break-pressure tank A way of reducing the pressure in, and thus the cost of, a water *main*. At vertical intervals of 100 m along the main, break-pressure tanks receive water from the main through a ball valve and deliver it to the downstream length of main. Breakages in pipelines at high pressure can be dangerous as well as extremely expensive.

breeching US term for a *flue*.

brewery wastes Although the waste liquors from *fermentation* have a high *BOD* and *suspended solids*, much of the *effluent* from a brewery comes from the washing processes. The BOD_5 of the effluent is about 400 mg/litre for beer brewing and 1000 mg/litre for cider brewing.

brine Often this means sea-water or a solution of common salt, NaCl. Ordinarily, sea-water has about 35 000 mg/litre (3.5%) of dissolved salts, including 20 000 mg/litre of *chlorides*, mainly common salt. Many other brines exist, some with as much as 300 000 mg/litre (30%) dissolved salts. *ASTM 31* defines brine as water with more than 30 000 mg/litre of dissolved matter. *See brackish water.*

briquettes of waste-derived fuel *See pelletiser.*

broad irrigation An old term for *land treatment.*

bromine, Br A *halogen* that, like *iodine* and *chlorine*, is used for *disinfection*. It has advantages over chlorine. Bromamines are more effective disinfectants than the chloramines discussed under *breakpoint chlorination*, but bromine is expensive. It is therefore unlikely to be used for disinfecting municipal water supplies, although it has been used to disinfect swimming pools. Sea-water contains only 50 to 60 mg/litre of bromine.

brown algae, phaeophytes, Phaeophyta Multicellular seaweeds which form large beds in coastal water and include *bladder wrack* and *kelp*.

Brownian movement Slow movement with random direction, seen under the *microscope* in small objects the size of *bacteria*. Some bacteria, in addition to Brownian movement, have their own motion.

brown smoke *Smoke* usually containing tar from coal, which can, like *black smoke*, be prevented by skilled operation of the furnace. Both types of smoke indicate incomplete burning of the fuel, which is easy to prevent. A particularly polluting but luckily rare brown smoke is caused by *nitrogen dioxide*, a poisonous gas occasionally vented from chemical works.

brucellosis, contagious abortion, Gibraltar fever, Malta f., etc. Disease in humans or cattle, caused by the bacterium *Brucella abortus*. It is conveyed to humans in unpasteurised milk from infected cows but may also be a *waterborne disease*.

Brugia malayi A parasitic *nematode* worm that causes *filariasis*.

brush aerator A *surface aerator* with a rotating horizontal shaft having metal blades attached to it. The original type was the *Kessener brush*, but this has been superseded by the *Mammoth rotor* and the *TNO rotor*. Brush aerators are most commonly used in *oxidation ditches*.

bryophyte, Bryophyta A plant *phylum* that includes liverworts and mosses, normally found in slightly polluted water.

BS(S) British Standard (Specification).

BSCP British Standard Code of Practice.

bubble gun A type of *air-lift pump*, used for *de-stratification of lakes and reservoirs*.

Buchner funnel test A rough assessment of the rate of formation of *sludge cake*, made by pouring *sludge* on to a filter paper over an ordinary ceramic Buchner funnel which is evacuated through a filter flask.

bucket latrine A system used in low-income communities, consisting of a bucket for excreta, which is frequently removed and emptied. If improperly maintained, it is smelly and may be dangerous to health.

buffering The property, possessed by some solutions, of resisting any change in *pH* value. Bicarbonate buffer systems protect natural waters from fluctuations in pH and thus help the water life. Buffers are usually a solution of a salt of a weak acid with a strong *alkali* or of a salt of a weak alkali with a strong acid.

builder Active *synthetic detergents* (syndets) compose only 15 to 35% of the commercial syndet. The remainder consists of builders that hold the dirt and include foaming agents (alkanolamides); carboxymethyl cellulose to prevent dirt from re-depositing during laundering; and various phosphates, usually *sodium tripolyphosphate*, or sodium perborate and sodium silicate. The large amount of polyphosphates in domestic detergents is responsible for up to 70% of the phosphorus in domestic *sewage*.

bulk density Weight per unit volume including voids—e.g. of the *medium* in a *trickling filter*, of the sand or anthracite in a *deep-bed filter*, etc.

bulking Inability of an *activated sludge* to settle and thus to separate from the *effluent*. The term covers a wide range of possibilities—e.g. severe growths of *filamentous organisms* such as *Sphaerotilus natans* or *Leucothrix*. The stringy growths may be not the fundamental reason for the bulking. Often the activated sludge loses its *flocculating* characteristics first and only then does the filamentous growth flourish. Bulking has been connected with low *dissolved oxygen* content, low *nutrient* content, high *sludge loading rate*, or complete mixing, but an exception can be found to most generalisations about it. *See flocculation.*

bund A wall built of earth or other material impervious to the fluid to be retained.

bunding of oil tanks Prevention of oil pollution by the building of a *bund* around an oil tank of such a height that the volume contained below the top of the bund is equal to the volume of the tank plus 10%. Then, if the tank bursts, all the oil will be held in. No drainage to domestic *sewers* should be provided from inside the bunded area and any rainwater pumped out must go to a treatment plant for removal of the oil.

buoyancy of stack gas The buoyancy of *flue gas*, in the chimney or afterwards, increases directly with the temperature and with the live steam content, because steam is about half as dense as air at the same temperature. However, if the steam temperature falls below the *dewpoint* and becomes water vapour, the flue gas may become denser and less buoyant than air. *See stack gas re-heating, wet scrubbing.*

burnout *Incinerators* should burn refuse until all or nearly all the carbon has gone. In the UK a maximum of 5% of residual carbon is allowed. If the residual carbon is below this level, it is unlikely that enough *organics* remain to cause smell. Most incinerators therefore have a final burnout zone or an *afterburner*.

buttermaking *See creamery wastes.*

by-wash channel A channel to reduce the amount of silt coming into a *reservoir* from direct *overland flow*. It is a *leat* or stream following a contour, which starts at the head of the reservoir, passes along the full length of the lake and discharges to the *spillway*. A leat may be cut along each of the two hillsides flanking the lake.

C

caddis flies, Trichoptera A *biological indicator of pollution*, found in clean or slightly polluted water. *See Hydropsyche.*

cadmium, Cd Like *lead*, cadmium in drinking water can be derived from pipes made of plastics stabilised with cadmium compounds. It is present also in waste water from cadmium plating. *See osteomalacia*; *see also* Table of Allowable Contaminants in Drinking Water, page 359.

caesium-137, ^{137}Cs An important *decay product* in *nuclear reactor wastes*.

cage bar, grate b., screen b. Thick, parallel steel bars that act as a *screen* in the *grate* under a horizontal-shaft *hammer mill*.

cage disintegrator A *mill* with a rotating cage having horizontal bars to reduce the size of brittle material. There may be one cage or several.

cage rotor A *TNO rotor*.

calcium, Ca A metal ion important in water because calcium salts are the commonest cause of *hardness*. *See* Table of Allowable Contaminants in Drinking Water, page 359.

calcium bicarbonate, $Ca(HCO_3)_2$ *See calcium carbonate, carbonate hardness*.

calcium carbonate, $CaCO_3$ Though regarded as insoluble, calcium carbonate dissolves in water to the extent of about 15 to 25 mg/litre, depending on the pH value. It forms the bulk of chalk and limestone and is the main cause of *hardness*. With *carbon dioxide* in water it dissolves as the bicarbonate:

$$CaCO_3 + H_2O + CO_2 = Ca(HCO_3)_2$$

Calcium carbonate re-precipitates inside water pipes or vessels when water is heated near boiling point.

calcium hydroxide, $Ca(OH)_2$ A chemical commonly used in the *coagulation* of water, in *water softening*, sewage sludge *conditioning, phosphorus* removal and *ammonia* removal from sewage *effluents*. It is usually added as a *slurry*, and raises the *pH* of the water. Commercial grades are known as *lime*—correctly as hydrated or slaked lime.

calcium hypochlorite, $Ca(OCl)_2$, bleaching powder A solid which liberates the *hypochlorite* ion, OCl^-, when dissolved in water. It can be used for *disinfection*.

calcium sulphate, $CaSO_4$ Calcium sulphate is a common cause of *non-carbonate hardness*. It is often dissolved from clays, but in industrial areas comes from the *sulphur dioxide* in the air. *Sulphates* in high concentrations in water are purgative.

Calgon *See sodium hexametaphosphate*.

calorific value, heating v., CV The amount of heat given off by a fuel when it burns completely. There are two values. The larger or gross CV is slightly too high because it includes heat from the steam, given up when the water formed in the burning is condensed. The lower or net CV gives a truer value of the heat that can be used. The net CV of *methane* is 46 000 kilojoules per kilogram; that of dry paper is about 15 000 kJ/kg; that of coal is about 30 000 kJ/kg.

CAMP   The Continuous Air Monitoring Program for measuring air pollutants throughout the USA.

can   A tinplate or aluminium container that can be opened only by tearing the metal. In the UK in 1978, 27 million tinplate cans were used daily, containing 0.5% tin. *See aluminium recovery, de-tinning.*

Candida utilis   Food yeast, a *single-cell protein.*

candle filter   *See filter candle.*

cannery wastes   The volume and quality of food cannery wastes vary with the product. Typical values of liquid canning wastes are: for peas *BOD₅* 400 to 4000 mg/litre, *suspended solids* 250 to 400 mg/litre; for potatoes BOD₅ 200 to 3000 mg/litre, suspended solids 1000 to 1200 mg/litre; for apples BOD₅ 1600 mg/litre, suspended solids 300 mg/litre. Canning is seasonal and this makes treatment design difficult.

capillary fringe   The ground immediately above the *water table* which contains water held by capillary rise. The lower part of the fringe is fully saturated. The fringe rises higher with increasing fineness of the soil and as the water table rises. It is the lowest part of the *zone of aeration.*

capillary suction time, CST   The time in seconds, under standard conditions, for water from a sludge to travel a given distance along a filter paper—a test developed by the UK Water Pollution Research Laboratory, and originally thought to relate to the *filterability* of the sludge. More recent tests have shown that CST and filterability can be related only for a given sludge and CST is not generally related to *specific resistance.*

Capitella capitata   A marine *polychaete* used in bio-assays of tidal mud and estuaries. It is more resistant to oil spills than other polychaetes.

Carbofloc process   A sewage *sludge* treatment process (1966) in which thickened sludge with *lime* added to it reacts with *carbon dioxide* gas to form *calcium carbonate*, that acts as a *chemical conditioner. Flue gas* from *incineration* can provide the carbon dioxide.

carbohydrate   Organic substances such as starch, *cellulose*, sugar that contain hydrogen, carbon and oxygen, with about twice as many atoms of hydrogen as of oxygen. Carbohydrates are widely distributed in plants and animals.

carbon   A non-metal that forms the bulk of most fuels, about 90% of anthracite and 70% of bituminous coals, and, being the basis of *organic compounds*, is essential to life.

carbonaceous BOD   The biochemical oxygen demand exerted in the *BOD* test, coming from the biochemical oxidation of carbon only. In domestic *sewage* 60 to 70% of the total carbonaceous BOD is measured in the 5 days.

carbonaceous oxidation   Oxidation of matter containing carbon, producing *carbon dioxide.*

carbon adsorption bed, c.a. column, c. contactor, activated c. filter   A tank used in *water treatment*, containing from 0.3 to 1 m depth of *granular activated carbon* through which water flows, usually downwards. Downflow contactors remove *trace organics* and other pollutants from drinking water that are difficult to remove in any other way, but they also filter out *suspended solids*, so they need *backwashing*. They can be designed for the spent carbon to be replaced while the water is flowing. Plants of up to 40 000 m³/day have been installed in the USA. They are designed for *surface loadings* of about 90 m³ m^{-2} day^{-1} to obtain greater than 80% reduction of CCE contaminants. *See adsorption.*

carbonate hardness, temporary h.   *Hardness* caused by bicarbonates (HCO_3) of *calcium* or *magnesium* or occasionally strontium or *iron in water.* On boiling, the carbonates are precipitated, producing the characteristic *scale* (fur) in hot water systems, kettles, etc., thus:

calcium bicarbonate, $Ca(HCO_3)_2 = CaCO_3 + H_2O + CO_2$

carbon–chloroform extract, CCE   A chemical procedure in which the *organic compounds* in water are adsorbed on *activated carbon* and then desorbed by the solvent chloroform. CCE indicates very diluted toxic organic compounds caused by low concentrations of *sewage* pollution. *See* Table of Allowable Contaminants in Drinking Water, page 359.

carbon dioxide, CO$_2$ A gas with high solubility in water—well water may contain 29 000 mg/litre of free CO_2—responsible for the *acidity* of many natural waters. CO_2 in water is either free, as carbonic acid (H_2CO_3), or combined with, e.g., *calcium* or *magnesium* as $Ca(HCO_3)_2$ or $Mg(HCO_3)_2$. Free CO_2 can be removed by a cascade which also oxygenates the water. CO_2 is formed by the oxidation of organic matter in fuels, soil and elsewhere. One method of making pure carbon dioxide gas for *stabilisation* of water is to burn propane or other clean hydrocarbon gas under water. Alternatively it can be bought in bulk from breweries or distilleries. *Flocculation* of water containing *carbonate hardness*, with aluminium or *ferrous salts* or ferric salts, releases CO_2 and some of this may remain for stabilisation, thus:

$$Al_2(SO_4)_3 + 3Ca(HCO_3)_2 = 3CaSO_4 + 2Al(OH)_3 + 6CO_2$$

The carbon dioxide content of the air (0.033% or 330 p.p.m.) has been steadily rising from some 318 p.p.m. in 1958 to 334 p.p.m. in 1976 (maximum summer values). Some scientists have stated that if the CO_2 content of the atmosphere continues to rise, a *greenhouse effect* will progressively warm the Earth's surface. The recent serious de-forestation of the Earth could raise the CO_2 content of the atmosphere and so could the increased consumption of fuels.

carbon filter   A *carbon adsorption bed.*

carbonic acid   The *acidity* formed by *carbon dioxide* dissolving in water.

carbonising   (1) The heating of coal, wood, etc., in the absence of air to produce coke, charcoal, etc. (2) In wool cleaning, the use of hot concentrated acids to char the vegetable particles in the wool and thus to loosen them.

carbon monoxide, CO   A poisonous, invisible gas that cannot be detected with certainty without instruments, because it has neither taste nor smell. It is formed when the *carbon* in fuel burns with too little air, in either furnaces or internal combustion engines. Internal combustion engines are responsible for 80 to 90% of the CO in the air but the *background level* has not been increasing, probably because microbes oxidise it to carbon dioxide, CO_2. Carbon monoxide is explosive in air and burns to form CO_2. The background level of CO in air is below 1 p.p.m. but in busy streets it reaches 15 to 20 p.p.m., occasionally exceeding 100 p.p.m. In the Blackwall Tunnel under the Thames it reaches 500 p.p.m. but in the centre of Hyde Park, London, only 1 p.p.m.

carbon:nitrogen:phosphorus ratio, CNP ratio   The average ratio of C:N:P in living matter is 106:10:1. For *biological treatment* the ratio is normally expressed as BOD_5:N:P and the minimum N and P contents should give a ratio of 100:5:1. *Nitrogen* deficiency can be corrected by adding an ammonium salt or *nitrates. Phosphorus* deficiency is corrected by adding phosphates.

carboscope   An optical instrument for estimating *dark smoke*, slightly different from the *telesmoke.*

carboxyhaemoglobin, COHb   A compound formed in the blood when its haemoglobin reacts with carbon monoxide in the air. COHb reduces the ability of the blood to carry oxygen.

Carchesium   A stalked *ciliate protozoon* often present in *trickling filters* or *activated sludge*, especially where *nitrification* is well established and a clean *effluent* is produced. It is also found in *sewage fungus.*

carcinogenic agent, carcinogen   Any cause of cancer. Carcinogens

46

are believed to include many *organic compounds* (e.g. polycyclic compounds from soot, smoke and automobile exhaust gases), *organo-chlorine compounds* and many naturally occurring organic or inorganic substances.

carpet manufacturing wastes Liquid wastes are discharged from the wool cleaning and synthetic textile industries as well as from mothproofing, mildew-proofing and laytexing. Some carpet makers still use *Dieldrin* for mothproofing, but many are using a *biodegradable* substitute. *See synthetic textile wastes, woollen industry wastes.*

carriage water The water in *sewage* (99.9% of the sewage) which carries it along the *sewers* to the treatment plant. It consists of the waste water from industry and dwellings plus some *infiltration*. In temperate climates it is often warmer than the outside air because of the hot water from houses.

carrier A person, animal, fish, bird, etc., that can harbour the *bacteria* or other cause of a disease for months or years in his body without harm.

Carrousel system An *extended aeration* activated sludge treatment for *sewage* from which only coarse solids and grit have been removed. It resembles an *oxidation ditch* but is larger and deeper. The *mixed liquor* is aerated and circulated by *cone aerators* at the ends of the channel.

carry-over In a boiler, the lifting of water with steam; in a *mechanical sorting plant*, the removal of unwanted material

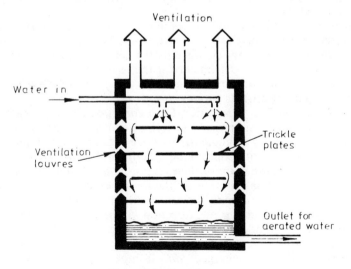

Figure C.1 Cascade aerator

47

together with the material being separated; in a furnace, the removal of ash by the *flue gases*.

car shredder A *fragmentiser*.

cascade aerator, free-fall a. An *aerator* of which two varieties exist, free-fall and bed types. The bed type is built like a *trickling filter* with air circulating up through the stone packing. In the free-fall type the water cascades over open aerated steps (*see Figure C.1*). As much as 90% of the *carbon dioxide* can be removed by a bed aerator.

casing, lining See *well casing*.

cast-iron pipes These old-established pipes are now of spun iron when in straight lengths, but bends, flanged pipes, hydrants and other special fittings are cast in vertical sand moulds. See *spun-iron pipe*.

catabolic metabolism, catabolism Transformation of food that provides living beings with the energy they need for building up their bodies, leaves, branches, etc. *Compare anabolic metabolism*.

catalyst A substance that speeds up a chemical reaction without itself being consumed. Platinum does this for the burning of fuel gas. *Enzymes* do it for biochemical reactions.

catchment, c. area, drainage area, gathering ground (1) The area of land that collects the water flowing to a given length of stream or to a *reservoir* or lake. (2) In the USA this land is called a *watershed* or drainage basin.

catchment yield The *runoff* from a *catchment*. See also *yield of a water source*.

catchwater A tunnel, pipe or *leat* dug or laid to bring water into a *reservoir* from a neighbouring *catchment*. The natural catchment of the reservoir is thus enlarged by the neighbouring area, which is called an indirect catchment for the reservoir.

cathode, kathode (adjective **cathodic**) The negative *electrode* in *electrolysis*. It attracts *cations*. See *anode*.

cathodic protection Protection from *corrosion* of the outside of underground or submerged metal structures such as pipes, by making the metal cathodic to an *anode* outside it, which may be consumable or not. Consumable (*sacrificial*) *anodes* are of zinc, magnesium, aluminium or their alloys. Non-consumable anodes are of graphite or platinum. In the impressed current system the pipe is made the *cathode* in a direct current circuit with a higher value than, and an opposite direction to, the estimated corrosion voltage. The anodes may need to be numerous so as to provide enough protection (BSCP 1021).

cation A positively charged *ion*. It moves to the *cathode* in *electrolysis*. Metals, ammonium and hydrogen all yield cations.

Compare anion.

cation exchange, base e. Replacement of one *cation* by another—e.g. *zeolites* replace *calcium* and *magnesium* by sodium in *water softening* by *ion exchange*.

cationic detergent *See synthetic detergents*.

caustic alkalinity *Alkalinity* caused by a strongly alkaline chemical—e.g. sodium hydroxide (NaOH) or *calcium hydroxide* ($Ca(OH)_2$). In water supply it is often defined as the alkalinity above *pH* 8.2, the so-called phenolphthalein alkalinity, because phenolphthalein changes from colourless at lower pH values to red at pH above 8.2. In natural waters any alkalinity above 8.2 is usually caused by carbonates or hydroxides.

caustic scrubbing The use of caustic soda, sodium hydroxide, NaOH, as the wash solution in a *wet scrubber* to remove *sulphur dioxide*.

cavitation Formation and rapid collapse of vapour or gas bubbles in low-pressure areas in a pumping system or near a ship's propeller, etc. It may occur where high velocity reduces the pressure, as at the tip of the *impeller* of a *centrifugal pump*. It often causes *corrosion*.

CCE *Carbon–chloroform extract*.

celerity The overall speed of a wave.

cell (1) In *controlled tipping*, the amount of refuse deposited in one day and enclosed all round with *daily cover*. (2) A minute unit in the structure of living matter, consisting of *protoplasm* surrounded by a membrane or wall.

cell age *See mean cell residence time*.

cellulose, $(C_6H_{10}O_5)_x$ The main constituent of the *organic* part of municipal refuse, including the bulk of paper, cardboard, string, straw, cotton, linen, silk, wood or flax, all of which decompose more slowly than *garbage*. At about 60% in the USA and 40% in the UK, cellulose is the bulk of municipal refuse. Some 7 million tonnes a year are thrown away in Britain and several schemes exist for utilising it. *Incineration* is practicable and on the Continent of Europe it is also profitable. *Fibre recovery* is practicable if there is a market for the product. Pulping in water segregates the cellulose fibres from the *plastics*, glass fragments, etc. Only half the fibres can be recovered. The remainder, even as a sludge of 40% water, can be burnt by *fluidised bed combustion*. *See acid hydrolysis*.

cellulose acetate A strong 'artificial silk', insensitive to moisture, and therefore used as a *membrane* for *reverse osmosis* and other *membrane processes*. One particular membrane is a film 0.2 μm thick of the dense *polymer*, resting on a coarser-pored base layer 100 μm thick of the same material. Production of cellulose acetate involves *alkaline* and acidic wash waters. *See acidity.*

cenosphere   A tiny hollow sphere of ash, usual in *flyash.*

Centralsug system   A Scandinavian method of refuse collection by suction, probably the earliest form of refuse transport by pipeline. This *pneumatic transport of solid waste* was originally designed for and used at hospitals because of its hygienic appeal. Refuse is dropped into the refuse chutes and remains there until the turbo-exhausters start up, several times daily.

centrate   Liquid removed by a *centrifuge*—e.g. the water from a *sludge.* The worst difficulty in the centrifuging of sludges is the disposal of the centrate, with its high concentration of suspended, non-settling solids. If the particle size is increased by *conditioning* the sludge beforehand with *ferric chloride* or *lime* or *polyelectrolyte*, some improvement may be expected. The centrate from a *raw sewage* sludge may have a *BOD$_5$* around 2000 mg/litre.

centre-pivot irrigation   The world's first successful automatic *irrigation* machine, likely to increase the demand for water considerably in warm dry regions. It can irrigate only circular fields, spraying water from a long perforated pipe, both pivoting and fed at the centre of the field and supported on wheeled carriages (towers) at suitable intervals. In the USA the largest units carry 800 m of pipe of 200 mm diameter on 20 towers. Towers are driven separately by an electric or hydraulic motor at speeds proportional to their distance from the centre, so as to keep the pipe straight and radial. Operation is continuous for about 2 months in spring and summer. The minimum time for a single rotation round a field is 12 h.

centrifloc   A circular, one-pass, *flocculator-clarifier.*

centrifugal   Description of the effect that tends to force whirling objects away from their centre of rotation.

centrifugal dust collector   A *centrifugal scrubber* or *cyclone* or *fan collector*, etc.

centrifugal pump, radial-flow p.   A rotodynamic pump in which a bladed *impeller* drives the water outwards so that it leaves the casing radially. The water enters the pump at the eye of the impeller. Pump casings may be either *volute* or *diffuser* types. Volute pumps are less efficient but can pump raw sewage or sludge. Diffuser pumps do not pass gross solids but are often used in multi-stage borehole or *turbine pumps*, in which the water from one stage flows into the eye of the next impeller, increasing the available head. Axial-flow pumps are not centrifugal. Mixed-flow pumps are partly centrifugal.

centrifugal scrubber   A *wet scrubber* in which the gas swirls because it has been injected tangentially or has been directed around by stationary or moving swirl vanes. Centrifugal scrubbers are best for collecting particles larger than 5 μm and have a gas pressure drop between 5 and 20 cm water gauge. The cyclonic spray scrubber is one type, with a spray either from the centre of a *cyclone* or from the walls, aimed inwards.

centrifuge A cylindrical vessel that by spinning separates a mixture of fluid and solid (or two fluids) within it, throwing the denser to the outside of the vessel. It is used both in the laboratory and for de-watering *sludges*, particularly in the USA. Modern centrifuges for *de-watering of sludge* are of two types, with either bowl conveyor or concurrent flow. In the first type (*see Figure C.2*) the bowl and the conveyor inside it are driven at different speeds. The helical conveyor pushes the sludge to one end. The *centrate* discharge is at the other end and the level of the centrate weir governs the volume of sludge in the centrifuge.

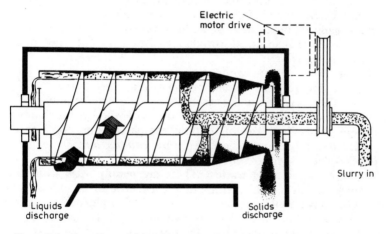

Figure C.2 Centrifuge with bowl conveyor (Pennwalt Ltd., Camberley, Surrey)

With the other type, using concurrent flow, the sludge and the centrate move in the same direction, the flow of wet sludge being adjusted by a skimmer device. The sludge discharged should have less than 80% water but the centrate may be high in fine suspended solids. See *dense-medium cyclone, disc centrifuge.*

CEQ The US *Council on Environmental Quality.*

cercaria The final, larval stage of flukes. *See larva.*

cesspit, cesspool A tank or pond that receives domestic waterborne waste without treatment, at sites where there is no *sewer* to remove it. The liquid overflows, and the *sludge* at the bottom is emptied every 6 to 12 months. Cesspits are required to be watertight in the UK but not in drier climates. In the USA the walls and floor have to be porous and may be built of stones laid dry without mortar so that the liquid can soak away, as in an *absorption pit*. *Septic tanks* have tended to supersede cesspits, which are often smelly. They should be located downstream of water supplies and should not be used for *sullage* disposal, since this less polluting liquid can be disposed of elsewhere. The UK Building Regulations require cesspits to be at least 18 m³ in volume.

Cestoda A sub-class of the *phylum* of Platyhelminthes, *flatworms*.

chain scraper A chain, rather than blades, that scrapes the sludge along the bottom of a *radial-flow sedimentation tank*.

chalk test, marble t. A crude measure of the undersaturation of *calcium carbonate* in water. The sample of water is mixed with chips of chalk or marble and the loss in weight of the chips is measured, indicating the amount dissolved. From this the *pH* may be calculated. *See Langelier saturation index.*

chalybeate A description of water, especially a mineral water, in which there is so much iron dissolved that it can be tasted. *See iron in water.*

channel flow *Open-channel flow.*

channel storage The volume of water in a watercourse.

char The black, usually crumbly, solid that remains of a substance after *pyrolysis*. The pyrolysis of wood yields charcoal.

charge box That part of a static compactor that contains the refuse during compression. *See static compaction.*

check valve A *non-return valve.*

cheese manufacturing wastes These are mainly *wash water* and whey, which is acidic, with a high concentration of milk sugars. Whey has a BOD_5 of about 30 000 to 40 000 mg/litre but this is diluted by the wash water to give an overall typical BOD_5 for cheese-making *effluent* of 2000 mg/litre.

chelating *Complexing* of a compound with a central atom (typically a metal ion) at two or more points, forming a ring structure—e.g. haemoglobin or *EDTA*.

chemical clarification The use of chemicals in *raw water* for *coagulation* and *sedimentation* of suspended material.

chemical conditioner A substance used in *chemical conditioning*.

chemical conditioning Mixing of a chemical with *sludge* to increase its *de-waterability* before *mechanical de-watering of sludge*. Typical conditioners used with sewage sludge, expressed as a percentage of dry solids, are as follows:

0.2 to 0.8% of *aluminium chlorohydrate*, $Al_2(OH)_4Cl_2$
20 to 50% *lime*, $Ca(OH)_2$

52

10 to 40% *copperas*, $FeSO_4.7H_2O$
1 to 7% liquid *polyelectrolyte* or
0.2 to 0.7% solid polyelectrolyte

Different sludges or mechanical de-watering techniques require different combinations of chemical. Ferric chloride is usually cohsidered too expensive in the UK. *Chlorinated copperas* and aluminium chloride have been used.

chemical fixation Converting a substance chemically into one that is insoluble, one way of *toxic waste disposal*. For example, the salts of aluminium, *cadmium, chromium, nickel*, tin and *zinc* can be precipitated as insoluble hydroxides, or they may be cast into solid concrete, which also is insoluble.

chemically precipitated sludge *Sludge* that is produced in a waterworks or *sewage* works when *coagulation* has been used to help the *sedimentation* of suspended matter.

chemical oxygen demand *See COD.*

chemical sludge *Chemically precipitated sludge.*

chemical toilet The simplest chemical toilet is a bucket containing a solution of *bactericide*. More elaborate systems may include flushing to a tank and re-cycling of the chemicals after settling. Portable chemical toilets as used in caravans, etc., should be designed to BS 2081, with disinfectant fluid, usually formaldehyde-based, to BS 2893.

chemical tracer *See tracer.*

chemical treatment Various water and waste water treatments involve the addition of chemicals—e.g. for the *coagulation* of *raw water*, occasionally of *raw sewage*; *pH* adjustment; *precipitation* of *heavy metals*, often as hydroxides; precipitation of phosphates as calcium phosphate; *disinfection* by *chlorine* or *ozone*; *conditioning* of sewage *sludge*; and *water softening*.

chemisorption *See adsorption.*

chemoautotroph A *chemotroph* that uses *carbon dioxide* as its carbon source—e.g. *nitrifying bacteria*.

chemoheterotroph A *chemotroph* that uses organic material as its carbon source—e.g. *Sphaerotilus natans*. Chemoheterotrophs include all animals and *fungi*.

chemosynthesis (adjective **chemosynthetic**) A method by which micro-organisms, including *chemotrophs*, obtain energy. They use the chemical energy of inorganic or *organic* materials (but not *photosynthesis*) to build up their organic matter.

chemotaxis *See taxis.*

chemotroph An organism that uses *chemosynthesis*, not *photosynthesis*, to build itself up, using the energy from chemical processes. Chemotrophs may be either *chemoautotrophs*, using *carbon dioxide* as a source of carbon, or *chemoheterotrophs*, using organic carbon.

CHESS Community Health and Environmental Surveillance System of the *EPA*. It includes the study of the long-term effects of human exposure to air pollutants.

Chilodonella A *ciliate protozoon* commonly found in *trickling filters* and in *activated sludge* plants that have a satisfactory *sludge*.

chimney (1) A pipe, usually vertical, of masonry, metal, fibre glass, etc., that discharges *flue gases, smoke* or other *air pollutants* at a level high enough for them not to offend people in surrounding buildings or on the ground below. (2) US term for a *back drop*.

chimney effect Hot air rises up a chimney because it is warmer and therefore lighter (more buoyant) than the air outside. This *natural draught* can occur away from chimneys—e.g. in lift shafts or stair wells, which in large buildings are subject to special fire regulations.

chipper A *shredding* machine used in the USA to reduce tree stumps to chips about 5 cm long and 1 cm diameter. The chips can be used as a mulch in gardens or for covering paths or for making paper or fibreboard.

chironomid, Chironomidae Non-biting midges whose *larvae* are *blood worms*. The larvae of *Chironomus riparius* live in sewage-polluted mud and can be a *biological indicator of pollution*.

Chlamydomonas A flagellate green *alga* usually found in *eutrophic* water, including *waste stabilisation ponds*, tinting it green. *See flagellum*.

chloramines Several compounds formed in *breakpoint chlorination* that make up the *combined residual chlorine* in a water.

Chlorella A unicellular *green alga* which is often the dominant alga in *waste stabilisation ponds*, since it tolerates temperature variations and can exist for limited periods in *anaerobic* conditions. *Chlorella* can grow well in the dark using *chemosynthesis*.

chlorides Compounds of *chlorine*—e.g. sodium chloride ($NaCl$, common salt). Chlorides are not removed from water or *sewage* by any treatment plant unless it includes *de-salination*. Since chloride in the form of $NaCl$ is eaten by humans and excreted in the urine to the extent of about 6 g per adult daily, the chloride content of water increases with each re-use, by about 50 mg/litre. Chloride can be tasted at 500 mg/litre, sometimes at half this level by those with a sensitive palate, but it is said not to harm healthy people even at 1500 mg/litre. In Britain rivers have

20 to 80 mg/litre but some African rivers have only 1 to 5 mg/litre. In natural waters chlorides can occur from the leaching of chloride-containing rocks. High chloride content may indicate also that fresh water is being polluted by *saline encroachment*.

Some water supplies may have 1000 mg of chlorides per litre without complaints arising, but *see* Table of Allowable Contaminants in Drinking Water, page 359; *See also brine*.

chlorinated copperas A mixture of *ferric chloride* and ferric sulphate, written as $FeClSO_4$, used as a *coagulant* for *raw water* and for *conditioning* sludge. It is usually made on site by oxidising ferrous sulphate (*copperas*) solution with the feed from a *chlorinator*. A disadvantage of its use in drinking water is the possibility of residual *iron in water*.

chlorinated hydrocarbons *Organic compounds* consisting only of *chlorine*, hydrogen and *carbon*—e.g. *DDT, polychlorinated biphenyls*. *See organo-chlorine compounds*.

chlorination Addition of *chlorine* to water, usually for *disinfection*. For drinking water enough is usually provided to leave 0.1 to 0.2 mg/litre of *free residual chlorine* in the supply as it leaves the works. A *solution* of chlorine in water is injected. This is made by adding to water the gas or one of its compounds—e.g. sodium or *calcium hypochlorite*. Chlorine is a strong oxidising agent. Although regularly used in waterworks, it has been little used for purifying final *sewage* effluents in the UK, because it can retard *self-purification* in the *receiving water*. Also, it converts phenols into chlorinated phenols, which have an objectionable taste even in minute concentrations, and non-toxic thiocyanate is converted by chlorine to highly poisonous cyanogen chloride. Chlorination is the last purification process in a waterworks, because *bacteria* and *viruses* can be shielded from chlorine by the tiny particles in a water that make it *turbid*. In clear water 0.2 mg/litre of free residual chlorine will kill 99.9% of bacteria in 30 min when the *pH* is less than 8.0, but some viruses may be more resistant. Taste troubles can occur with as little as 0.05 mg/litre of free chlorine, but most waters can take 0.2 mg/l without difficulty. In an emergency 2 mg/litre is allowed to remain in a drinking water. Chlorination is accepted practice in US sewage treatment, for reduction of smell in pumping stations, for grease removal before *pre-aeration*, for oxidation of *ferrous salts* to ferric (mainly the sulphate), for prevention of *ponding* at *trickling filters* and for disinfection of *stormwater* or final *effluent* or *raw sewage* discharges. Dosage rates for disinfection are 6 to 25 mg/litre for raw sewage, 5 to 20 mg/litre for *settled sewage* and 2 to 10 mg/litre for an effluent after *biological oxidation*. High dosage rates are required partly for the solids content of the water and partly because of the reactions with *ammonia*. *See also breakpoint chlorination, chloramines, combined residual chlorine, dechlorination, hypochlorination, pathogen, superchlorination*.

chlorinator Equipment used at waterworks for dissolving *chlorine* in water and adding it in the right proportion to the water supply.

chlorine, Cl_2 The commonest *halogen*, a poisonous gas used in *chlorination* because it is cheap and easily available.

chlorine demand The amount of *chlorine* consumed by a certain volume of water in a certain time.

chlorine dioxide, ClO_2 A poisonous green gas that is an even stronger oxidising agent than *chlorine*. It is used for removing taste from drinking water, especially that of *chlorophenol*. Chlorine dioxide has further advantages over chlorine, being less affected by high *pH*, and it does not react with *ammonia*, but it is much more expensive.

chlorine residual, available c. Chlorine that exists in water either as *hypochlorite* (free residual chlorine) or in combination with other substances—in particular, *ammonia*—as *combined residual chlorine*. *See breakpoint chlorination*.

Chlorobiaceae (also **Chlorobacteriaceae**) *Photosynthetic*, green *sulphur-oxidising bacteria* that use *carbon dioxide* as their carbon source. They may oxidise sulphur or sulphides in *anaerobic* conditions in sunlight but they do not release *oxygen*. They proliferate in stagnant pools and waters containing *hydrogen sulphide*. Water containing them looks green.

chlorofluorocarbons, $CFCl_3$ and $CFCl_2$ *Aerosol propellants* that may be used also as refrigerants.

chloro-organic compounds *Organo-chlorine compounds*.

chlorophenol, chlorphenol Any compound of *phenol* and *chlorine*. Chlorophenols are often toxic but have a medicinal taste, even at one part in 500 million (2 μg/litre).

Chlorophenothane Another name for *DDT*.

Chlorophyta The *green algae*.

chocolate mousse Water-in-oil *emulsion* naturally emulsified at sea from viscous oil. It has up to 70 or 80% water and is very stable, less *biodegradable* than oil, not easily removed by *emulsifiers* and impossible to burn, even with a flame-thrower. On the beach it forms tarry lumps as solids are picked up and water slowly evaporates.

cholera A serious intestinal disease that can kill in a few hours, caused by the bacterium *Vibrio cholerae*. As with other *waterborne diseases*, healthy *carriers* of the disease may number as much as 9% of the population in warm climates. It is usually transmitted by *faecal* contamination of water, possibly of food.

Chromatiaceae, Thiorhodaceae *Photosynthetic*, purple, *sulphur-oxidising bacteria* that obtain their carbon from carbon dioxide, although they do not release *oxygen*. They may oxidise sulphur or sulphides in *anaerobic* conditions in sunlight. They occur in similar conditions to *Chlorobiaceae* and may give a strong red colour to water.

chromatography Any one of several ways of analysing or separating mixtures of liquids, gases or solutions by selective adsorption on to, e.g., activated carbon, alumina, silica gel, chalk, etc. *See gas–liquid chromatography.*

chrome tanning *See tannery wastes.*

chromium *See hexa-chrome; see also* the Table of Allowable Contaminants in Drinking Water, page 359.

chronic Long-lasting.

chronic toxicity A danger from long-lasting but low levels of pollutants that may not kill but nevertheless may cause lasting harm. *Compare acute toxicity.*

Chrysophyta A *phylum* of yellow-green *algae* of many diverse forms that includes *diatoms*.

chute-fed incinerator An *incinerator* in a US apartment block, provided with both a *flue* and a separate chute, so called to distinguish it from the *flue-fed incinerator.*

cilia (singular **cilium**) Minute hairs on micro-organisms that enable them to move gently.

ciliate protozoa, Ciliaphora *Protozoa* with hairs (*cilia*) that help them to move. They are common in *trickling filters* or 'healthy' *activated sludges*. Unsatisfactorily activated sludges contain few ciliates. They may swim freely in the liquid or they may be stalked (peritrichs), with a stalk attached to solid matter in the liquid. Peritrichs often form colonies.

circular sedimentation tank A *radial-flow sedimentation tank* or *upward-flow sedimentation tank.*

cistern (1) A water tank that may be open but should be covered to keep its contents clean. Usually installed in the roof space or attic of a house in Britain, it is unusual in the USA or the European Continent. In the UK all the water for baths, WCs and hot water systems must come from the cistern. Cisterns reduce the problem of *back siphonage* and give the consumer a reserve supply of non-drinking water, although at greater cost and complication of household fittings than the direct mains supply used on the Continent. Typical domestic storage cisterns contain 220 litres of water. (2) A *flushing cistern.*

clack valve A *non-return valve.*

Cladophora, blanket weed Stringy green *algae* commonly found in *nutrient*-rich waters. They mat into a green blanket in the water.

clarification Removal of tiny suspended solids from a water or *sewage*, in small quantities, often as little as 100 p.p.m., frequently by *coagulation*. *Sedimentation* is clarification of dirtier water, though still with relatively small quantities of dirt. *See* below.

clarifier In the UK, usually a tank for settling solids out of water or *sewage* after chemical *coagulation*. In the USA it is often synonymous with *sedimentation tank*. In *water treatment* it is common to settle a coagulated water before *sand filtration*. Clarifiers may be divided into simple *flocculator-clarifiers* and *solids contact clarifiers*.

clarify Of water, to remove *turbidity* (therefore suspended matter) and to make transparent.

class A division in *taxonomy* between phylum and order.

classification (1) Of living creatures, *taxonomy*. (2) Of a bulk material, its division into different particle size grades with a *screen, trommel* or other *classifier*.

classifier Any device used in *mineral dressing*, or the cleaning of gravel, grit, sand, etc., that separates the material into grains of two or more sizes. The simplest is a dry *screen* (sieve), but many are wet—e.g. *screw classifiers, wet cyclones*. In *sewage treatment* a classifier may be a device that separates *organic* solids from grit. A common method used in conjunction with a Detritor is a *reciprocating-arm grit washer*. In the mechanical sorting of refuse, *trommels* are very often used for primary classification, with holes of about 20 cm. *See also air classification*.

Claus process Extraction of the element sulphur from the gas *hydrogen sulphide* (H_2S) by passing it mixed with air at about 150°C over an iron or bauxite *catalyst*. The exhaust gas (tail gas) consisted of SO_2, H_2S, S, COS (carbonyl sulphide) and CS_2 (carbon disulphide), corresponding to some 15 000 p.p.m. of SO_2, but it has recently been reduced to about 250 p.p.m., with an increase in the recovery of sulphur.

Clean Air Acts 1956, 1968 The main UK laws governing air pollution, the declaration of smokeless zones, adequate heights for industrial chimneys, etc., preceded by the Alkali Act 1906 and the Public Health Act 1936, and followed by the *Control of Pollution Act 1974*. Before any prosecution can be considered, the offender must be warned in writing not more than 2 days after the offence. *Smoke* emissions by industry diminished by 96% between 1956 and 1973. Ground level concentrations of *sulphur dioxide* in towns diminished also, even though total emissions both there and in the country may have risen.

Clean Air Act 1970, Public Law 91-604   This US clean air act has rather more teeth than the British ones. Repeating offenders may incur a fine of up to $50 000 per day. The Act requires new generating stations either to burn low-sulphur coal (less than 0.7% S) or to install *flue-gas desulphurisation* units. The *particulate emissions* allowed under the Act are 1.24 kg per megawatt-hour for oil and 1.86 kg/MWh for coal. For a coal with 3% sulphur (high-sulphur coal) flue-gas desulphurisation can add 15% to the cost of the plant.

Clean Air Regulations 1971   For solid fuel furnaces these UK regulations under the Clean Air Acts can be summarised as follows: allowable emissions of dust and grit vary uniformly from about 0.5 kg/h for a furnace of 293 kW continuous rating to 114 kg/h for one of 169 000 kW. Calculated per megawatt-hour (1000 kWh), this is a variation from 1.7 kg/MWh at the lower end of the furnace range to 0.7 kg/MWh at the upper end. Emissions allowed for oil-fired furnaces are much smaller, about a quarter of those for solid fuel, but are still larger than what is allowed on the Continent.

clear water storage   Water storage in a *service reservoir*.

cleated conveyor   A rubber belt conveyor to which upstanding cross-strips (cleats) of metal or rubber are riveted or welded, enabling it to carry material up steep slopes. An ordinary smooth rubber belt cannot carry material up a slope steeper than 18° to the horizontal. A cleated belt can occasionally raise material up a slope as steep as 60° to the horizontal, but some material will be spilt at each end of the belt.

cleated-wheel tractor   A *compactor* (1) for use in *controlled tipping*.

climax community   The final stable community that remains after many developments in its *ecology*.

clingage   Oil that clings to the sides, top and bottom of the bunkers and oil tanks of ships, and is washed out with *oil tanker washings*.

clinker   Stony, usually dense and sharp-edged ash that has partly melted, unlike *slag*. 'Clinkering' may mean either the formation of clinker in the furnace or its removal. After cooling and sizing, clinker can be sold as builder's fill or for roadmaking or as *cover* for *controlled tipping*.

clinoptilolite   A natural *zeolite* used at *Lake Tahoe*, USA, for *ion exchange* of *ammonia* as a cold-weather alternative to the *ammonia stripping* tower. The zeolite adsorbs ammonia and can be regenerated by passing a saturated solution of *lime* plus sodium chloride over it. If clinoptilolite is used as the main method of ammonia removal, the ammonia can be extracted from the *regeneration* liquor in a small stripping tower. Traces of ammonia in the *effluent* from the clinoptilolite may have to be removed by *breakpoint chlorination*.

closed-recirculation system An industrial cooling system like that of the car engine, which should need no water once the system has been filled and consequently wastes less water than either the *open-recirculating system* or *once-through cooling*. It also cannot cause *thermal pollution* of *receiving water*.

closed water circuit A chemical plant, coal washery, etc., with *zero-discharge layout, closed-recirculation system*, etc.

Clostridium A genus of anaerobic *bacteria*, present in *anaerobic sludge digestion*, some of which can cause such diseases as botulism and tetanus. *See* below.

Clostridium perfringens, C. welchii An anaerobic *bacterium* of *faecal* origin that has *endospores* that survive even 5 h boiling and outlast *Escherichia coli* in water. In the dry state the *spores* last indefinitely. Their presence in water almost certainly indicates faecal pollution at some time or place. *See coliform group*.

cloth binding The clogging by fine material of a *screen* or *filter* cloth.

cloth filter A *fabric filter*.

CNP ratio, C : N : P ratio The *carbon : nitrogen : phosphorus ratio*.

coagulant A substance that helps *coagulation*; often *lime, alum, copperas* or *alumino-ferric*.

coagulant aid Many substances, such as activated silica, stone dust, *polyelectrolytes*, clays, are added as suspensions or *solutions* in water to help *coagulation*. Since coagulation of drinking water is always followed by *filtration*, the distinction between these and filter aids is not sharp. Aids are a small fraction of the main coagulant. The US Food and Drugs Administration publishes lists of such substances that are approved for use in drinking water. In the UK similar lists come from the Committee on New Chemicals and Materials of Construction for Use in Public Water Supply and Swimming Pools, of the Environmental Protection Directorate, Department of the Environment.

coagulation The addition of a chemical to water or *sewage* so as to precipitate, usually, a metal hydroxide that catches and so removes from the water most of the tiny suspended particles. The chemical lessens the *surface charge* of the suspended matter, unlike *flocculation*. Many *colloidal* substances, *bacteria* and *algae*, have a net negative charge. Therefore the *coagulant* is often a metal *ion* of the form Al^{3+} or Fe^{3+}, although it is the *hydrolysis* products of these metals that act as coagulants, not the free metal ions—e.g. the ion: $[Al(H_2O)_6]^{3+}$. Coagulation of *raw water* before flocculation and settling is a common *water treatment* process, whereas coagulation of sewage is relatively uncommon. Any coagulant added must help to settle the solids and 'form a readily settleable *floc* itself, so that subsequent *sand filters* will work properly. The most common coagulant for raw water is *alum*, which precipitates as aluminium hydroxide under alkaline conditions. *Alkalinity* is normally maintained by adding *calcium hydroxide. Ferrous salts* and ferric salts are less common for coagulating raw water, because of the hazard of leaving residual *iron in water. Polyelectrolytes* also may be added in small quantities.

coal carbonisation waste   *Coke plant wastes.*

coal industry waste water   Two main sources of water pollution are *acid mine drainage* and coal washery wastes. *Deep-cone thickeners* and other devices developed by the UK National Coal Board have greatly reduced the bulk of the washery wastes and enabled washery water to be re-circulated. In new NCB washeries all the dirty water is clarified and re-circulated in the plant, except for the small proportion that leaves in the form of moisture on the coal or on the reject to the tip. Black water drained from colliery tips has been greatly reduced by grassing the tips and by maintaining gentle surface slopes not steeper than 1 in 6 or so to the horizontal.

coarse-bubble aeration   *Aeration* through pipes with large holes or an open end, producing large bubbles, generally above 4 mm diameter. The pipe is nearer the surface of the water than with *fine-bubble aeration* and a low-pressure blower can be used. Even in de-oxygenated water, only about 5% of the oxygen supplied is dissolved. This means that a supply of 100 m^3 of air at 760 mmHg dissolves only about 1.5 kg of *oxygen* in de-oxygenated water and even less in *mixed liquor*—about 0.8 kg. *See Inka process, oxygenation efficiency.*

coarse filter   A *roughing filter.*

coarse screen   In *sewage* works, a *screen*, with spaces at least 30 mm wide, that protects *centrifugal pumps* from excessively large solids. At *raw water* intakes the spaces are similar. Coarse screens are often hand cleaned.

coarse-to-fine filtration   *Filtration* of water through a sand *filter* or *multi-media filter* bed in which the water approaches the coarse grains first. Coarse solids are thus filtered early from the water by the coarse sand, while further into the filter the finer solids are filtered out by the finer sand. Most of the bed is used efficiently, instead of only the upper part. The same end is achieved by *upward-flow sand filters* or *dual-media filters*.

coastal waters According to the UK Public Health Act 1936, this means the area up to 3 miles from low-water mark of spring tides—derived from the ancient idea of the length of a cannon shot.

coccus A spherical *bacterium*.

COD, chemical oxygen demand, dichromate value The COD of a natural or waste water is measured by a test involving a strong oxidising agent. A boiling mixture of concentrated sulphuric acid and potassium dichromate ($K_2Cr_2O_7$) is the standard reagent, together with a *catalyst*, silver sulphate. Mercuric sulphate is also added to combine with *chlorides* which otherwise would precipitate the silver catalyst as silver chloride. The test can be completed in 2 h—much more convenient than the 5 days of the *BOD* test. The COD value is normally higher than the BOD value because more organics can be oxidised chemically than are *biodegradable* in the BOD test. Typically, for domestic sewage the ratio COD : BOD_5 is from 1.5 to 2. A high COD : BOD_5 ratio for an industrial waste may indicate the presence either of organics that are hard to biodegrade or of toxic material inhibiting the BOD results. Acidic *potassium permanganate* ($KMnO_4$) can also be used as the oxidising agent but this gives a much reduced value. *See permanganate value.*

coefficient of haze, COH, smoke shade A US way of measuring the smokiness of air. It is estimated from the darkness of a 1 cm spot on a filter paper through which 304.8 m (1000 ft) of air have been drawn.

coefficient of uniformity *See uniformity coefficient.*

co-enzyme Any one of many complex, non-protein, organic substances essential to the activity of an enzyme.

COHb *Carboxyhaemoglobin.*

coil filter *See vacuum filter.*

coke plant wastes The distillate from the making of coke separates into two layers: a dense tarry layer from which many useful materials are made; and a lighter layer, the *ammoniacal liquor*, which is highly polluting. The tar is separated and cooled. The cooling water contains *phenols*. The quenching of red-hot coke after it has been pushed from the oven produces a water with coke particles that are recovered by *sedimentation. See spent liquor.*

cold digestion *Anaerobic sludge digestion* in a tank that is not heated. It is suitable only for small *sewage* works, because of the unavoidably long *detention periods*. The gas is not collected and the tanks may be open.

cold-end corrosion, low-temperature corrosion *Corrosion* of boiler metal at temperatures below about 230°C, caused by acid *condensation* in the smoke passages. *See incinerator corrosion.*

Coleoptera The beetles.

Coleshill AWT plant A physico-chemical *advanced wastewater treatment* plant at Coleshill, Warwickshire, UK, with a capacity in 1976 of 450 m³/day, which can evaluate processes that need large pilot testing before the expense of building a full-size plant can be risked. Investigations at Coleshill started in 1975, and processes such as *chemical clarification, multi-media filtration* and *activated carbon* adsorption have been investigated as well as the use of chemical clarification before *biological treatment.*

coliform count An estimation of the *bacterial purity* of a water by the number of *bacteria* from the *coliform group* per ml. The three current methods of laboratory testing are: the *most probable number*; *plate counts*; and *membrane filtration.*

coliform group, coli-aerogenes g. *Bacteria* that exist in topsoil and in the intestines of animals and birds. Each person excretes many billions of coliform organisms daily, most of them harmless. Coliforms are used as an indicator because their presence is taken to show that disease-causing microbes also may be present, even though the coliforms may have come from a bird colony. *See Escherichia coli.*

coliphage Any *virus* that reduces *Escherichia coli* (q.v.).

collecting lorry (in USA **sanitation truck**) A vehicle that collects refuse. Modern types compress it about 2.5 to 1 or more. Discharge may be by floor tipping, by moving floor (conveyor floor), compaction plate, etc. *See compactor truck.*

collecting sewer A *lateral.*

collector system *Underdrains* or proprietary systems of perforated pipes, ducts or plates set at the bottom of *filters* or *trickling filters.*

Collembola *Springtail.*

colliery spoil, minestone Stone or washery waste from coal mines, a material that can safely be used to *cover* solid refuse at the end of each working day. *See spoil heap.*

colloids (adjective **colloidal**) Particles smaller than 0.002 mm but larger than 0.000 001 mm (1 nanon). Colloids may be removed from water by *coagulation*. Colloidal systems include some *emulsions. See surface charge.*

colon bacillus *Escherichia coli.*

colour of water True colour in natural water is caused by large *organic* molecules and may be removed by *coagulation* and *sedimentation*. Most other colours are bleached by *disinfection*, especially by *ozone.*

Colpidium A free-swimming *ciliate protozoon* found in *trickling filters* and *activated sludge* plants, especially those with poor *effluent*.

combined residual chlorine The weak disinfectants, NH_2Cl (monochloramine), $NHCl_2$ (dichloramine) and NCl_3 (trichloramine or *nitrogen* trichloride), that are formed in water by *chlorination* when chlorine reacts with *ammonia*, NH_3. Whether one of these is formed rather than another depends on *pH*, with NCl_3, for example, existing significantly only at pH below 4.0. At high pH only monochloramine is formed and at pH 7.0 monochloramine and dichloramine are found in about equal proportions. *Compare free residual chlorine.*

combined system, c. sewer In the combined system of *sewerage* one *sewer* pipe carries both the rainwater and the *sewage*, domestic or industrial, unlike the *separate system*, with its two sewers. The storm flows may reach 100 times the *dry-weather flows* and exceed the capacity of any treatment works. Consequently, combined sewers are provided with *stormwater overflows* that send diluted *raw sewage* into a *stormwater sewer* and thence into the nearest river. With the separate system the stormwater discharged directly into the river is less polluted. The combined system is, however, cheaper to build. Sewers laid to a fall that gives a *self-cleansing velocity* when flowing part full in dry weather should not become foul.

combined water Water chemically held in a compound—e.g. the water in alum crystals. *See bound water.*

commensalism The steady co-existence of various species together in particular locations in which one receives benefit but the other is not harmed. For example, *bacteria* can exist on the human body without causing disease, although occasionally they may do so. *Compare symbiosis.*

commercially dry, air d. Referring to *sludge*, means that it has no more than 10% water.

comminution Size reduction, usually by a *mill*.

comminutor In sewage *preliminary treatment* a combined shredding *mill* and screen (*Figure C.3*) that reduces the size of solids without removal from the *sewage* flow. It thus has the advantage that operators do not have to handle the *screenings*. Much grit can quickly blunt the cutting edges of comminutors and they may cut rags into long stringy pieces that cause problems in later treatments. *Compare screenings disintegrator.*

common water flea *Daphnia magna.*

community medicine A modern name for *public health*.

community water supply The US *EPA* defines this as a public drinking water system that delivers to at least 15 service

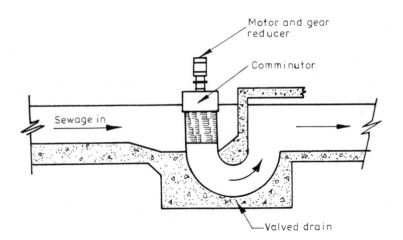

Figure C.3 Comminutor

connections the year round or to 25 permanent residents. Sampling conditions are stricter than for non-community supplies.

compaction of landfill All *controlled tips* settle because of the conversion into *carbon dioxide* or *methane* of part of their *organic* material. They also compress under the weight of the material over them. Deep tips therefore settle more than shallow ones. Usually 90% of the ultimate settlement takes place in the first 5 years, most of the remainder in 20 years. Concentrated loads should not be placed on a tip, even an old one. Concrete piles or other structures designed to carry heavy loads through a tip are liable to be severely corroded by acid *leachate*.

compaction ratio Of municipal solid waste, its original volume divided by its final volume after compaction.

compactor (1) Or steel-wheeled compactor. Several types of wheeled bulldozer with large steel-tyred wheels carrying radial cleats or teeth, resembling civil engineering rollers for compacting earthen banks, and claimed to compact to densities above 1 tonne/m³. The steel-wheeled vehicle is specialised for *landfill* sites and may not travel under its own power on public roads. Its teeth are claimed to exert pressures up to 14 megapascals (2000 psi). Some 18 t compactors can be made 2 t heavier by filling their steel wheels with water. (2) *See static compaction.*

compactor truck A refuse-collecting lorry. Those used in the USA may have a capacity between 11.5 and 30.5 m³, sometimes with an auxiliary power supply to drive the compactor mechanism while the lorry is moving. The auxiliary power supply is needed less in rural areas than in cities.

compensation water The amount of water that has to be discharged below a reservoir to maintain the river flow at an equitable level for downstream landowners with *riparian rights*. It is often realistic to assume that the compensation water need not be greater than the flow which is ordinarily exceeded for 90% of the time. The minimum river flow is called the *prescribed flow*. In modern large schemes the amount and type of compensation water have to be agreed and may take the form of continuous flow, of intermittent *freshets*, or of some combination of them.

complete-mixing system, completely mixed process Any treatment process in which the *influent* is evenly distributed throughout the tank as it enters. This system may be used in *activated sludge* treatments to distribute the influent *settled sewage* throughout the *aeration tank* and to protect the sludge from shock loadings. Recent research has indicated that completely mixed activated sludge plants may be more likely to produce *bulking* of the sludge. The primary digester in *anaerobic sludge digestion* is often operated as a completely mixed system.

complexing, sequestering Formation of a complex *ion* by the union of one ion with another atom or molecule—e.g. copper with ammonium, forming $Cu(NH_3)_4^{2+}$. *See chelating*.

composite sample, composite A combination of two or more *samples*, made so as to obtain a representative sample of water over the period of flow in question. The composite may be flow-weighted, which means that the ratio between the volumes of each sample in the composite is equal to the ratio between their flows at the time of sampling.

compost Rotted leaves, plant stalks, bark, twigs, etc., a material of low bulk density that improves poor soils, whether clays or sands. Like a sponge, it can hold 30 to 35% water by weight, and its *organic* content encourages earthworms that further improve the soil. Where there is a market for it, it can be made from municipal refuse, mainly the *garbage*. If compost contains high contents of *heavy metals*, they may harm the plants.

composting Conversion by *bacteria* or *fungi* in air of that part of solid refuse that rots quickly, mainly *garbage* and cotton or woollen rags, and, more slowly, paper and cardboard. There is no agreement about the point at which garbage becomes *compost* (see *curing*), although there is agreement about the process, that it is *aerobic* and *thermophilic digestion*. Further degradation must not take place in *anaerobic* conditions during storage. Anaerobic composting is conceivable but would be smelly, even

66

slower than aerobic composting and therefore much more expensive, although the Beccari system is an exception, being partly anaerobic. To ensure that air is present, the water content should not exceed about 65%, but it also should not fall below about 45% and preferably should exceed 50%. During average *mechanical composting*, the temperature exceeds 60°C for at least 96 h of aerobic digestion. The process is helped by grinding the material below about 25 mm maximum size. The three main operations in composting are: mechanical preparation, including separation of metals, bottles, etc., *shredding* and wetting; then aerobic digestion in a composting machine for several days or by *windrowing* for several weeks; then final upgrading for the market—e.g. removing the rest of the glass, wood, plastics bags, etc. Compost is matured and stable when it has lost the smell of sewage or refuse. *Sludge* has a carbon : nitrogen ratio of about 10:1. With a C : N ratio in the compost higher than 80, little digestion will take place. If the C : N ratio is lowered, by adding sludge, to about 30 or 35%, the digestion period falls to 10 days or less. The maximum particle size should be between 10 and 140 mm.

compost toilet   A receptacle for excreta as well as for domestic refuse. *Aerobic* and *anaerobic sludge digestion* take place, helped by an air vent.

compression feeder   A device that squeezes refuse before forcing it into a *mill*.

compression vehicle   A *collecting lorry* on which, as in most modern ones, there is a mechanism that compacts the refuse (US *compactor truck*).

compressive settling   *See settling regimes.*

CONCAWE foundation   Conservation of Clean Air and Water, Western Europe, the oil companies' international study group, based in The Hague, Holland.

concentrating viruses   The difficulties of detecting *viruses* are increased by the fact that, apart from being exceedingly small, they are scarce, even when dangerously infectious. Consequently, an important aspect of virology is merely finding the viruses, usually by concentrating them into smaller volumes of water. Some of the many methods are: *adsorption* on to solid particles, or on to various types of filter, including some made of insoluble *polyelectrolytes*; also *hydro-extraction, phase separation* and *ultra-centrifuging.*

concrete degradation   Concrete pipes may corrode inside and outside: (a) Their outsides react with *sulphates* in the soil to form calcium sulpho-aluminate, which occupies a greater volume and spalls off. Pipes break up quickly in *groundwater* containing

50 mg/litre of sulphates. *See* **sulphate-resisting cement.** (b) Inside, in warm climates, sulphuric acid attacks *sewers* by *sulphide corrosion*, but any other acid also may be a hazard.

condensate Liquid formed as a dew by release (condensation) from a gas or air during cooling. It is common in chimneys or elsewhere at temperatures below the *dewpoint*.

condensate polishing In high-pressure steam power stations, cooled steam (condensate) forms 98% of the feed water and make-up water only 2%. Consequently, the quality of the condensate is important and must be maintained. The condensate must be polished to bring its *conductivity* below the acceptable maximum of 0.1 microsiemens/cm. This usually means passing it through *mixed-bed ion-exchange resins* and the removal of the *suspended solids* by filtering through *filter candles*. The suspended solids are mainly oxides of metals from the turbines or tubes.

condensation *See condensate.*

conditioning (1) Of *sludge*, any treatment that makes de-watering easier—e.g. adding minerals or using *mechanical flocculation* or *elutriation* or *chemical conditioning*. (2) In *mineral dressing*, adding *flotation* agents to a mineral *pulp* to ensure that the valuable mineral floats well and that the unwanted *gangue* sinks.

conditioning tower An *evaporative tower*.

conductive induction The operating principle of an *electrostatic separator*. An earthed rotor feeds mixed refuse over its outer surface into an electric field near a charged *electrode*. The conducting materials are attracted to the electrode and are thus separated from the non-conductors, which move towards the rotor.

conductivity, specific electrical conductance of water *See electrical conductivity*.

conduit Any channel or pipe that carries a liquid flow. It may be an open conduit or a closed pipe. Pipes may be of circular or other cross-section. If a pipe flows full, it is under pressure; if partially full, it is under *open-channel flow*.

cone aerator A vertical-shaft *surface aerator* for *activated sludge*, which agitates the *mixed liquor* and throws it outwards in a low trajectory above the surface—e.g. the *Simplex system* and *Simcar aerators*. A powerful upflow forces liquid up from the bottom of the tank (*see Figure C.4*). A *draft tube* may be fitted below the cone in the liquor to help circulate it upwards. The *aeration* motor may have from 1 to 75 kW output. Aerators dissolve about 2.5 kg of *oxygen* in de-aerated water per kWh (*see oxygenation efficiency*). A 75-kW motor will circulate the liquid in an aeration 'pocket' 5 m deep and 15 m square.

68

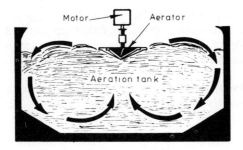

Figure C.4   Cone aerator

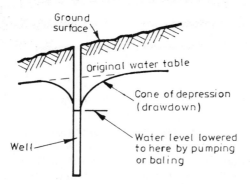

Figure C.5   Cone of depression about a well

cone of depression   The theoretically conical depression of the *water table* around a well or borehole from which water is being pumped or baled (*Figure C.5*).

confined groundwater   *Groundwater* that is overlain not by unsaturated ground, like **unconfined groundwater**, but by impermeable soil or rock. Consequently, it can be under a pressure higher than atmospheric and is able to rise in a pipe above the bottom of the impermeable bed that confines it. At any free edge it may be unconfined. *Artesian wells* are the typical confined groundwater. *See piezometric surface.*

coning   Behaviour of a plume of smoke that is expanding uniformly as it leaves the chimney.

conjunctive use of resources   The integrated use of resources of water. One example is the combined use of water from uplands and from boreholes to provide a water supply.

connate water   Water in deep sedimentary strata that has been there since they were deposited.

conservancy system   The domestic use of buckets, earth closets,

69

etc., instead of WCs, so that no *carriage water* is needed.

conservative pollutants Pollutants that do not decay, that are *non-biodegradable*—e.g. *heavy metals* and many *pesticides.*

consolidation tank A *sludge thickening* tank that uses gravity alone.

constant-velocity grit channel A unit for removing grit from *sewage,* usually with a *standing-wave flume* (a Venturi section) downstream, and a cross-section that ensures a velocity in the

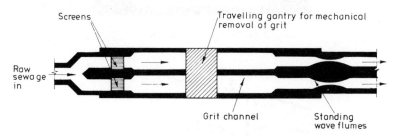

Figure C.6   Constant-velocity grit channel (plan view)

channel of 0.3 m/s, whatever the depth of flow. The *organic* solids are carried in suspension at this velocity and the grit is deposited and then (usually) removed mechanically. (*See Figure C.6.*) Constant velocity is assured because the channel cross-section is so shaped that its resistance to flow is proportional to the depth of water in it.

consumptive use of water Loss of water in use, either by *evaporation* or by absorption into vegetation. It includes the *evapo-transpiration* from a crop.

contact aerator A variety of *submerged-bed aeration* with a swinging *diffuser* for air and no solid packing.

contact bed, c. filter A forerunner of the *trickling filter.* It was a watertight tank packed with coarse stone, coke, brick or clinker of about 80 mm size, alternately filled with *sewage* and allowed to remain quiescent for 2 h, then emptied and left empty for 2 h.

contactor A *carbon adsorption bed* or other reactor.

contact stabilisation An *activated sludge* treatment in which the *settled sewage* is aerated for only 30 min in a contact tank with *return activated sludge.* This is followed by *sedimentation,* after which the top water flows away to the river. The return sludge is aerated in a re-aeration tank for 3 to 6 h before it returns to the contact tank. The *organic* pollutants in the sewage are adsorbed (*see adsorption*) on to the activated sludge in the contact tank and then metabolised by the bacteria during the re-aeration (stabilisation) stage. (*See Figure C.7.*) The sludge loadings are

70

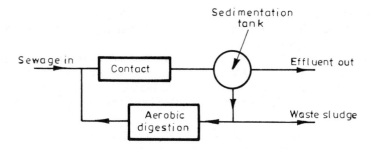

Figure C.7 Contact stabilisation (flow diagram)

similar to conventional activated sludge but the *aeration tank* (contact tank) is much smaller. Problems may arise in the settleability of the sludge after the contact tank, because activated sludge usually settles better after stabilisation. Stabilisation is incomplete because of the short detention period in the contact tank. *Primary sedimentation* is sometimes omitted. The method may be used in *packaged sewage treatment plants*, from which the excess sludge may go to a separate tank for *aerobic digestion*.

contagious abortion *Brucellosis*.

container compaction *See static compaction*.

contamination In some US states 'contamination' implies a hazard to health, as opposed to 'pollution', which is impairment of water quality without health hazard.

continuous filter A *trickling filter*.

continuous-flow stirred-tank reactor, CFSTR A chemical engineering description of a tank with a *complete mixing system* of flow.

continuous ion exchange Equipment which uses three *ion exchange* columns. One tank contains resin that is treating the water, another resin undergoing *regeneration* and the third contains resin being rinsed or stored before return to the treatment tank. The resin beads are easily transported between the tanks in a current of water. Continuous ion exchange is appreciably more complicated than fixed-bed equipment.

continuous phase *See emulsion*.

contraries Substances in municipal refuse that hinder a later process and should be removed for either disposal or other treatment. BS 3440 defines pernicious contraries in paper as vegetable parchment, wet-strength resins, and adhesives that are not water-soluble and prevent pulping for papermaking. Non-pernicious contraries are easily detectable, and include string, rags, metal and wood.

71

controlled-air incinerator   A *two-stage incinerator*.

controlled tip (USA **sanitary landfill**)   A refuse pile correctly built (by *controlled tipping*).

controlled tipping   The cheapest method of disposing of solid waste, used for 80% of disposals in Britain, the USA and probably most other countries, is to bury the refuse by piling it in layers 50 cm thick or so on suitable land and to run *compactors*

Figure C.8   Controlled tip during construction

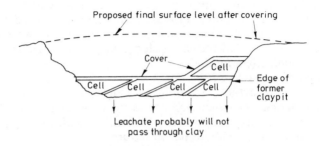

Figure C.9   Controlled tip (longitudinal section)

or bulldozers over it (*see Figure C.9*). Each daily *lift* of several 50-cm layers is overlaid with *daily cover* some 15 cm thick, completely enclosing it to form a cell 2 to 3 m high. The word 'controlled', like 'sanitary', implies that the tipping is inoffensive and that the *leachate* does not pollute rivers, lakes or *groundwater*. Tip sites should be free of ponds or streams and if necessary be culverted (*see culvert*). They should be at least 300 m from the nearest house and with good access roads. Earth, chalk, furnace ash or other *cover* must be available. To help in compacting the refuse, the tip layout may be such that the vehicles bringing the refuse have to drive over what has already been covered but not over uncovered refuse. They are often laid out in parallel fingers, 12 m wide, between which are gaps that are filled in windy weather. After tipping at the bottom of the finger, the refuse is pushed up the slope by a steel-wheeled compactor or whatever tractor is available. The finger layout enables the *aerobic* bacteria to flourish at least at the edges. The gaps are closed preferably during wind, because paper and plastics then blow off the tip face less easily. After tipping, the refuse is broken down by soil microbes, mainly *bacteria* and *fungi* that reduce it in volume and partly gasify it to *methane* and *carbon dioxide*. The latter makes the leachate weakly acidic. Because the air in the fill is quickly used up by the *aerobes* and because they degrade the fill more quickly than the *anaerobes*, steps are often taken to ensure that air is not cut off from the centre of the tip. The use of fingers helps. In the USA it has proved feasible to blow air into a tip, reducing the likelihood of dangerous concentrations of methane arising above it. About 50 to 60% water is said to be ideal for effective biodegradation. However, if the water content rises much above 60%, it pushes out the air and stifles the aerobes. In the USA it is recommended that the water table should be at least 1.2 m below the bottom of the tip, preferably more, to ensure that the *capillary fringe* is not in contact with any part of it. *See balefill, biodegradable, compaction of landfill, final cover, landfill, methane recovery from landfill, tip temperature, trench method, venting of landfill.*

Control of Pollution Act 1974   A UK law giving wide new powers to local authorities concerned with waste disposal, street cleaning, water pollution, consents and their publication, noise on streets or construction sites and from machinery, as well as *air pollutants*—e.g. from motor vehicles and the burning of insulation off electrical cables. Because of the government finance needed for full enforcement, much of the Act was not in operation even in 1980.

convection   Transfer of heat in a fluid by movement of the fluid. Usually the warmed fluid rises and the cold part descends. When the sun heats the ground, this warms the air above it, which rises by convection, mixes with other air and reduces pollution.

convection dryer   A *dryer* that works by the *convection* of hot gas, usually air.

cooling tower   A tower at a power station or other industry with large boilers or generating capacity, used for the air-cooling of hot water by evaporation (*Figure C.10*). *Effluent* from *sewage treatment* works has been used to cool UK power station water, resulting in further aeration of the effluent. *See dry cooling tower.*

cooling water   Some 70% of industrial water is used for cooling and very large quantities are needed for electrical generation and other industry. *See thermal pollution.*

copper, Cu   Compounds of copper are rare in natural waters but they can enter drinking water, either from copper tubing by

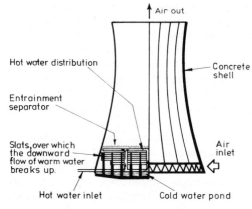

Figure C.10 Cooling tower

cuprosolvency or from the dosing of an *impounding reservoir* with 0.5 mg/litre of copper sulphate (0.2 mg/litre Cu) to reduce *algae*, followed by similar additions daily for a few days. Copper salts are very toxic to fish; 1 mg/litre can stain sanitary fittings and much lower contents can accelerate *corrosion* of other metals. In some waters only 0.02 mg/litre will pit aluminium. *See* Table of Allowable Contaminants in Drinking Water, page 359.

copperas   A commercial grade of *ferrous sulphate*, $FeSO_4.7H_2O$. Used as a *coagulant* in conjunction with *lime*, it forms a precipitate of ferric hydroxide. *See also chlorinated copperas.*

copper sulphate, $CuSO_4$ A chemical sometimes added to drinking water in small quantities to eliminate nuisance *blooms* of *algae*. It was used as early as 1905. Care must be exercised, because the highest desirable level of *copper* in drinking water is 0.05 mg/litre. At a level of 0.14 mg/litre of $CuSO_4$ trout are killed and 2 mg/litre will kill virtually all water life. In addition, since it kills *protozoa* as well as algae, the bacterial population increases after a *reservoir* has been dosed with copper sulphate, although the protozoa quickly recover.

corona discharge   A luminous electrical discharge, seen as a bluish glow from high-voltage conductors. At the hanging wires in an *electrostatic precipitator*, corona charges the dust, ionises it and so causes it to migrate to the other *electrodes*, the plates that collect the dust. *See flashover.*

corralling an oil slick   Encircling an oil *slick* with a floating boom.

corrosion   Attack on the surface of metal or other solid. In water or *sewage* there are three main causes: (1) scouring of the surface by the water or sewage, (2) presence of dissimilar metals close to each other, (3) chemical or biochemical attack. *See* below; *see*

74

also cathodic protection, cavitation, concrete degradation, coupon test, cuprosolvency, de-zincification, galvanic corrosion, incinerator corrosion, Langelier saturation index, plumbosolvency, sulphide corrosion, tuberculation.

corrosiveness of water   Water that is slightly *alkaline* and deposits a thin protective *scale* of *calcium carbonate* is considered non-corrosive. Lack of corrosiveness is measured by the *Langelier saturation index*—an index of the likelihood that a water will deposit a scale. Apart from acidity (*pH*), *chlorides* in water also may be corrosive. Aluminium, copper and many steels, including some *stainless steels*, are affected, because the inert oxide layer on the surface of the metal is attacked by the chloride ion. *Polyphosphates* (0.5 mg/litre) may be added to a water to protect iron or steel by forming a film of iron phosphate that prevents further corrosion. If polyphosphate is added, it is important to use *ring mains* everywhere and to eliminate dead ends where *bacteria* might flourish. For steam boilers the requirements are quite different, although some *alkalinity* may be desirable. There must be no *dissolved oxygen*, which is highly corrosive. High-pressure boilers need exceptionally pure *feed water*. Low-pressure and medium-pressure boilers are much less demanding and their solids are usually limited by intermittent or continuous *blowdown*. *See also* above.

cotton textile wastes   Wet processing of cotton includes de-sizing (removal of starches used in processing raw cotton), scouring or kiering (caustic washing), bleaching, dyeing and printing. The mixed waste waters are *alkaline*, having 300 to 600 mg/litre *BOD$_5$*, mainly soluble, plus *detergents, hypochlorite* ions, and waste dyes that may contain *chromium*.

Coulter counter   An instrument that measures the size distribution of particles between 0.6 and 300 μm in size. A dilute suspension of the particles in an electrically conducting liquid is forced through an opening with *electrodes* each side. When a particle passes through an opening, it initiates a voltage pulse proportional to the size of the particle.

Council on Environmental Quality, CEQ   A committee of three scientists in the US President's Office, established in 1970 under the *National Environmental Policy Act* to advise him on the quality of the environment and action to be taken.

countercurrent, counter flow   Description of a flow process in which two fluids or a solid and a fluid, etc., flow in opposite directions—e.g. in an *ammonia stripping* tower the flow is countercurrent because the water trickles down and the air is blown upwards. The opposite of countercurrent is concurrent—flowing in the same direction.

75

counter ion   An *ion* of opposite charge to that of the ion under discussion. Thus, a *cation* is counter to an *anion*.

coupon test   Insertion of a piece of weighed metal in a water *main*, with re-weighings at intervals. This aims at being a comprehensive test of the *corrosiveness of water* but is not yet widely used. It measures the effects of all the factors involved in *corrosion*, including water speed and turbulence.

cover   In *controlled tipping*, the layer of topsoil, chalk, ash, minestone or other fine material laid over refuse. There are at least two types of cover: *daily cover* and *final cover*, used after the tip is completed. Final cover must be topsoil, at least 23 cm deep for playing fields and 60 cm deep for farming. Old tip material, if it is at least 5 years old, can be dug out for use as cover but should first be screened to remove bottles, metal, etc. Covering is an essential part of controlled tipping, and after compaction by the vehicles at the tip, it seals the surface against rats and other *vectors of disease*. It also reduces smell, prevents refuse becoming windborne and reduces fire risk. Clay, except for sealing against exit of gas or entry of rain, is undesirable, especially in wet weather, when lorry tyres cannot grip it. Ashes or clinker are better. The amount of cover to be allowed for in design is about 1 m^3 per 4 to 6 m^3 of refuse. As little as 15 cm of well-compacted cover prevents the emergence of fly larvae, but 50 cm of uncompacted cover does not.

coverload pressure   *Overburden pressure*.

cowshed wastes   The manure from cow byres may be spread on land or treated in *anaerobic lagoons* at a volatile *suspended solids* loading of 0.1 kg *BOD$_5$* per m^3 of lagoon per day. The BOD from one cow is about equal to that of 20 people.

Coxsackie virus   Waterborne viruses of many types, some of which cause heart disease.

cracking of an emulsion   *See emulsion*.

creamery wastes   In making butter or margarine, water is used both for cooling and for washing. The waste water has a *BOD$_5$* of up to 1000 mg/litre, a nearly neutral *pH* and 200 mg/litre of fats and oils. A *grease trap* is essential before discharge to the *sewer*. The waste readily submits to *biological treatment*. *See dairy and milk bottling wastes*.

critical concentration range   In a *bio-assay*, the range of concentrations of pollutant between the highest at which all test animals survive for 48 h and the lowest at which all die within 24 h (ASTM 31).

critical depth   The depth of liquid flowing at the *critical velocity* in an open channel.

critical particle size of a cyclone   The smallest size of particle that

76

can be caught with 100% efficiency—i.e. at which all larger particles are caught.

critical velocity   In an open channel, that velocity at which the *Froude number* is 1. In a pressure pipe the critical velocity is at the change point from *laminar flow* to *turbulent flow* or vice versa. In practice this is at a *Reynolds number* of about 2000. *See subcritical flow, supercritical flow.*

crocidolite   The most dangerous type of *asbestos.*

cross-connections   (1) Connections between the pipes of a drinking water supply and a water supply that is doubtful or not drinkable are forbidden in the UK and many other countries. Consequently, water from a *cistern* is not allowed to flow back into the main, and this is prevented usually by *air gap protection* or a *non-return valve* or both. (2) Connections between *pressure zones* of a water supply, such that water can flow either way and one zone may augment the supply in the neighbouring zone.

crown corrosion of sewers   *See concrete degradation, sulphide corrosion.*

crown of a sewer   The top of a sewer.

crude sewage   *Raw sewage.*

crustaceans   Mainly water animals that use *oxygen*, consume *organic* substances, and have a hard body or crust. They are an important food for fish, are not normally found in *biological treatment* processes and are an indicator of clean water. They include barnacles, crabs, lobsters, shrimps and prawns, and are important *predators* on micro-organisms. *See Cyclops, freshwater shrimps.*

cryofragging, cryofragmentising, cryogenic fragmentation   A process designed to help the *de-tinning* of *cans* and used in the UK at Hartlepool. Bales of cans are dipped in liquid *nitrogen* at $-195°C$ which cools them below $-120°C$. They are then shredded in a *hammer mill*. While still cold, the metal is separated magnetically from the dirt to achieve a dirt content of only 0.7%. Aluminium also is recovered, by differential magnetic separation. *See below.*

cryogenic processing   One of the many cryogenic processes is used for breaking scrap tyres so as to obtain their rubber and steel. Using liquid *nitrogen* (*see cryofragging*), the tyres are cooled to $-190°C$ and shredded in a hammer mill to form 'rubber crumb' of any desired size. The rubber is not contaminated by steel, but the steel contains some rubber and fibre. The cost of the nitrogen is 70% of the process cost.

cryogenics   The science of the behaviour of matter at very low temperatures.

cryology   The study of snow, ice, hail, sleet—in fact, of any water at low temperature.

cryptic growth Growth of *bacteria* based on the death and break-up of other microbes. This occurs during *endogenous respiration*.

CSTR A continuously stirred tank reactor—i.e. a *complete mixing system*.

Culex fatigans A *mosquito*, the main *vector* of *filariasis*. It thrives in polluted water in the tropics, breeding in pit latrines, *septic tanks*, pools, rivers containing sewage, etc. Tropical areas affected by this mosquito are increasing because of the growth of towns.

Culicidae The *mosquitoes*.

cullet Clean broken glass of one colour. Cullet is often worth the trouble of salvage, because it requires less fuel to make bottles from cullet than from the original minerals.

cultural eutrophication US term for *eutrophication* accelerated by human activity.

culture A culture of *bacteria* consists of micro-organisms and their nutrient *medium*, in a suitable container such as a test-tube, cultivated under specific conditions of temperature, light and food—*incubation*.

culture medium See *medium*.

culvert A covered channel or large pipe to take a *sewer* or stream under a road, railway or *controlled tip*.

culverting Building a *culvert*.

cumec A unit of flow, 1 cubic metre per second.

cupola A furnace in which iron is melted to make iron castings. Cupolas emit grit, ash and sometimes oil vapour. The *carbon monoxide* that flowed from their stacks is now reduced by an *afterburner* in the *stack*, which may be no more than a gas flame that eliminates risk of explosion. Following the afterburner and any quenching device used, the *flue gases* pass to *dust arrestors*. See *shaft furnace*.

cuprosolvency Drinking water that will dissolve *copper* is cuprosolvent. Cuprosolvency occurs in hot waters with low *pH* and free *carbon dioxide*, and is stopped by raising the pH. High *chloride* concentrations also increase cuprosolvency, as with *dezincification*.

cup screen A rotating, drum-shaped *screen* made of fine stainless steel wire mesh through which the influent *raw sewage* passes outwards, leaving solids inside. Buckets (cups) scrape them from the inside, lift them out and throw them into a solids hopper. The direction of flow is thus opposite from that of a *drum screen*.

curing, maturing Storage of *compost* after it has passed through the stage of digestion, whether by *windrowing* or in a *mechanical* *composting plant*. During the 10 to 60 days of curing, *aerobic digestion* continues, the colour darkens and the temperature remains near 60°C. There is no unpleasant smell, because the pile remains *aerobic*. The product looks better and is more marketable, but the main reason for curing is to ensure that the compost is completely stabilised and so will not rob the soil of *nitrogen*.

cusec A unit of flow, 1 cubic foot per second.

cut In *mineral dressing*, a particle size or a specific gravity that describes the performance of a *classifier* or *separator*. In a *cyclone* dust arrestor the cut is the *critical particle size*. In *dense-medium treatment* the cut is the specific gravity at which separation takes place. All particles lighter than the cut float and all those denser sink.

cutwater The upstream end of a bridge pier or of any other wall that divides a flow channel. It is sharp-pointed to provide the least resistance to flow and to catch as few solids as possible.

cyanide compounds Cyanides (CN^-), although *biodegradable* by certain *bacteria*, are poisonous and should be chemically removed before they are discharged into rivers. Fish are killed by them, electroplating *effluents* being their normal origin. One way of testing for them is to run the suspected water through a fish tank. *See* Table of Allowable Contaminants in Drinking Water, page 359; *see also hexa-chrome*.

cyanide wastes *Cyanide compounds* in *metal-plating wastes, ammoniacal liquor*, etc., can be removed by *chlorination* at *pH* 11, which oxidises the cyanide to cyanate. At lower pH values, extremely toxic cyanogen chloride is formed.

cyanophage A *virus* that attacks *blue-green algae* but is harmless to humans. Cyanophages are more resistant to *chlorination* than coliform bacteria and have therefore been used as indicators of the presence or absence of viruses.

Cyanophyta The *blue-green algae*.

cyanosis, methaemoglobinaemia An illness that can be caused by high content of nitrates in drinking water, especially for infants less than 6 months old, whose haemoglobin is attacked. *See* Table of Allowable Contaminants in Driking Water, page 359.

cycle of stabilisation The cycle of *self-purification*.

cyclone (1) Or depression (weather). A region of low atmospheric pressure, with its minimum at the centre. North of the equator the winds blow counter-clockwise round the centre; south of the equator, clockwise. Cyclones usually accompany rain and storms, unlike *anticyclones*.

(2) The original 'straight-through' cyclone is a straight tube for cleaning gas which does so by the swirling of the gas throwing the

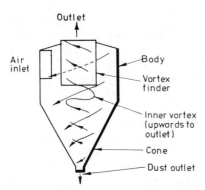

Outlet

Air
inlet

Body

Vortex
finder

Inner vortex
(upwards to
outlet)

Cone

Dust outlet

Figure C.11   Cyclone for gas cleaning

dirt particles to the wall of the tube, from which they are removed by 'bleeding' off a small proportion of the gas flow. This has been developed to the much commoner reverse-flow cyclone (*see Figure C.11*), usually with tangential inlet, a vessel shaped like a kitchen funnel, cylindrical above and conical below, with the point (apex) of the cone downwards. The exit from the cyclone (*vortex finder*) is a central tube at the top, next to the tangential inlet. The gas flow reaches it by swirling first round the outside wall down to the apex, where its diameter of swirl is greatly reduced. It is forced upwards and passes up and out through the vortex finder. Cyclones for gas cleaning exist with flow rates from 2 to 100 m³/min and a maximum gas speed around 20 m/s. For sawmills, cyclone sizes vary from about 10 to 200 cm diameter, the diameter being 3 to 5 times that of the inlet duct. For flue gases, cyclones built of ceramic can function even at 1100°C. They can catch dust from 2 mm down to 10 microns or smaller. If there is much coarse dust, larger than about 50 μm, an *expansion chamber* should be inserted ahead of the cyclone to catch it and thus reduce erosion of the cyclone wall. Cyclones do not have the 99.9% efficiency of electrostatic precipitators or fabric filters, but they are inexpensive, extraordinarily versatile and, being without moving parts, usually trouble-free, except for blockages, which can occur with any equipment.

In a *dense-medium cyclone*, fluid speeds are much lower than in dust arrestors, water being 800 times as dense as air, and the pressure losses are correspondingly higher. At the inlet to a dense-medium cyclone, the centrifugal force on the fluid and particles is about ten times the force of gravity. At the throat, near the point of the cone, it can reach 70 *g*—one reason for the sharpness of *cut* that is possible with these units. In dust collectors a much higher centrifugal force can be reached, about

2500 *g*, to catch particles of 1 μm or smaller. *See also axial-inlet cyclones.*

cyclonic spray scrubber A *centrifugal* type of *wet scrubber.*

cyclo-nitrifying filter A *trickling filter* that nitrifies the *effluent* from *activated sludge* treatment. After *nitrification* some of the effluent is returned to the *aeration tank* for mixing with the settled sewage to inoculate the mixed liquor with *nitrifying bacteria.*

Cyclops A *crustacean* about 1 mm long that is common in *zooplankton* and may be used in *bio-assays.* It is the *vector* of *Guinea worm.*

cyst A protective capsule or sac that can be formed by some *protozoa* and a few species of *bacteria* around a 'seed', ensuring its survival. It does not germinate until water and nutrients become available. Some protozoan cysts and all bacterial cysts are formed as a 'resting' cell in a different environment. Other protozoa form a cyst as an integral part of their life cycle and these reproductive cysts have thinner walls than the protective cysts. *See also Entamoeba histolytica, endospore.*

cytology The study of living *cells.* Since cells are the smallest units of life, they are seen only through the microscope.

cytoplasm The contents of a *cell*, apart from its nucleus.

D

daily cover, primary c. Earth, demolition refuse, *clinker*, ash, gravel, etc., about 15 cm thick, that completely encloses the refuse laid each day in a *controlled tip* to form one *cell.*

dairy and milk bottling wastes The *wash water* from dairies contains some whole milk, separated milk and buttermilk. The waters are slightly *alkaline* but may become acid because of the *fermentation* of milk sugars. The BOD_5 of milk bottling wastes is about 500 mg/litre. *See also cheese manufacturing wastes, creamery wastes.*

damsel flies *See dragon flies.*

Daphnia magna, common water flea A swimming *crustacean* up to 5 mm long. Like other *Daphnia*, water fleas are useful in testing for the toxicity of substances in water, because they are small, visible, easy to cultivate, have a life span of 2 months at 25°C and have offspring in their first week of life. Their *median tolerance limits* to many toxic substances have been published.

Darcy's law For the velocity of percolation of water in a saturated soil:

velocity = coefficient of *permeability* × *hydraulic gradient*

dark smoke *Smoke* as dark as grade 2 of the **Ringelmann chart**, the level at which prosecutions start under the UK **Clean Air Acts**. Continuous dark smoke for more than 4 min is an offence unless it is caused by soot blowing. It is also an offence to emit dark smoke on separate occasions within 30 min that total more than 2 min. Fines can reach £400 per day.

daughter product A *decay product*.

DDT, dichlorodiphenyltrichloro-ethane, dicophane, chlorophenothane, etc. An *insecticide*, now forbidden in many parts of the world because it is *non-biodegradable* and therefore may continue undesirably as a poison. It had been in use for 30 years before this fact was discovered, but, according to WHO in 1972, there was then no evidence that DDT harmed humans. It is toxic to fish at very low concentrations (less than 0.03 mg/litre) and may also affect many other forms of life. It temporarily eliminated some diseases (*malaria*) in hot countries. *See biomagnification, organo-chlorine compounds, vector control.*

dead storage In a *reservoir*, the volume of bottom water that cannot be withdrawn, either because it is below the outlet or is dirty, or for other reasons. *Compare live capacity.*

dead storage In a *reservoir*, the volume of bottom water that cannot be withdrawn, either because it is below the outlet or is dirty, or for other reasons. *Compare live capacity.*

de-aeration Feed water for steam boilers should have all its *dissolved oxygen* (DO) removed to stop corrosion. The water is passed through a vacuum chamber or heated under pressure to above 100°C. The DO can be reduced below 7 μg/litre. Chemical de-aeration also is possible, by adding sodium sulphite (Na_2SO_3) or *hydrazine* (N_2H_4), both of which react rapidly with dissolved oxygen.

decanting valve A valve that can be opened to withdraw sludge or the cleaner top water from a *sedimentation tank* or thickening tank. For selective withdrawal, a floating outlet pipe may be used. *See telescopic valve.*

decay product, daughter p. A substance that is formed by *radioactivity*.

decay rate The rate at which the concentration of a pollutant or the activity of a *radioactive isotope* falls, expressed by stating its *half-life*.

de-chlorination Removal of excess chlorine after *chlorination*, usually by adding *sulphur dioxide*:

$$HOCl + SO_2 + H_2O = H_2SO_4 + HCl$$

Where sulphur dioxide is not available, in small works, *sodium thiosulphate* can be used.

82

declining growth A *growth phase* that follows the rapid, *logarithmic growth* phase and precedes the *endogenous phase of growth*. The quantity of food is limited, so the micro-organisms cannot reproduce at the maximum rate.

decontamination filter A filter of steel wool, clay and *activated carbon* or resins, which can remove 99.99% of radioactive materials from water.

de-duster, de-dusting plant *Dust arrestors.*

deep-bed filter *Dual-media filters, multi-media filters, rapid gravity sand filters,* etc.

deep coarse filter bed A *filter* bed used in the USA for purifying sewage *effluent*, some 2 to 3 m deep, with rounded, uniformly sized sand grains from 1 to 6 mm diameter, which achieves in-depth filtration. The filter bed expands by only 10% during *backwashing*, using air with a little water to flush out the solids. These beds nitrify sewage effluents or act as *anaerobic filters* in *de-nitrification*.

deep-cone thickener A thickening tank that replaces *lagoons* for settling *tailings* from a coal washery. It thickens a 95% water *slurry* quickly to 40 or 50% water with the help of a *polyelectrolyte*.

deep-shaft system An *activated sludge* system for treating domestic *sewage* or strong industrial waste, in which air is blown into a shaft or borehole of 0.4 to 10 m diameter, sunk 60 to 300 m into the ground. The depth and consequent high pressure are an advantage, because for every 10 m depth the pressure rises by 1 atm, which correspondingly increases the *solubility* of oxygen. Air is injected into the outer upflow annulus to start the circulation and in the central downflow tube for the whole time. Once circulation has started, it becomes automatic, because gas comes out of solution in the upflow annulus with falling pressure as the liquid rises, as in an *air-lift pump*. The *mixed liquor* flows from the shaft to a *flotation tank*. The froth (*sludge*) from the flotation tank is returned to the shaft.

deep-well disposal *See disposal well.*

degree Clarke An old British measure of the *hardness* of water. One degree Clarke was 1 grain of *calcium carbonate* per gallon (14.5 mg/litre $CaCO_3$).

degree of hardness (1) *See* degree Clarke. (2) One French degree of *hardness* is equivalent to 10 mg/litre as $CaCO_3$. (3) One German degree of hardness is equivalent to 10 mg/litre as CaO, which is equivalent to 18 mg/litre as $CaCO_3$.

de-inking Removal of dirt, staples, printing ink, etc., from scrap paper so as to enable it to be re-used. A flotation agent resembling soap solution is added to the *pulp* of old paper. Air blown into the bottom of the liquid carries off the ink in a froth. About half the cost of a de-inking plant is spent on its final water purification plant.

de-ionised water *De-mineralised water*.

demersal Concerned with the sea bed, as opposed to *pelagic*.

de-mineralised water, de-ionised w. Very pure water, from which both *suspended solids* and dissolved solids have been removed. It is essential for the *feed water* of high-pressure boilers, for the cosmetics and other industries. At sea *distillation* has been used longest, but ashore the use of *ion-exchange resins* is more than competitive with it in both cost and quality of water. Power stations, which have the largest de-mineralising plants, use only ion exchange. All processes make a waste *brine* that may be hard to dispose of.

de-mister An *entrainment separator*.

demography The study of the statistics of human population.

Dendrocoelum lacteum A free-living *flatworm* (*planarian*) which is more tolerant than others of organic pollution and is found in mildly polluted water. It may be a *biological indicator of pollution*.

dengue, breakbone fever An acute, influenza-like disease of many Asian cities, caused by a *virus*, which may lead to fatal bleeding. It is spread by *mosquitoes*, especially *Aëdes aegypti*.

de-nitrification The opposite of *nitrification* in water, the removal of oxygen from *nitrates* in *anoxic* or *anaerobic* conditions, resulting eventually in gaseous *nitrogen*, which bubbles off and is thus removed. Many *heterotrophic* bacteria found in *aerobic* biological processes can adapt to using the oxygen that is combined in nitrates if the *dissolved oxygen* is very low or zero (*respiration* under anoxic conditions). The *sludge* in the bottom of a *secondary sedimentation* tank has little oxygen and with a *nitrified effluent* it may rise to the surface because of de-nitrification, as the nitrogen bubbles lift the sludge. In *activated sludge* treatment, if the sludge is well nitrified, some de-nitrification can be achieved by not fully aerating the *return sludge* in the first compartment of the *aeration tank*. The return sludge is then efficiently mixed with the incoming effluent from the *primary sedimentation* tanks. Some 50% de-nitrification has thus been regularly achieved. Alternatively, a reduced number of air *diffusers* can be used instead (one-third of the number ordinarily used for *aeration*) with equally good results. *Sewage* may also be de-nitrified by locating an anoxic activated sludge plant after secondary sedimentation. Methanol may be added to the secondary effluent, as it enters the anoxic mixing tank, to act as an *organic* carbon source for the de-nitrifying *bacteria*. There

84

is a further stage of sedimentation, from which the sludge is returned to the anoxic mixing tank. Alternatively, an *anaerobic filter* may be used to de-nitrify the effluent from secondary sedimentation. The same process has been observed in polluted rivers when a batch of well-nitrified effluent meets some *raw sewage*. The nitrogen bubbles off. Appreciable losses of nitrate by de-nitrification have been noticed from effluent left for a few days in a lagoon for tertiary treatment.

de-nitrifying filter   An *anaerobic filter*.

dense medium   *Carbon dioxide* is a dense medium compared with air, and water is dense compared with oil, but in commercial *mineral dressing* a dense medium is any fluid denser than water—e.g. a suspension of magnetite in water. *See* below.

dense-medium cyclone, heavy-medium c.   A *wet cyclone* through which there is a circulation of *dense medium*, used in *mineral dressing* to separate coal or metallic mineral from stone.

dense-medium treatment   *Float and sink* treatment, the use of *dense-medium cyclones*, etc.

dense mineral   *Garnet, ilmenite*, magnetite are used in *dense media* as well as in *multi-media filters*.

densifier   A *pelletiser* (briquetting unit) for *waste-derived fuel*.

density current, gravity c., suspension c., turbidity c. (also **internal flow, layered f., subsurface f., stratified f.**)   With two similar bodies of water in contact, if one is warmer than the other, the warmer water rises and flows horizontally over the colder. Similarly, clean, fresh water entering a body of salty or dirty water at the same temperature travels over the *brine* or dirty water, because it is lighter. Likewise, silt-laden water or brine entering fresh water at the same temperature travels along the bed, because it is denser. In *sedimentation tanks* such density currents can be caused by a temperature difference of only 0.2°C. They create large stagnant areas in the tank and reduce the actual *detention period* in the tank by up to 25%. *See de-stratification of lakes and reservoirs, thermal stratification*.

density separation   *Gravity separation, skimming*, the use of a *centrifuge* and, occasionally, *flotation*.

de-oxygenation   Depletion of *dissolved oxygen*, desirable in the *feed water* of a high-pressure boiler but undesirable in most *biological treatment* of sewage. In *sewage* it is caused by the activity of *aerobes* that use the dissolved oxygen to oxidise *organic* materials to *carbon dioxide*, water and *nitrate*. *See de-aeration*.

Department of the Environment   The UK government department responsible for *pollution control* and for the water industry in England and Wales. In Scotland some of its duties belong to the

Scottish Development Department.

deposit gauge A container for measuring *dust fall*, which sometimes includes liquid *particulates*. The types are described in BS 1747. *See also directional dust gauge.*

de-salination, de-salting Removal of dissolved salts from water. The term is usually confined to any of the processes for making drinking water out of sea-water (3.5% salts or 35 000 mg/litre) or from *brackish waters*). Drinking water should have less than 500 mg/litre of total dissolved solids. If the wastewater from de-salination becomes saturated with a salt, it may precipitate out and block pipes or create other problems, particularly in *membrane processes*. In connection with the recovery of heat from the *incineration* of refuse, de-salination is a convenient way of using surplus heat, because the water (and thus the energy spent in de-salting it) can be stored. In other uses of heat from incineration the heat cannot be stored—e.g. electrical generation or district heating. *See de-mineralised water, saturated solution.*

de-salination factor The concentration of dissolved substances in the *influent* water of a *de-salination process*, divided by their concentration in the de-salted water.

de-salination processes *Distillation, multi-stage flash distillation*, and the less common *membrane processes of electro-dialysis* and *reverse osmosis*. Membrane processes are more expensive than distillation, but this situation may alter.

de-silting Removal of *silt*—a preliminary treatment for *raw water*, with *sedimentation* for only 0.5 to 1 h, used for very *turbid* waters (more than 10 000 mg/litre of *suspended solids*) if no other storage is available for the water before *clarification*.

de-sliming Removal of particles smaller than about 0.5 mm from a mineral pulp so as to make it more easily treated in a *wet cyclone* or other *mineral dressing* unit. Fine material is the most difficult and expensive both to extract and to dry afterwards.

de-sludging Removal of *sludge* from a *sedimentation tank, septic tank*, etc. *See telescopic valve.*

desorb, desorption Removal of adsorbate after *adsorption*.

de-stabilisation of particles The removal of electrical charge from particles of *colloidal* size or smaller. This is never a simple process, but *coagulation* may achieve it, enabling the particles to coalesce and sink, so that they can then be removed as sludge.

de-stratification of lakes and reservoirs The summer *thermal stratification* of lakes or *reservoirs* into an *epilimnion* layer above, separated by a *thermocline* from the cold *hypolimnion* below, prevents mixing of the waters even by the summer winds, and reduces the quantity of good *raw water* that can be taken for treatment. Consequently, some water engineers have, since 1965, pumped the hypolimnion up to the epilimnion. *Bubble guns* are often used. Another method is to inject the incoming river water by jets near the bottom of the reservoir, either horizontally or upwards at 20 to 45° to the horizontal. When the river water is dirty, the jets are by-passed and the influent comes in slowly, which allows the solids to settle. The *dissolved oxygen* content of the hypolimnion has been improved by the mixing, which also has reduced the temperature gradient in the water. Sulphide concentrations up to 10 mg/litre that occur in midsummer in the bottom are eliminated by the mixing. The *ammonia* and *manganese* contents also are reduced. As long as pumping continues, the *plankton* content drops rapidly, and this reduces the solids loading on the *filters* at the waterworks, improving their efficiency. These improvements have been observed in *eutrophic* water, ideal for the growth of *algae*.

Desulphovibrio desulphuricans A *sulphate-reducing bacterium* that exists in *anaerobic* conditions and forms sulphides. It is often present in *septic sewage*. It is *heterotrophic*, consuming *organic* carbon which is oxidised as the sulphates are reduced to sulphides.

de-sulphurisation *See flue gas de-sulphurisation*.

detention period, retention p., hydraulic residence time The time during which *sewage* or *raw water* is kept in a tank or *reservoir*, so as to enable it to undergo its due treatment. The theoretical detention period is the volume of the tank divided by the flow rate through it. The actual detention time may be significantly less, because of poor distribution of flow or *density currents*. The longer the detention period, the larger and the more expensive must the tank be. *See also mean cell residence time*.

detergents *Synthetic detergents, dispersants, emulsifiers*.

de-tinning Removal of the tin coating from the tinplate of which most *cans* are made. Scrap steel for steelworks is regarded as of good quality and is relatively highly priced if it contains less than 0.03% tin.

de-toxification The elimination of danger from poisons. *See toxic waste disposal*.

Detritor, Dorr–Oliver Detritor A square settling tank for removing grit from *raw sewage*. The inlet is arranged so that the flow is evenly distributed across the tank, with a velocity of 0.3 m/s at maximum flow. The grit is removed and sent to a *classifier*. *Surface loading rates* are typically 200 m^3 m^{-2} day^{-1} at maximum flow. If operated correctly, Detritors can produce a very clean grit.

detritus In *sewage treatment*, has two senses: (1) Abrasive material, such as sand or cinders, removed in a *grit chamber*. (2)

Accumulated, dead or dying vegetation or animals.

detritus chamber A *grit chamber*.

de-waterability The ability of a *sludge* to lose water; effectively the same as *filterability*.

de-watering of sludge Removal of water from *sludge*. Various processes may be used, including *mechanical de-watering of sludge* or *sludge drying beds*. The typical 65 to 80% water content of de-watered sludge varies with the method of de-watering and the nature of the sludge. *See also conditioning, sludge thickening.*

dewpoint The highest temperature of a particular gas or air sample at which it deposits moisture—a measure of its moisture content or *relative humidity*. The higher the dewpoint the higher the moisture content. *See acid dewpoint.*

de-zincification Corrosion of brasses and other copper–zinc alloys by the removal of zinc, often caused by corrosion from the *chlorides* in the water. It can occur with a hot or cold water of *pH* above 8.2, and at pH less than 8.2 in hot water only, when the chloride content is not low enough in relation to the accompanying *carbonate hardness*. River waters with as much as 60 mg/litre chloride are now softened by sodium carbonate alone so as to remove the *non-carbonate hardness* and to keep the carbonate hardness as high in relation to the chlorides as it should be. Meringue de-zincification is a name for the fluffy white corrosion product from zinc. Some 0.2% arsenic in brass prevents de-zincification provided that there is less than 35% zinc in the brass. *See cuprosolvency.*

diagonal flow pump A *mixed flow pump*.

dialysis (verb **dialyse**) Separating a *colloid* from a *solution* by a *semi-permeable membrane* through which the solution diffuses. Dialysis in *hydro-extraction* helps in *concentrating viruses*, and in *electro-dialysis* purifies water.

diatomite, diatomaceous earth, kieselguhr, moler earth, tripolite A soil composed of the tiny siliceous skeletons of *diatoms*, therefore mainly insoluble, and used in *pre-coated filters*.

diatomite filter A *pre-coated filter* with a slurry of *diatomite* as the pre-coat, laid on *filter candles* to collect the *suspended solids*. Diatomite filters do not need *coagulants*, but the filter candles need re-coating with new diatomite after *backwashing*. This low-cost, compact alternative to *sand filters* is not recommended for polluted waters with much suspended solids or *bacteria*.

diatoms, Bacillariophyta Unicellular *algae* with a silicified cell wall that divides the cell in two. They often occur in large masses and are common in sea-water or fresh water.

dichromate value, dichromate oxygen demand *See COD.*

dicophane Another name for *DDT*.

Dieldrin An *organo-chlorine compound* used as a pesticide. In an investigation into its toxicity to fish it was introduced into a stream at concentrations between 0.06 and 0.6 parts per thousand million, with two effects. It poisoned the *larvae* eaten by the fish and thus reduced the food supply but it also sickened the fish and the fish population fell. Dieldrin may nevertheless still be used for mothproofing carpets, but it has been banned in many countries.

differential medium A *selective medium.*

diffused-air system An *activated sludge* treatment with air introduced near the floor of the *aeration tank* through *diffusers* that form coarse or fine bubbles. *See coarse bubble aeration.*

diffuser (1) Or air diffuser. In the *activated sludge* process, a device for dissolving air in the *mixed liquor.* Swinging pipe diffusers have the advantage that they can be swung out of the tank, and may produce fine or coarse bubbles. For diffusers in the bottom of the tank, access is possible only by emptying the tank. Fine bubbles of 1 to 3 mm diameter are produced by plate, dome or tube diffusers made of ceramic materials or of fused silica or alumina. Coarse-bubble diffusers are often perforated metal tubes. Fine-bubble diffusers must use filtered air or they clog. See *oxygenation efficiency.* (2) On a sewer *outfall,* the perforated end of the pipe, which spreads the effluent effectively, merely by the shape and layout of the holes. (3) In a *centrifugal* pump, the guide passages around the *impeller,* which enable it to lift clean water through high heads.

digested sludge Sewage *sludge* which has undergone *anaerobic sludge digestion* or passed through an *aerobic digester* should have a fibrous structure and a peaty smell which is not

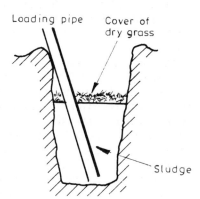

Figure D.1 Emergency pit digester for digestion of sludge in a warm climate (Reproduced from 'Water Treatment and Sanitation; sample methods for rural areas', Intermediate Technology Publications Ltd.)

unpleasant, unlike *raw sludge*, but it still may contain organisms that cause disease. Digested sludge is usually more difficult to dewater than raw sludge, although many texts state the opposite. *See Figure D.1*; *sludge disposal*.

digester, digesting, digestion *See aerobic digester, anaerobic sludge digestion, composting*.

digester gas *Sludge gas*.

dilution The usual ultimate method of disposal of *sewage* effluent into a river or the sea or a lake, or of many pollutants into the air, although many of them are washed out of the air by rain. *See Royal Commission effluent*.

dilution factor The *available dilution*.

Diphyllobothrium latum A broad *tapeworm* transmitted by food—e.g. infected freshwater fish and some *crustaceans* that have received the infection from the faeces of infected humans, dogs, etc. The disease can be reduced or eliminated by proper cooking. The worm grows to enormous lengths in the intestine and causes anaemia.

diplo A twin. All rod-shaped *bacteria* are diplos just before they divide.

dip pipe An outflow pipe for *effluent* from, e.g., a *septic tank*, which dips about 0.45 m below the water surface to avoid removing *scum*.

Diptera The main *insect* inhabitants of *trickling filters*, usually flies with one pair of wings, but the group includes midges, *mosquitoes*, gnats and some *filter flies*.

direct filtration The use, in raw *water treatment*, of small quantities of *coagulant*—e.g. less than 10 mg/litre of *alum*, followed by filtration without settlement. It is possible with clean *raw water* only.

direct incineration *Incineration* of all the refuse received, including, often, much that does not burn. Unlike *separation and incineration*, there is no systematic mechanical sorting beforehand, although some bulky wastes may be removed or sheared.

directional dust gauge A *deposit gauge* described in BS 1747, consisting of a tubular collecting head with vertical slots pointing in different directions, indicating where the dust comes from.

direct-supply reservoir An *impounding reservoir* which supplies *upland catchment water* to a city through a necessarily expensive *aqueduct* (*Figure D.2*). To maintain the quality of the water, the *reservoir* is not open for public recreation. A *river-regulating reservoir* has multiple uses and is more modern in conception.

disc centrifuge A common industrial *centrifuge* which produces a thick slurry rather than a *sludge cake*. The liquid moves up

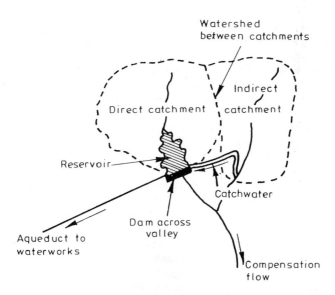

Figure D.2   Direct-supply reservoir

through a stack of discs, the solids moving to the outside of the centrifuge.

disc filter   A type of *vacuum filter*.

discharge   A volume of flow per unit of time, often expressed in *cumecs*.

discharge prevention   An *acid mine drainage* pollution control method aimed at holding all contaminated water within the mine workings. The sulphides are submerged and their rate of oxidation and production of acid are thereby reduced. This involves the heavy expense of building watertight dams underground and injecting the rock around them with cement mortar. Such dams have been built in coal mines in eastern USA as well as at sulphide mines.

disc mill   A *mill* with horizontal grinding discs. The lower, stationary disc has holes in it through which the material passes when it has become small enough. The upper disc has knobs that strike the refuse. The lower disc also may rotate (multiple-disc mill) but in the opposite direction to the upper one. The exit for the crushed refuse then is at the rim, not through holes in the lower disc.

disconnecting trap   An *intercepting trap*.

discrete settling   *See settling regimes.*

disc screen   A circular *screen* used in *water treatment*, which rotates in its own plane, perpendicular to the line of flow.

91

disinfection   Destruction of germs that cause human disease. It is not sterilisation, which is the destruction of all microscopic life. For drinking water, ordinary *water treatment*, of chemical *coagulation, sedimentation* and *filtration*, can reduce *bacteria* and *viruses* by greater than 90%. The disinfection is usually by *chlorination*, occasionally by using *ozone* or other chemical disinfectants, but *ultra-violet radiation* also has been used. It has often been shown that chlorination of *turbid* water that has not been well coagulated, settled and filtered will give poorer disinfection because bacteria and particularly viruses enmesh themselves in a protective coat of solid particles. These ordinary pre-treatments are therefore an essential part of disinfection. It is not customary in the UK to disinfect *sewage* effluent, because this can impair the *self-purification* of the *receiving water*. In the USA chlorinated *effluents* are often de-chlorinated before discharge into the river. Boiling a polluted water disinfects it but, of course, does not remove poisons.

disintegration product   A *decay product*.

disintegrator   (1) In dust collection many types of disintegrator exist but all have a rapidly turning wheel with water flowing on to it. The movement of the wheel breaks up the water into minute drops projected into the dusty gas, which help to catch the dust. (2) A *screenings disintegrator*.

disintegrator scrubber   A *wet scrubber*.

dispersal   Spreading out—*dilution*.

dispersant   A *synthetic detergent* or *emulsifier* used for washing oil from polluted beaches, sometimes also for emulsifying the oil before it has beached. These substances can be poisonous to *shellfish* at trace concentrations. Their purpose is to produce oil-in-water *emulsions* that spread the oil and thus enable micro-organisms to reach the oil and destroy it. Some biologists believe that dispersants make the oil more toxic. *See chocolate mousse, scavenging*.

dispersed-air flotation   A *froth flotation* method in which 1 mm diameter bubbles are blown into sewage by *diffusers* or *spargers* so as to remove grease and oils before *sedimentation*.

disperse phase   Part of an *emulsion* or *dispersion*.

dispersion   A gas, liquid or solid in which one substance, the disperse phase, is distributed throughout the other, the continuous phase. *Emulsions* and *solutions* are dispersions in this sense, but many others exist. (Not the same as *dispersal*.)

dispersoids   Liquids dispersed as droplets in air or other space as clouds, *mist, fog, smog*, spray, etc.

disposal well, injection w.   A well or borehole into which corrosive or otherwise dangerous liquids can be injected and left for disposal. The method has long been used in Texas for disposal of **brine** pumped up with the oil, into old oil wells, but it can pollute **groundwater** and even encourage earth slips. In Iran wells used for disposal of **raw sewage** into the pervious ground have seriously affected the drinking water because of **chloride** from the sewage.

dissolved-air flotation, saturation f. A *flotation* method in which *sewage* or *sludge* is first saturated with air and then flows to a tank where the air comes out of *solution* in tiny bubbles of 70 to 90 μm that float the sludge and enable it to be skimmed off (*see Figure D.3*). *Coagulants* may be added to help *flocs* to form.

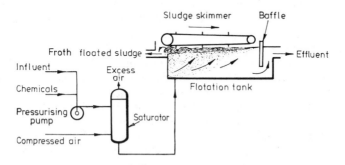

Figure D.3 Dissolved-air flotation for treatment of sewage, raw water or sludge

Two methods of dissolved-air flotation have been used. One of these is vacuum flotation, in which a short period of **aeration** at atmospheric pressure is followed by application of vacuum in a separate tank. For **sludge** thickening it is commoner to use a **saturator** in which the sludge is aerated for some minutes at 3 to 4 atm pressure and then flows to a *flotation tank* at atmospheric pressure. The dissolved air coming out of solution attaches itself to the flocculated particles and floats them. A typical design target for a mixed **primary sludge** and **secondary sludge** is 100 kg of dry solids per m² of tank per day to produce a sludge with 5 to 7% solids. For surplus activated sludge (*see **surplus sludge***) alone this loading should be halved, but only 4% solids can be obtained. The minimum depth of the flotation tank is about 2 m.

dissolved oxygen, DO Oxygen dissolved in *sewage* or water is needed for the existence of *aerobic* water life, but the *solubility* of oxygen in water is low. The maximum at 0°C is only 14.6 mg/litre at sea level, falling to 10 mg/litre at 16°C. With *chlorides* reaching 5000 mg/litre (0.5%) the solubility falls to 9.5 mg/litre at 16°C; with 10 000 mg/litre of chlorides, to

93

9 mg/litre. All this applies at atmospheric pressure. But when the external gas pressure increases manyfold, as occurs in the *deepshaft system*, the solubility of oxygen rises and much more oxygen dissolves. In steam boilers dissolved oxygen causes *corrosion* and must be removed from *feed water* before it enters the feed pump. The damage done by oxygen can be especially serious in boilers operating at pressures above about 17 atm. DO is measured by *dissolved oxygen electrodes* or by the *Winkler test*.

dissolved oxygen electrodes For measuring *dissolved oxygen* in water, a constant voltage is applied across *electrodes* covered by a *membrane*. The oxygen diffusing through the membrane causes a current to flow that is directly proportional to the oxygen concentration.

dissolved oxygen sag curve The *oxygen sag curve*.

dissolved solids *See total dissolved solids*.

distillation Boiling followed by *condensation* of the vapour, the commonest method of *de-salination* of sea-water, used at sea for many years. It could be used on solar stills in hot countries, but these are not so cheap as might be expected, because the glass must be cleaned and, if need be, replaced. Stills heated by the burning of fuel provide water that becomes cheaper as their output increases but are not economic for general use on land. *See multi-stage flash distillation*.

distillery wastes The liquid wastes after producing whisky, etc., from grain are high in dissolved or suspended organic substances, but much of the solid matter is recovered for sale as animal feed. The remaining liquid has a *pH* of 4.5, a *BOD$_5$* of about 500 mg/litre and *suspended solids* up to 1000 mg/litre, but is readily *biodegradable*.

distribution reservoir A *service reservoir*.

distributor A device that spreads the *effluent* from *primary sedimentation* over the upper surface of a *trickling filter*. Distributors may be electrically driven or operated by a *dosing siphon*.

district heating *See incineration with energy recovery*.

divide North American term for a *watershed*.

DO *Dissolved oxygen*.

dome diffuser *See diffuser*.

domestic waste, house refuse (USA **trash**) The amount of house refuse collected in the UK was about 20 million tonnes in 1978, roughly 1 kg per person per day or 350 kg per person per year. The amount collected is slowly increasing in weight but more rapidly increasing in volume. *See solid waste analysis*.

Donaldson sludge density index The *sludge density index*.

Dorrclone classifier A cast-iron *wet cyclone* used in *sewage treatment* to wash grit by centrifugal force after it has been removed from the *grit chamber*.

Dortmund tank, Kniebühler t. An *upward-flow sedimentation tank* with an elaborate inlet distributor and a grid of surface *weirs*. Many so-called Dortmund tanks have, however, mixed radial and upward flow.

dosing siphon A *siphon* that empties a *dosing tank* when it is full (*Figure D.4*).

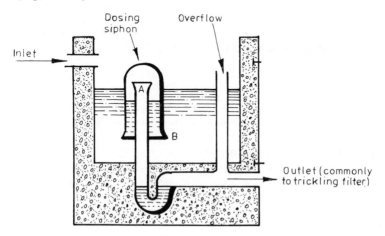

Figure D.4 Dosing siphon (condition of rising water level). As soon as the sewage level rises to the top of the pipe A under the dosing siphon bell, it flows out through A continuously until the level drops to B, admitting air to the bell

dosing tank A tank in a *sewage* works in which *effluent* collects until there is enough in it to operate the next treatment process. Where *trickling filter* influent is spread by *distributors* turned by the reaction of the water flowing out of them, dosing tanks ensure that the flow is fast enough to rotate the distributors. When the dosing tank is full, it empties quickly through the *dosing siphon*. There is thus no flow except when there is enough influent to rotate the distributor. *See tipping trough.*

DOT label A hazard sign slightly resembling the diamond hazard sign of *Hazchem*, used in USA and named after Department of Transport.

double-bed filter A *dual-media filter*.

double shredding *Two-stage shredding.*

double side-weir overflow A *side-weir overflow* with *weirs* each side of the *sewer* (*see Figure D.5*).

doubling time *Generation time.*

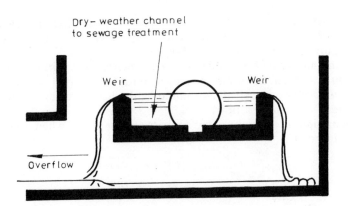

Figure D.5 High double-side weir, a common storm overflow in vertical section). The weirs are higher than the centre line of the sewer. In a 'low double-side weir' the weir crests are below the sewer centre line. The whole structure may be covered with a manhole cover, or, if it is in a sewage works, permanently open

downdraught, downwash On the lee side of a building downward-eddying wind may occasionally bring smoke near the ground. To prevent this, UK power station chimneys are ordinarily at least 2.5 times as high as the nearest tall building, usually the boiler house. Similar rules apply to other industrial chimneys.

downflow contactor A downward-flow *carbon adsorption bed*.

down time, outage Time during which equipment is out of action, especially for such reasons as maintenance and repair.

DPD, *N,N*-diethyl-paraphenylene diamine A substance used to test for the available *chlorine* in drinking water. It produces a red colour with *free residual chlorine*.

Dracone A synthetic rubber tube, a 'sausage' built to contain fluid and to be carried or towed—e.g. containing drinking water for a small island. Dracones have been adapted for use as floating booms.

dracontiasis, dracunculosis, dracunculiasis A disease caused by *Guinea worm*.

draft US term for flow rate.

draft tube (1) A tube, connected to the outlet from a water turbine, which has a gradually varying diameter to vary correspondingly the speed and the pressure head of the water. (2) A vertical tube resting on legs on the floor of an *aeration tank*, directly below a *turbine aerator*. *Cone aerators* often have no draft tube, since the aerator itself can produce the required flow.

dragon fly, damsel fly, Odonata Large flying insects. The *nymphs* mostly inhabit sluggish reaches of rivers and can tolerate reduced *dissolved oxygen.*

drag-out Unwanted dip paint, electroplating solution, etc., the liquid that drips off an object after it has been removed from a dip tank. Similarly, with *magnetic separators*, drag-out may be paper, rags and dirt that are dragged out with the metal.

drainage area The area draining to a *gulley, sewer*, river, etc. It may be different from the natural *catchment area.*

drainage basin North American term for a *catchment area.*

drain rods Flexible rods about 1 m long, usually in sets of ten, made of cane or of tightly coiled steel springs with threaded ferrules at each end, which screw together and can be pushed to and fro in a drain to unblock it (rodding). The rods are inserted by a person in a *manhole* or at an *access eye.*

draught stabiliser (USA **barometric damper**) A vertical metal plate pivoted horizontally and located in the wall of a *flue* from a furnace, which ensures that the *natural draught* to it is not excessive. When the draught is too high, the plate is sucked inwards by the draught, admitting cool air to the flue and reducing the draught through the furnace. It automatically closes again when the draught diminishes, ensuring correct draught at all times.

drawdown, dropdown Lowering of a water or *groundwater* level—e.g. by withdrawal of water from a *reservoir. See cone of depression, specific capacity of a well.*

Dreissena A *freshwater mussel.*

drift loss Water lost from *cooling towers* as drops and water vapour carried out by the rising air stream.

drinking water, potable water Water that can be drunk is supplied usually from a *main* after treatment at a *waterworks* to bring it up to the high level of *drinking water standards.* Untreated but drinkable water can be found in wells or mountain streams. Even untreated lowland river water can be drunk if people are used to that particular water. For example, some people from Cairo dislike tap-water and prefer Nile water because of its 'pleasant taste'.

drinking water standards, water quality standards Quality in drinking water implies the absence of suspended matter, excess salts, unpleasant taste and all harmful microbes, but it does not imply purity. A water with as much as 300 mg/litre of *total dissolved solids* or even more can be an excellent drinking water. Quality limits for each contaminant are often published at three different levels: the desirable minimum, the permissible and the absolute maximum levels. In a well-operated plant, desirable levels should be reached. Absolute maximum levels are those above which, except in emergency, water is undrinkable. The standards are stated by such bodies as the *World Health Organisation*, the UK Ministry of Health or the US *EPA*. The EPA seriously adopted standards, and its list published in 1976 resembles that published by the US PHS in 1962. *See Escherichia coli, irrigation, swimming water*; individual pollutants by name—e.g. *cyanide compounds*; and the Table of Allowable Contaminants in Drinking Water, page 359.

drip irrigation *Surface irrigation* through pipes made of plastics, which deliver the water drop by drop to the plant through tiny holes in the pipes. This method reduces waste of water compared with other methods of spray or surface *irrigation*, and has been used with great success for irrigating orchards, strawberry beds, sugar cane plantations, etc., but it is unsuitable for closely planted crops such as wheat. The system brings the soil to *field capacity* near the plants and not away from them, so weeds are discouraged, but a major problem is clogging of the holes.

drop A liquid particle in air, larger than 0.2 mm (200 μm).

drop box In USA an open-topped, but screened, steel skip or similar container up to about 35 m³ capacity, in which people can place their refuse.

drop connection, d. manhole A *back drop*.

droplet A liquid particle in air, smaller than 0.2 mm (200 μm).

drought In the UK a drought is defined as 15 consecutive days with less than 0.25 mm of rain each day.

drum pulveriser, wet p., rotary-drum p. Many makes of *mill* with a horizontal, cylindrical, hexagonal or octagonal steel drum that slowly rotates on its axis at speeds from 2 to 12 rev/min, according to the make. They may be continuous or batch machines. Water is added with the refuse, but for eventual *composting*, sewage *sludge* may replace water. In some machines

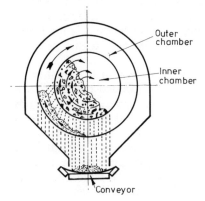

Figure D.6 Drum pulveriser (cross-section) (Dano Ltd, Switzerland)

a central shaft rotating in the opposite direction from the drum carries arms that help to break up the material. In other types the milling is done by the hard objects in the refuse (*see Figure D.6*). Most types embody an internal *trommel* through which the material drops out as soon as it is small enough. Tyres, carpets and large metal objects pass out as oversize. The refuse after processing may be denser than dry shredded refuse, but this can be caused entirely by the weight of the added water. For fast composting, wet pulverising is essential, but the weight of the added water has to be paid for as a transport load even if it is merely tipped. The slow-moving drums of this process demand less energy than do *hammer mills*. Power costs are therefore lower and less maintenance is necessary. Since drum pulverisers cannot crush bulky wastes, these should be removed beforehand for treatment in a *trade-waste incinerator* or *alligator shears* or other powerful machine. A stated maximum volume of water to be added to UK refuse is 360 litre/tonne, bringing its moisture content between 40 and 50%, enough to soften cardboard and paper. *See enriched pulverised refuse.*

drum screen   A cylindrical drum of fine stainless steel wire mesh, rotating in the raw influent *sewage*, separating the coarse solids from it (*see Figure D.7*). The sewage passes into the drum

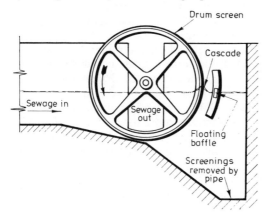

Figure D.7   Drum screen (outlet axial), cross-section

radially and out axially, unlike the *cup screen*. The *screenings* wash off the drum as it rotates and they settle into a sump for removal by a pump or bucket elevator.

dry catch   Particles from *flue gas* that are caught on a filter paper when the gas is at 121°C—a way of estimating *particulates*, in use by the US *EPA*. 'Wet catch' particles pass through the filter and may amount to about 10% of the dry catch, but are caught only at lower temperatures.

dry cooling tower   A type of *cooling tower* used in regions that are short of water. Its water circulates through cooling tubes without evaporating, but it can be four times as expensive as a wet cooling tower of the type usual in the UK. In cold weather, when they are not working, they can be damaged by frost unless they are drained, but this has not prevented them being used in Hungary and the USSR.

dryer   Equipment for drying solids—*band dryers, flash dryers, fluidised beds, rotary kilns, spray drying units*, etc. A UK working party in 1974 recommended the following maximum dust from dryers, in summary. The allowable emission at 15°C varies from 0.45 kg/h for a *flue gas* flow of 7.1 m³/min up to 116 kg/h for a flow of 8500 m³/min. For each 28.3 m³/min above this level, the allowable dust rises by 0.39 kg/h. The *grit* content may not exceed 20% of the total mass of the dust.

drying bed   A *sludge drying bed*.

dry-plant design   A method of dealing with metal-finishing *effluents* that involves piping each effluent separately to its own disposal or *neutralisation* point, enabling valuable chemicals to be reclaimed more easily. New metal-finishing plants should be laid out in this way, but it is often too expensive to modify existing plants to achieve such ideal complete separation.

dry pulveriser   A *hammer mill*.

dry riser   An empty vertical pipe in a high building, with hydrant outlets at each floor. Fire engines connected to the street *fire hydrants* can pump water into the dry riser so that fires can be fought at any level from inside the building.

dry-weather flow, DWF, dwf   The average daily flow rate of *sewage* on days when not more than 2.5 mm of rain has fallen in the previous 24 h. The dwf is often equal to the water consumption minus the *evaporation* plus the *infiltration* or minus the exfiltration. Although evaporation in the UK is considered zero, in hot dry climates it may reach 40% of the water consumption. Infiltration may contribute from 0 to 30% of the dwf.

dry well   The part of a pumphouse containing the pumps. It is separated by a wall or floor from the *wet well* containing the liquid to be pumped. The suction pipes to the pumps pass through this dividing wall. The dry well allows work to be done on the pumps without draining the wet well and the pumps are self-priming.

dual flushing system   A WC flushing system that can release either 4.5 litres or 9 litres at will. Normal closet flushing in the UK is at

9 litres and this alone is thought to consume some 3 million m³/day, so dual flushing could save water.

dual-media filter, double-bed f., two-layer f. A unit for downward, *coarse-to-fine filtration* of drinking water, in which the upper 600 mm consists typically of grains of anthracite, of 1 to 2 mm size, lying above 150 mm of fine sand of 0.5 mm size. Although the coarser anthracite particles are less dense than the sand, the two layers remain in position after *backwashing*. *Polyelectrolyte* added beforehand enlarges the size of the *floc*; otherwise the fine floc passes through the anthracite layer, while if the floc is too large, the anthracite layer will be overloaded. Compared with conventional *sand filters* at loading rates of 4 to 6 m³/m² of surface area per hour, the cycles between backwashings can be 1.5 to 3 times as long. Alternatively, loadings of up to 17 m³ m⁻² h⁻¹ can be achieved, although 6 to 9 m³ m⁻² h⁻¹ is commoner. *See multi-media filter.*

dual supply system Two water supplies, one for flushing WCs and another for drinking and cooking, often used in countries where drinking water is scarce.

dumb well An *absorption pit*.

Dumping at Sea Act 1974 Under the Oslo Convention this law prohibits *ocean dumping* without a licence from the UK Minister of Agriculture, Fisheries and Food, except what is needed for the ordinary working of ships.

dust According to BS 3405, dust includes solids between 1 and 75 μm diameter. *See also flyash, fume, grit, particulates.*

dust arrestor, d. collector Any device for catching *dust*, usually from *flue gases*; consequently much of the dust caught is *flyash*. It can include *air filters, cyclones, electrostatic filters, electrostatic precipitators, expansion chambers, fabric filters, fan collectors, gravel-bed filters, wet scrubbers,* etc. Dry types, such as fabric filters, electrostatic precipitators or *inertial separators,* cannot pollute water. Any wet cleaner produces acid-polluted water that is difficult to clean. *See Figure P.1.*

dust bin According to BS 792, a container not larger than 115 litres. Formerly always of galvanised steel, dust bins are now more often of polypropylene, which is very much lighter and more resistant to *corrosion* but no stronger. In the 1960s dust bins began to be replaced by disposable sacks or by the hinged dust bins of *dustless loading.* Sacks greatly reduce the weight lifted by dustmen and halve the distance they must walk, since there is no empty bin to be returned to its owner. Dust bins should not have to be carried more than 25 m.

dust burden The mass of dust in *flue gas*, air, etc., usually now expressed in g/m³ of gas, or in g/m² of deposit per day. *See dust* *fall.*

dust collector A *dust arrestor.*

dust deposit *See dust fall.*

dust-extraction hood, suction h., exhaust h., fume h. A metal canopy leading into ducting that is connected to an exhaust fan sucking air from a tipping point for *collecting lorries* or from an area with poisonous gas (steel furnace). The fan sucks the air away from the dusty area but may also deliver it to *dust arrestors* and must do so if the air is re-circulated into the work area. The advantage of re-circulating air into the work area is that it saves a considerable amount of money in heating the building in cold weather. The fan power needed is large, rarely less than 20 h.p.

dust fall, d. deposit, fallout The amount of *dust* and *grit* that falls on an area, expressed in g or kg m^{-2} day^{-1} or in tonne m^{-2} month^{-1}. In British towns a typical figure is 0.1 g m^{-2} day^{-1}, which is 30 tonne km^{-2} month^{-1}. Particles larger than 5 μm form the bulk of the deposit. Dust fall is measured in a *deposit gauge.*

dustless loading A system of refuse collection involving the use of *dust bins* with hinged lids. On the lorry is a device that hoists the dust bin, inverts it over a special close-fitting opening and empties it. The use of disposable sacks of paper or plastics instead of this expensive dust bin may be equally dust-free and easier for the dustmen.

dust nuisance Investigators for the UK Central Electricity Generating Board found that *dust* and *grit* from 50 to 100 μm in size gave rise to most complaints of *nuisance*. It is deposited mainly within 1 km from a chimney of medium height. At an ambient annual average of 80 μg/m^3 of ordinary non-poisonous dust, people over 50 are liable to die early; at 100 μg/m^3 children suffer for 3 or 4 days, people fall ill, some have to go to hospital and many older people cannot go to work.

dust removal *See dust arrestors.*

Dutch rasp A *rasp mill.*

dwarf tapeworm *Hymenolepsis nana.*

DWF, dwf *Dry-weather flow.*

dyeing wastes When wool, cotton or synthetic fibres are dyed, the spent dye liquors contribute 15 to 30% of the *BOD* load from the various textile processes.

dye tracers *See tracers.*

dynamic precipitator A *fan collector.*

dysentery Formerly any disease like diarrhoea, aggravated by blood flowing with the stools; now means that caused by *Bacterium dysenteriae* and the amoebic dysentery caused by *Entamoeba histolytica. See shigellosis.*

dystrophic Description of a lake water like that in a peaty district, rich in humic acid causing a low *pH*, containing undecomposed plant fragments but with few *nutrients*.

E

easement A legal right to move past or over land—e.g. the *riparian right of* an upstream landowner to discharge water downstream where neither bank of the watercourse belongs to him.

easy-open tin cans Tinplate *cans*, with an aluminium top that can be torn open merely by pulling an aluminium ring, contain by weight about 92% steel, 0.4% tin, 1.5% lead (in solder), 3.7% aluminium, and 1.8% lacquer, paint and paper. This mixture is unacceptable to the steel industry as steel scrap, aluminium being the worst contaminant, but tin and lead are also undesirable. If these three metals can be removed, a prime scrap results. Aluminium easy-open cans are acceptable as scrap to the aluminium industry with no treatment except crushing.

EC *Electrical conductivity.*

ECHO virus Many types of Enteric Cytopathogenic Human Orphan virus have been thought to cause *waterborne disease* and *gastro-enteritis*.

E. coli *Escherichia coli.*

ecological pyramid *See trophic level.*

ecology (adjective **ecological**) The study of the relationships between living organisms and their surroundings, including the materials, *nutrients* and details of the plants, animals and micro-organisms, their variations, scarcity, abundance, etc.

ecosystem An *ecological* system, one in which there is a cycle of interchange of materials between living organisms and their environs. In a tropical rain forest, for example, the dead leaves and branches, built up by *photosynthesis*, drop to the ground, where they rot and provide the materials needed by the roots. The forest flourishes even though the soil is poor.

ecosystem development strategy A planning strategy (e.g. for a lake) that would reduce the *nutrient* content of *sewage* and any other *effluent* flowing into it. It might include *aeration* and stocking with fish. Elimination of one nutrient is easier than eliminating several and it is possible to reduce the *phosphorus* content of sewage effluent by more than 90%. The land next the lake and streams should be re-forested and kept free of fertilisers. Soil *erosion* should be controlled by conservation methods and single-crop agriculture eliminated where possible.

ED *Electro-dialysis.*

eddy-current separator, linear-induction s. A device that can electrically extract non-magnetic materials such as non-ferrous metals from refuse. Magnetic fields are induced in aluminium pieces by a coil of copper wire through which an alternating current passes.

eddy flow *Turbulent flow.*

EDTA, ethylenediamine tetra-acetic acid A *chelating* agent which forms complex compounds with many *cations*. It is used to determine the concentration of *calcium* or *magnesium* ions, and thus the *hardness* of the water.

eductor A device fitted to a hose near its mouth, enabling a chemical to be sucked in, in the correct proportions.

EEC, enteropathogenic *Escherichia coli* Types of *E. coli* discovered in the 1960s, responsible for waterborne epidemics of infantile diarrhoea.

EEM *Ether extractable material.*

effective height of a chimney The true height of a chimney plus the distance above it of the vapour plume. It depends on the *efflux velocity.*

effective migration velocity, EMV The rate, usually between 4 and 17 cm/s for UK electrical generating boilers, at which the *dust* in the *flue gas* approaches the collecting plates of an *electrostatic precipitator* (ESP), measured perpendicular to the plates. The migration of the charged particles from the charged electrodes towards the earthed plates causes a slight current to flow. An ESP is therefore not truly electrostatic and has sometimes been called an electro-filter or electro-precipitator.

effective particle size, D_{10} size The grain size that is larger than 10% by weight of the soil particles—the same as the size that is smaller than 90% of the particles. Effective size is useful for describing sands used in *filters* as well as in estimating *permeability* and for calculating the *uniformity coefficient.* Permeability in cm/s is roughly equal to 100 times $(D_{10})^2$ when D_{10} is in cm.

effluent Description of fluid that flows out. In *sewage* works it usually means the treated sewage that flows out of the works to the river, but in 'stack effluent' it means gases from a chimney. *Compare influent.*

effluent standards *See sewage effluent standards.*

effluent stream A stream that receives *groundwater* from its banks or bed, unlike an *influent stream*, which loses water to the ground.

effluvium (plural **effluvia**) *Effluent*, what flows out, usually applied only to offensive smells.

efflux velocity The speed at which *flue gas* leaves a chimney.

Where dry *dust arrestors* are used, the efflux velocity should, according to the *Alkali Inspectorate*, be at least 15 m/s at full load so as to maintain a negative pressure in the chimney. If a *wet scrubber* is used, the efflux velocity should not exceed 9 m/s, so as to avoid the removal of large drops of liquid from the walls of the chimney. To achieve velocities above 15 m/s, *forced-draught fans* and usually also *induced-draught fans* are needed.

egg-shaped sewer A *sewer* shaped like an egg with the point down. It has the advantage of slightly higher speed at very low flow, which discourages the deposition of solids in the sewer. It is, however, more expensive to build than a circular sewer, and this often outweighs its advantages.

Eichhornia crassipes *Water hyacinth.*

ejector Venturi scrubber, jet ejector. A *Venturi scrubber* in which the motive power for the high-energy scrubbing is provided by pumps rather than fans (*Figure E.1*). The water consumption is high, equivalent to about 5% of the flow of gas, but most of it is re-circulated. The amount of make-up water needed increases with the amount of solids removed from the *flue gases*. The power input is as high as that to a *Venturi scrubber*—about 4 kW

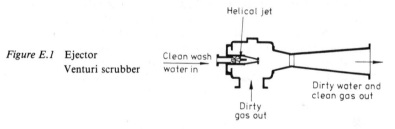

Helical jet

Figure E.1 Ejector Venturi scrubber

Clean wash water in

Dirty gas out

Dirty water and clean gas out

per 1000 m³/min of gas per mm of mercury pressure drop, and the pump pressure is about 5 atm. Scrubber water is highly corrosive and wears away even stainless steel, so *pH* control by adding *lime* or other *alkali* may be needed. Raising the pH of the water also raises its uptake of the pollutants hydrochloric acid, hydrofluoric acid and *sulphur dioxide*. Part of the scrubber water can sometimes be used for quenching furnace ash, thus reducing water treatment costs.

electrical conductance The reciprocal of electrical resistance (1/ohm), the unit being the siemens (S) or mho, commonly millisiemens (mS) or microsiemens (μS).

electrical conductivity The ability of a conductor to pass electric current. It is the *electrical conductance* of a conductor 1 cm long and 1 cm² in cross-sectional area, stated commonly in micromho/cm (microsiemens/cm, μS/cm). For water the value in μS/cm is roughly proportional to the concentration of

105

dissolved solids. Thus 150 μS/cm corresponds to about 100 mg/litre of *total dissolved solids*.

electrical precipitator   An *electrostatic precipitator*.

electro-chemical series   The *galvanic series*.

electro-chemical treatment of sewage   *Electrolytic treatment of sewage*.

electrode   A terminal of one of the two electrical conductors leading current to an *electrolyte*, electric furnace, *electro-dialysis stack*, etc. One is called the *cathode*, the other the *anode*.

electro-dialysis, ED   A method, formerly used only for making *brackish water* drinkable, developed since 1975 in Japan for *desalination* of sea-water (*Figure E.2*). Discussed in 1928, it could not become practicable until about 1950, when the first *ion-exchange membranes* were made. When salt water is electrolysed, the *cations* travel to the cathode and the *anions* to the anode, leaving theoretically pure de-salted water between cathode and anode. This does not in fact happen, but in ED it is encouraged by dividing the electrolytic cell alternately with cation-selective and anion-selective membranes that are permeable, only to cations and anions, respectively. When a direct current voltage is applied to the *electrodes*, the various ions move towards their electrodes. But no cation can pass an anion-permeable membrane and no anion can pass a cation-permeable membrane. The water in neighbouring compartments is alternately more and less salty than the *raw water*. ED does not remove *bacteria, viruses* or *organic* impurities, and the raw water

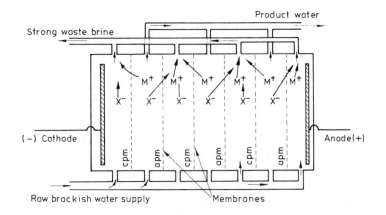

Figure E.2   Electro-dialysis, some cells of an electro-dialysis vat
apm = anode permeable membrane
cpm = cathode permeable membrane

must be carefully filtered beforehand so as to avoid clogging the pores of the *ion-selective membranes* through which it must flow. The lower limit of *salinity* for the purified water is about 500 mg/litre, because below this value the electrical resistance and the power demand increase disproportionately. *See also dialysis* and below.

electro-dialysis stack An *electrolytic* vat containing water to be de-salted by *electro-dialysis*, with several hundred parallel, alternating cation- and anion-selective *membranes* forming separate cells. The membranes are very thin, about 0.2 to 0.3 mm thick overall, and are only about 1 mm apart, separated by spacers of about the same thickness. Many *polymers* are available to make membranes. *See de-salination.*

electro-dynamic Concerned with flowing currents of electricity and the magnetic fields they cause.

electro-dynamic separator A device to separate paper from *plastics*, an electrically earthed rotating drum on to which refuse is loaded. Near the drum is an *electrode* at high voltage. Paper is attracted to the electrode and plastics to the drum. Moisture content is important. At 15% moisture the separated paper was devoid of plastics but the plastics contained some paper. At 55% moisture, separation in one pass was satisfactory. Similar equipment has been used for separating electrical conductors from non-conductors.

electro-flotation A *flotation* method by which *electrodes*, inserted into *sewage*, release gases when current passes through them (*Figure E.3*). The gases form small bubbles 0.1 to 0.01 mm

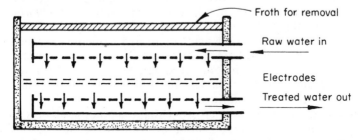

Figure E.3 Electro-flotation tank (cross-section)

across and do not break up the crust of *scum*. The method is used for treating industrial *effluents* or for thickening *activated sludge*. Typically a waste activated sludge of 0.5% solids (5000 mg/litre) requires a current of 40 A per m² of tank to thicken it to 5% solids, at a maximum surface loading of 15 m³/day per m² of tank.

107

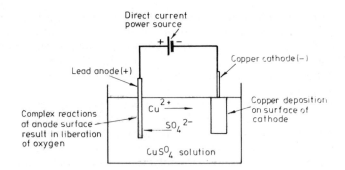

Figure E.4 Electrolysis, making pure copper from a solution in water of $CuSO_4$

electrolysis (adjective **electrolytic**) The conduction of electricity through an *electrolyte*, involving decomposition of the dissolved or molten substance, with *cations* migrating to the cathode and *anions* to the anode. Metal may be deposited at the cathode (*see Figure E.4*).

electrolyte A substance which dissociates into *ions* when molten or dissolved in water and will then conduct electricity by *electrolysis*. A conducting solution is also an electrolyte.

electrolytic corrosion *Galvanic corrosion*.

electrolytic treatment of sewage, electrochemical t. of s., Føyn's process In Norway, where electricity is relatively cheap, *electrolysis* has been used for the final purification of biologically treated sewage *effluent* after mixing with sea-water. The *anode* is below in the denser sea-water and *chlorine* is liberated there. Above, at the *cathode*, in the lighter effluent, a *flocculent* precipitate of magnesium hydroxide and magnesium ammonium phosphate ($MgNH_4PO_4$) floats off with the hydrogen bubbles that are also released. Some 90% of the phosphate can be removed in this way and the water is disinfected by the chlorine.

electron microscope A microscope that examines small objects by electrons in a vacuum instead of by light, as in the *optical microscope*. It has a much higher magnification, some 50 000 to 500 000 times, because electrons have a shorter wavelength than visible light. Vision is possible on a fluorescent screen and good photomicrographs can be taken. The cell structure of *bacteria* and *viruses* can be studied, since the *limit of resolution* is about 0.5 nm, appreciably better than the optical microscope, with its limit of about 200 nm.

electromagnet A powerful magnet for lifting iron and steel or for

108

extracting them from other material. Inside a coil of copper wire through which direct current passes is an iron core that is strongly magnetised by the current. If the current is switched off, the magnet drops its load. At its working temperature it is warm, can take less current than when cold and has less lifting power. Electromagnets should therefore be tested only when warm.

electrophoresis   Migration of suspended *colloidal* particles in an *electrolyte*, caused by an electrical potential across immersed *electrodes*. The method is used for the dip painting of car parts.

electrophoretic mobility   The rate of movement of a particle to an *electrode* under a given electrical potential. It may be used to compare the *surface charge* characteristics of *sludges, algae*, etc. Surface charge tends to change with *pH*. The pH at zero charge is the *iso-electric point*.

electro-plating wastes   *See hexa-chrome, metal-plating wastes*.

electrostatic   Concerned with electrical charges, not with the flow of current.

electrostatic filter   A type of *air filter* that attracts dust particles to its charged *filter* medium. Dust masks that use it may keep in good condition for years.

electrostatic precipitator, ESP   Probably the most versatile of all dry *dust arrestors*, the ESP can remove particles down to 0.01 μm at any temperature below 450°C when as usual it is built of mild steel, and at any efficiency, depending on the customer's requirements. Increase of efficiency from 99 to 99.99%,

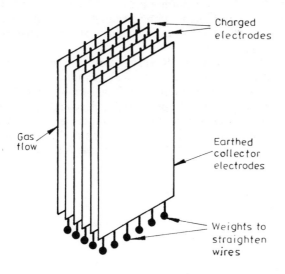

Figure E.5   Electrostatic precipitator, plate type

109

however, means doubling the size and cost. ESPs are often used with municipal incinerators. The gases pass between rows of *electrodes*. Connected to a d.c. voltage supply of some − 30 000 to − 50 000 V, the charged electrodes (corona electrodes) usually consist of a strong wire carrying a weight at the foot, hung midway between the other (earthed) electrodes, which are steel plates about 30 cm apart (*see Figure E.5*). The dust collects on the earthed plates, which are periodically rapped to remove the dust. If the voltage exceeds − 50 000 V, *flashover* between electrodes may prevent dust being collected. The gas should contain water vapour in order to reduce the resistivity. ESPs work best at about 200°C and are popular because they do not cause a high pressure drop (3 to 20 mm water gauge), nor do they have a high power demand, about 0.7 to 1.4 kW per 100 m³/min of gas treated. But they are expensive to install and they require a rectifier and a voltage regulator. Typical dimensions of ESP are: *flue gas* speed in precipitator 0.9 to 1.8 m/s (in flue 10 to 20 m/s); plate height 3.5 to 10 m; plate length from 30 cm to 1.5 times the height; *detention period* 3 to 6 s. To avoid *corrosion* and bridging of particles over the exit into the dust hopper (caused by *condensation*), hoppers are always insulated and sometimes heated in cold weather to keep their temperature above the *dewpoint* of the flue gases. Small quantities of *ozone* are produced in ESPs when the corona electrodes are negatively charged. This can be overcome by operating at a positive d.c. voltage. But the maximum positive voltage without arcing is substantially less than with negative voltage; hence, the efficiency of the ESP is then smaller. *See effective migration velocity, tube precipitator*.

electrostatic separator Several devices exist that can electrostatically separate electrical conductors from non-conductors—e.g. aluminium from glass or paper (*Figure E.6*). There are many ways of applying the electrical charge—e.g. by *conductive induction*. Repeated contact between grains of material shaken together also can create a charge unintentionally. If the materials are not dry, the charge is lost and separation

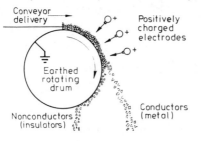

Conveyor delivery

Positively charged electrodes

Earthed rotating drum

Nonconductors (insulators)

Conductors (metal)

Figure E.6 Electrostatic separator

is impossible.

elemental sulphur   The uncompounded element *sulphur*.

elephantiasis   Great swellings of the legs as a result of many worms in the blood, usually from *filariasis*.

elute   To rinse. In the sense of the regeneration of an *ion-exchange* resin, the regenerating *brine* is the eluant; after it leaves the ion-exchange tank, it becomes the eluate.

elutriation   (1) Separation of mixed particles of different sizes or densities by a rising current of air or water—in an *air column separator* with air or in a *rising-current separator* with water. Many uses of the principle exist. (2) Washing of *sludge*, usually with final *effluent* to enable it to be de-watered more easily. The sludge is mixed with 5 to 10 times its own volume of effluent and settled. Soluble substances are removed (phosphates and bicarbonates) and non-settleable fine solids. The chemical sludge conditioners work more effectively and the dry solids content of the sludge improves to 6%. The amount of *chemical conditioning* should be smaller and the *filterability* of the sludge better. The main problem is the quantity of solids that are re-circulated to the treatment system in the elutriation water. Losses from *digested sludge* have reached 2.5 to 10% of the solids. *See de-watering of sludge.*

emergency by-pass   A pipe or channel that avoids a treatment tank that is not in use.

emergency water supply   A *water treatment* to prevent *waterborne diseases* can quickly be created as shown in *Figure E.7*.

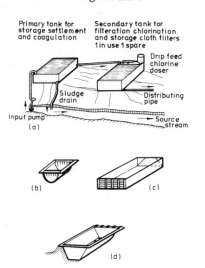

Figure E.7   Emergency water supply for a country district
(a) *General layout.* All tanks should be covered, for example with plastic sheet.
(b) *Cloth filter.* A wooden frame supports 2 or 3 layers of cloth without joints or seams. It removes suspended matter only.
(c) *Emergency tank made with timber.* This has wooden sides with uprights at 1 m spacing or less, lined with plastic sheet. Max. water depth, 1 m.
(d) *Emergency tank made without timber.* This is an excavated pit lined with plastic sheet. Corners are made by folding. Stones or sacks of earth are laid on the bottom and rim to maintain the shape (reproduced by permission from 'Waterborne Treatment and Sanitation, simple methods for rural areas' published by Intermediate Technology Publications Ltd, London).

111

emission What is sent out—emitted: not only pollutants, but also the scent from a rose, or radio waves in broadcasting.

emission factor, emission rate The amount of pollutant given off by a fuel. *Flyash* emissions are very variable and depend heavily on the *dust arrestors* but without them emissions could reach or exceed 100 kg/t—10% of the fuel.

emission standard A permissible level of pollutant sent out in water or air. *See dust fall, particulate emission*, and above.

emitter A source, usually of pollution.

emulsifier An additive that helps to stabilise an *emulsion* by reducing the *surface tension* in the continuous phase and preventing coalescence between the particles in the disperse phase.

emulsion One type of *dispersion*, a permanent mixture of two or more liquids that do not dissolve in each other, one of them usually being in tiny bubbles. In an oil-in-water emulsion the oil is the disperse phase (in bubbles) and the water is the continuous phase. When the oil has to be recovered from the emulsion, the emulsion has to be 'cracked' or 'broken'.

emulsion breaking, e. cracking Separation of the two parts of an *emulsion* by adding chemicals that encourage the droplets of the disperse phase to join up to form larger drops.

EMV *Effective migration velocity.*

enamelled steel sheet *See white goods.*

encapsulation *Toxic waste disposal* by casting the substance into something like concrete which is insoluble and thus inaccessible to water and other agencies that might bring out the poison. Some *radioactive wastes* are encapsulated into glass (see *FINGAL*), although they may still have to be stored in a safe place where they can be artificially cooled.

enclosed aerated filter A *trickling filter* that may be unusually deep but is completely enclosed, and equipped with a fan blowing air continuously through it. *Organic loadings*, measured per m³ of medium, have been claimed to be double those of conventional trickling filters.

encroachment *See saline intrusion.*

Endamoeba histolytica *See Entamoeba histolytica.*

endemic disease Disease that constantly recurs in a particular locality.

endobiotic Description of something living within another cell or plant.

endogenous phase of growth A growth phase of a population of micro-organisms in which there is no new input of food from outside. Individual microbes use the *nutrients* from dead cells and the number of living cells decreases. *See autolysis, growth* *phases of bacteria.*

endogenous respiration *Respiration* of *cells* in the *endogenous phase of growth.*

endospore A distinctive type of dormant cell that is formed among some types of *bacteria.* Endospores can have no detectable life for many decades but still be able to germinate in a suitable environment. They resist heat, ultra-violet light, desiccation and many toxic chemicals. Some species of *Bacillus* and *Clostridium* form endospores. *See cyst.*

endotoxin A *toxin* released by the degeneration or death (lysis) of bacterial *cells.*

energy accounting Estimation of the cost of things in terms of the energy needed to make them rather than in money.

energy analysis The questioning process on which *energy accounting* is based. It usually does not consider human energy.

energy line, e. grade line, e. gradient A line representing the total *head* (total energy) *of water* or other fluid flowing in an open channel or pipe. It is the sum of the potential energy (head or altitude), pressure energy and velocity head (kinetic energy). It is therefore the *hydraulic gradient* plus $v^2/2g$, where v is the velocity and g is the acceleration due to gravity. In *open channel flow* the pressure energy is always zero. (*See* diagram, page 210.)

energy recovery from solid waste This can mean many things—e.g. *incineration with energy recovery* or the production of *waste-derived fuel* in some form or another, even the gases or oils obtained from *pyrolysis.* One form of energy recovery, not always recognised as such except by *energy accounting*, is the *reclamation of solid waste.*

enriched pulverised refuse, EPR A *waste-derived fuel* made in a *drum pulveriser* using, instead of water, a liquid such as waste oil sludge. After leaving the drum, it is screened at 38 mm and the undersize is used as fuel. Most of the iron and steel are in the +38 mm size and are magnetically extracted. Textiles and plastics that are +38 mm are reduced to −38 mm. EPR is claimed to be a suitable supplementary fuel for use with coal on a chain-grate stoker in the proportions 22% EPR, 78% coal by weight.

enrichment culture The use of a *selective medium* to produce a microbial *culture* that is enriched with one group or species of micro-organism. For example, 5000 mg/litre of sodium sulphate in an *anaerobic* culture will enrich it in *sulphate-reducing bacteria.*

enrichment of lakes *See eutrophication.*

Entamoeba histolytica A parasitic *amoeba* that causes *dysentery.* Its *cysts* can be killed by heavy *chlorination*, such as 60 min exposure to 8 mg/litre of chlorine at 10°C. The cysts pass the illness to people, and may be carried by flies or on the fingers or food or occasionally by water. The cysts may persist for more than 6 months, even in clean natural waters.

enteric bacilli *Bacteria* that can live in the intestine.

enteric disease Any disease caused by microbes living in the intestine. The microbes pass out in the *faeces* of the infected person. Such diseases include *infectious hepatitis, poliomyelitis, shigellosis, typhoid fever*.

enteric fevers *Paratyphoid, typhoid fever* and food poisoning only, as opposed to the more general *enteric diseases. See Salmonella*.

enterobiasis Infection by *pinworm*.

Enterobius vermicularis Pinworm.

entrainment Removal of gas bubbles or of solid or liquid particles by a fast-moving fluid in contact with them. Sea spray and sandstorms are examples of entrainment by wind.

entrainment separator, mist eliminator, de-mister A unit that follows a *wet scrubber* or other wet treatment of gas, removing droplets of wash water in a 10 to 15 cm thick bed of knitted fine stainless steel or plastics thread. The pressure drop is usually below 25 mm water gauge and the de-mister may help to catch more dust. One good reason for mist elimination is to increase the buoyancy of the *flue gas*. The *Alkali Inspectorate* require de-misters to be fitted to *cooling towers* at power stations.

enumeration of bacteria *See bacteriological examination*.

environment Surroundings.

environmental control engineering (in UK **public health engineering**) The design, construction, operation and often also the management of equipment for water treatment and supply, sewerage networks, sewage and other waste treatment, effluent quality control and atmospheric pollution. It therefore covers many branches of engineering—chemical, civil, structural, mechanical, electrical and water engineering—as well as the sciences of chemistry, biology, hydrology, microbiology and subjects such as building law and public health.

environmental engineering In general, any engineering techniques concerned with the human environment. In the UK the term may have the restricted sense of the heating and ventilation of buildings—i.e. air conditioning. In the USA it is usually synonymous with *environmental control engineering*.

environmental health The interests and duties of *environmental health officers* include housing, food, health and safety at work, leisure, entertainment, health education, pest control. They are interested in the health implications of *environmental control* *engineering*, including water supply, *sewerage, sewage disposal,* refuse and its treatment and disposal, *nuisances*, infectious diseases, *pollution control* and noise.

environmental health officer, EHO An officer of a UK district council, who has had training in *environmental health*. Some 4800 EHOs are employed by the 450 or so district councils. *See Royal Commission on Environmental Pollution.*

environmental impact US term for any change in the environment for better or for worse, especially the effects on air, land or water of solid or liquid or gaseous wastes, smells or noise.

environmental impact statement, EIS In the USA the *National Environmental Policy Act 1969*, followed by the Environmental Policy Act 1970, requires a statement of its effect on the environment to be published for every major federal development scheme. At least two alternatives to the project must be published with the EIS: (a) what happens if nothing is done (the null alternative) and (b) any alternative schemes to the chosen one. In the UK an EIS is not obligatory but the *Department of the Environment* recommends planning authorities to ask for one in the exceptional instance of an important industrial development.

environmental lapse rate The measured change (usually drop) in air temperature with height. *See adiabatic lapse rate, inversion.*

Environmental Protection Agency *See* EPA.

environmental quality standards *See air quality standards, water quality standards.*

enzymatic hydrolysis All biodegradation, whether *aerobic, anaerobic* or simple human metabolism, involves some *hydrolysis* of complex *organic* compounds into simpler molecules by the action of *enzymes* of microbes. In the context of the treatment of solid wastes the term can mean conversion of waste paper to glucose and ethanol by enzymes, using less expensive reactors than *acid hyrolysis* but with much longer reaction times of around 20 h. *See biodegradable.*

enzyme *Bacteria* and other living *cells* produce enzymes—*catalysts* with a *protein*-based structure—in extraordinary variety. Enzymes accelerate biochemical reactions without alteration in themselves, although they may pass through an intermediate stage in combination with a substance, converting it to an end product that finally releases the enzyme. Enzyme names usually end in *-ase*—e.g. diast*ase*, oxid*ase*, reduct*ase*, ure*ase*. They exist in enormous variety throughout life and each one has its specific biochemical function, either outside a cell (extracellular enzymes) or inside one (intracellular enzymes). Their action is strongly influenced by temperature, *pH* or toxic substances.

enzymology The study of *enzymes*.

EPA, Environmental Protection Agency A powerful US federal agency formed in 1970 to attack pollution of air, water and land.

Ephemeroptera *Mayflies*.

epibenthic organisms Life on the sea bed, including fixed oysters, slow-moving hermit crabs, gasteropods and some fish that enjoy shallow water.

epibiotic Living on the surface of another organism.

epidemic A disease that spreads quickly.

epidemiology The medical study of *epidemics*.

epilimnetic Concerned with the *epilimnion*.

epilimnion The warmer upper layer of water in a lake, sea or *reservoir* which occurs in temperate climates in summer months, It is usually well mixed by wind action and the resulting turbulence. It consequently is uniform in temperature and well aerated. If the *nutrients* are available, *algal blooms* may occur. In a deep alpine lake of about 140 m total depth the epilimnion in September is often 10 m deep and in one instance at least was at 20.1°C throughout this depth. *See thermal stratification.*

Epistylis A stalked *ciliate protozoon* common in polluted streams or *activated sludge* or *trickling filters*.

equivalent per million, e.p.m. The concentration of a substance in mg/litre, divided by its *equivalent weight*. This is the same as its *milli-equivalent per litre* if the solution has a *specific gravity* of 1.

equivalent weight, equivalent Of an element or compound, the weight that will combine or react with or replace one atomic weight of hydrogen. For an acid the equivalent weight is the molecular weight divided by the number of hydrogen atoms in the acid that react with an *alkali*.

Eristalis tenax A species of *Diptera* whose *larva* (the rat-tailed maggot) can exist in de-oxygenated (*see de-oxygenation*) water and muds, breathing through tubes that communicate with the air above. It indicates gross *organic* pollution.

erosion, scour Removal of topsoil by heavy rain, or removal of material from stream and river beds or banks by fast currents. The first can harm bacterial and other life in lakes by burying it in sediment but this can be prevented by soil conservation in farming practice. Earth-moving work also can severely pollute water by encouraging scour of topsoil into it.

Escherichia coli, E. coli, **colon bacillus** (formerly *Bacillus coli, Bacterium coli*) A *bacterium* of exclusively *faecal* origin which forms 90% of the *coliform group* of bacteria in the human intestine. It is distinguished from other coliforms by its ability to grow at 44°C in *MacConkey's broth*, although in tropical climates other bacteria also grow at 44°C. Its presence in water indicates faecal pollution. It is normally thought to cause no illness in humans, although some strains may cause *gastro-enteritis*. *E. coli* survives longer than most bacteria from human faeces which might cause illness. For drinking water, no sample should contain any *E. coli* for 95% of the year, and no 100 ml sample should contain more than two *E. coli*.

ESP *Electrostatic precipitator.*

ether extractable material, EEM Oil, grease and tar, which are determined in *sewage* or *sludge* by extraction with petroleum ether. EEMs are harmful to *biological treatments* and may also foul *electrodes* or float systems in *wet wells*, causing faulty operation of pumps. Disposal of sludge to farms may be undesirable if much EEM is present. *Primary settlement* scum may be 45% EEM by weight. *See greases and fats.*

eucaryote A higher *protist.*

Euglenophyta, phytoflagellate protozoa A group of unicellular *algae* with *flagella*, mostly inhabiting fresh water—e.g. *Euglena*, which can be abundant in stagnant, *nutrient*-rich waters such as *waste stabilisation ponds.*

euphotic zone The water in a pond, lake or sea, down to the effective depth of penetration of light for *photosynthesis.*

eutrophic (from Greek, corpulent) Description of a water that is rich in *nutrients*. Eutrophic waters have plenty of life, resulting in *turbidity* and possible algal *blooms* in summer. The colder, lower layer of a eutrophic lake becomes depleted of oxygen in summer because of the descent of decaying *organic* matter from the warmer upper layer. *See oligotrophic.*

eutrophication The process of becoming *eutrophic*, which may take many thousands of years in natural conditions. An originally *oligotrophic* lake becomes enriched with *nutrients* brought from weathered rocks dissolved in the runoff from the surrounding *catchment*, which stimulate the life in the lake. This process can be greatly accelerated by humans with their inputs of *sewage*, fertiliser washed off farmland, etc., resulting in complete eutrophication in 50 to 100 years. The key nutrient that can limit eutrophication is thought to be *phosphorus. Nitrogen* can be fixed from the air by some *algae* and carbon enters the system by *photosynthesis*. Hence, man-induced eutrophication can best be limited by phosphorus removal in the sewage works.

evaporation In the *hydrological cycle*, loss of water to the air as vapour.

evaporative tower, quenching t., spray t., Equipment that cools the gases from an *incinerator* before they enter a gas cleaning plant without heat recovery. (In an incinerator with heat recovery the *flue gases* are cooled by a boiler that recovers their heat.) Water is sprayed into a chamber through which the gases pass. There are two main types—wet-bottom and dry-bottom systems. The wet-bottom system costs less but with coarse sprays uses more water and pollutes it, which results in a highly acid waste water that needs treatment. In dry-bottom systems the sprays are finer and consequently the pressure loss is higher, but all the water is evaporated. The spray volume must be adjusted accurately to ensure this, and the water must be filtered or the jets will block. With either system the water uptake increases the mass of flue gases but reduces their volume by about 40%. For a typical refuse burned with 100% excess air, the amount of water is 2 to 2.5 kg per kg of refuse. Failure of the spray system can result in burnout of the *electrostatic precipitator*, a very expensive unit.

evapo-transpiration The combined loss of water vapour to the air that results from *evaporation* and *transpiration* in the *hydrological cycle*.

excess air Air that is theoretically not needed for burning, yet passes through a furnace. In efficient furnaces (e.g. boilers) it should be kept as low as possible to maintain the efficiency and to reduce the number and mass of dust and grit particles carried off. If too much excess air is used, the *flue gases* are cooled and heat is lost up the chimney. For efficient burning of coal there should be not more than 50% excess air. This may mean 12% *carbon dioxide* in the flue gases. *Incinerators* have been operated with 300% excess air, which resulted in unnecessarily large and expensive *dust arrestors*. Chemically reducing conditions near incinerator *grates* have been blamed for high-temperature *incinerator corrosion*. Therefore, to provide the oxygen to eliminate the reducing conditions, 100 or even 150% excess air may be the target of operation for an incinerator.

excess-lime softening Unlike *calcium carbonate*, magnesium carbonate is soluble and magnesium is therefore removed in *water softening* by precipitating magnesium hydroxide. This is formed only at *pH* above 11.0, so lime in excess of that needed for normal *lime softening* must be added to obtain the required *alkalinity*.

excess sludge *Surplus sludge*.

exfiltration Leakage outwards, the opposite of *infiltration*.

exhaust ventilation, e. system Ventilation by a fan which sucks the foul air towards itself, as with a *dust-extraction hood*.

expansion chamber, settling c., subsidence c. A chamber of much larger cross-section than the *flue* to which it is connected, which

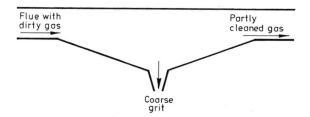

Figure E.8 Expansion chamber (vertical section)

therefore slows down the *flue gases* so that they drop the larger dust particles (*see Figure E.8*). About 50 μm is the smallest size that can be deposited. The gas speed should never exceed 3 m/s and ideally should drop to 0.3 m/s. The widening of the flue as it joins the chamber should not be steeper than 30° with the axis of the flue, so as to make sure that the gas does not eddy. Other *dust arrestors*, such as *fabric filters, electrostatic precipitators* or *cyclones*, must be used for catching the fine dust.

extended aeration An *activated sludge* process for *sewage* that uses a long *aeration* period of 24 h or more, as with the *oxidation ditch*. The main advantage is a low *sludge production* (less than 0.5 kg dry sludge per kg BOD_5 applied); therefore it can operate in small rural communities where sludge is removed every 3 months. Extended aeration plants do not need *primary sedimentation*. They may be designed to a *volumetric loading* of 0.25 kg BOD_5 per m³ of *aeration tank* capacity or a *sludge loading rate* of 0.05 to 0.15 kg BOD_5 per kg of sludge per day. The *effluent* is well nitrified. *See oxygenation capacity.*

extended filtration The use of a *high-rate trickling filter* in such a way that some of the *humus sludge* returns to mix with the *influent* raw or *settled sewage*. This increases the rate of organic *adsorption* by the microbes. The *effluent* from the *humus tank* passes through a *pebble-bed clarifier* to complete its *biological treatment*.

F

fabric filter, bag f. A highly efficient *dust arrestor*, the oldest known type, used in many industries and in many sizes, having long, vertical, cylindrical textile or felt tubes (bags) through which the dust-laden gas is blown, usually outwards (*see Figure F.1*). Bag filters remove at least 99% of the dust if no bags are broken and function in the same way as a household vacuum cleaner. They can, however, cause up to 170 mm water gauge

119

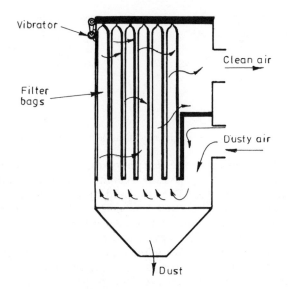

Figure F.1 Fabric filter

loss of gas pressure and are easily blocked, especially if the gas is cooled below its *dewpoint*. Bags have the following maximum working temperatures: paper 50°C; wool or felt 90°C; nylon 200°C; glass fibre 265°C if siliconised and graphited. Woven stainless steel bags are said to tolerate temperatures above 400°C. Bags have been impregnated with sodium bicarbonate so as to react with *sulphur dioxide* and extract it from the *flue gases* but this is unusual. Maintenance is slow, dirty and expensive. Both working costs and energy demand are generally lower than for good scrubbers but more than for precipitators. *See bag cleaning of fabric filters, baghouse.*

facultative anaerobic bacterium, f. anaerobe A *bacterium* that can grow in the presence or absence of *dissolved oxygen*. Consequently, it is either *anaerobic* or *aerobic* according to the conditions around it, unlike *obligate anaerobes* or *obligate aerobes*. Facultative bacteria usually function better as aerobes.

facultative lagoon, aerobic–anaerobic l., hetero-aerobic l. A *waste stabilisation pond* which is *anaerobic* in the bottom layers and bed and *aerobic* in the top water (*see Figure F.2*). It has been successful in warm countries where the land is available. In the delicate relationship between the two layers *photosynthesis* occurs by day in the top layer, with the *algae* evolving oxygen and consuming the *carbon dioxide* given off by the *bacteria* that exist much deeper. The algae are limited to the top 30 to 50 cm.

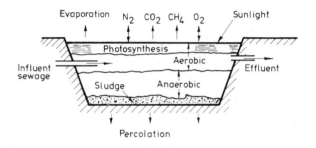

Figure F.2 Facultative lagoon

Chlorella is often the dominant alga but *Chlamydomonas* and *Euglena* also occur. Aerobic bacteria with the help of this oxygen oxidise the waste in the upper layers. Below 10°C the anaerobic activity in the lower layers is negligible, but above 23°C it is so intense that *methane* may lift the *sludge* off the bottom. The middle layers of the pond are partly aerobic and partly anaerobic. The water is 1 to 1.5 m deep. If it is shallower, plants grow from the bottom, making the lagoon into a swamp; if deeper, the lagoon is predominantly anaerobic. There are daily variations in dissolved oxygen and *pH*. The *dissolved oxygen* in the upper layer may reach 15 to 30 mg/litre by the afternoon and fall nearly to zero before dawn. The pH rises with photosynthesis, as the algae use up the carbon dioxide. Typical *BOD* loadings are 15 to 40 g of BOD_5 per m^3 of lagoon per day. A lagoon may work more efficiently if the dissolved oxygen from the upper layer is sometimes mixed with lower layers by wind or a re-circulating pump, encouraging the bacteria below. Facultative ponds are de-sludged every 10 to 20 years. To obtain a good *effluent*, a facultative pond should be followed by two or more *maturation ponds*.

faecal coliform bacteria *Bacteria* that inhabit the intestines of humans and animals, including *Escherichia coli. See coliform count*.

faecal streptococcus *Streptococcus* bacteria that commonly occur in faeces. *See Streptococcus faecalis*.

faeces (adjective **faecal**; USA *feces, fecal*) Solid waste excreted by humans, animals, birds, etc.

fallout, dust fall, particle fall *Dust* that falls from the sky, often in rain or snow. Whether coarse or fine, it may originate from chimneys, volcanoes, sandstorms, forest fires or nuclear explosions.

family A division in *taxonomy* between order and genus.

fan collector, impeller c., turbocollector, dynamic precipitator A

121

dust arrestor consisting of a *centrifugal* fan that throws the dusty part of the *flue gas* to a slot in the outside of the fan casing. The slot leads either to a dust hopper or to *cyclones* that further clean the gas. The flow passing through the cyclones may be 10 to 15% of the total, but the arrangement allows them to be smaller, more efficient and less expensive than if they were to treat the whole flow. Fan collectors have temperature limitations because of their gas seals and bearings, which must not be overheated.

fanning plume A smoke *plume* that occurs in stable air, when the gases quickly reach their highest level and then travel horizontally, remaining visible for a long distance—consequently, with little *dilution*. Fanning may thus produce a concentrated pollution on distant hills or tall buildings.

farm wastes *See cowshed wastes, piggery wastes, poultry house wastes.*

fauna The animals of a region or period.

feces American spelling of *faeces*.

feed lot A large field near a US city, where many thousands of animals are fattened for some months before slaughter. The feed lot drainage should be led to a *lagoon* with a volume at least equal to the rainfall *runoff* of the worst 24-h storm that occurs in 20 years.

feed water Water that is pumped into a steam boiler under pressure. Most of the feed water is condensed steam; the rest is make-up water. For electrical generating station boilers in which the steam pressures are around 140 atm, the total ionised solids should be below 0.05 mg/litre (5 parts per hundred million) and silica should be below 0.02 mg/litre, as otherwise turbine blades could be damaged or *scale* in the boiler could reduce heat transmission. With these severe requirements, it is essential to filter and de-mineralise even the returned *condensate*. Low-pressure boilers do not need such pure water, but it must not be corrosive. Chemicals may be added to it to improve the water quality and the sludge can be removed in *blowdown*. *See de-aeration, corrosiveness of water.*

fellmongering The preparation of hides, usually sheepskins, before tanning. *See tannery wastes, beamhouse wastes.*

fermentation *Anaerobic* biodegradation that transforms a large molecule to smaller ones with the extraction of energy. Thus sugar is converted into alcohol and *carbon dioxide* by yeasts; *methane* and carbon dioxide are produced from sewage *sludge* in *anaerobic sludge digesters*.

ferric chloride, $FeCl_3$ A *coagulant* or *chemical conditioner* that is expensive compared with the alternatives. *See ferrous salts.*

122

Ferrobacillus ferro-oxidans   An early name for *Thiobacillus ferro-oxidans*. *See Thiobacillus.*

ferrous metals   Iron and steel.

ferrous salts   Ferrous iron has a valency of two, compared with the ferric valency of three. Ferrous salts are cheaper and more easily available, but readily oxidise to the ferric form. For drinking water treatment, however, alum is preferred. *See copperas, iron in water.*

ferruginous discharges, red water   Discharges of *iron in water*, common in *acid mine drainage*. The *acidity* is caused by the natural oxidation of iron pyrite to sulphuric acid and iron sulphate by the *sulphur-oxidising bacteria* that occur in mine workings. They also de-oxygenate the water. In a stream that receives it, red ochre (ferric hydroxide) is deposited as soon as some *oxygen* enters the water. Fish and all other life can be killed when such water joins their stream.

ferrule   A screwed connection—e.g. into the wall of a water *main* for joining a branch pipe on to it.

feuillée   A latrine consisting of a shallow hole in the ground, where the *faeces* are lightly covered with earth after each use.

FGD   *Flue gas de-sulphurisation.*

fibre recovery   Obtaining *cellulose* fibres from old paper, etc., or textile fibres from old clothes, so as to use the fibres.

field capacity, field moisture c., moisture-holding c.   The amount of water in the topsoil that can be held in it against the force of gravity. It is the ratio between the weight of water held by the soil after free drainage and the weight of the dry soil. It varies from 5% for sands to about 24% for clay loams. Soil at field capacity is in the ideal state for crops to flourish. *Compare wilting coefficient.*

field drain   A *land drain*.

filamentous organisms   Wiry or stringy organisms: some *fungi, bacteria* or *algae,* which may occur in *bulking* sludge, *sewage fungus,* etc.

filariasis   A disease caused by a tropical *parasite* that is believed to be increasing; it is estimated that there are already some 250 million sufferers from it. It is caused by the *nematode* worms *Wuchereria bancrofti* and *Brugia malayi*, which release vast numbers of their *larvae* into the blood. The main *vector* of the disease is *Culex fatigans*, a *mosquito. See elephantiasis.*

fill and draw tank   A *sedimentation tank* of an old type that was alternately filled and emptied, now superseded for most purposes by continuous-flow tanks.

filter   A true filter separates solids from a fluid by straining them from the flow through it—e.g. an *air filter* or a *deep-bed filter*.

The grains of sand, anthracite, etc., form pores between one another that allow the water through but hold the solids. Filters also remove particles smaller than the pore size, because of their **adsorption** to the surface of the medium. In **water treatment, suspended solids** are removed usually by a deep-bed filter. Another true filter, the **membrane filter**, can be used for obtaining high-purity water by **ultra-filtration.** In **sewage treatment**, sludge is de-watered by the **filter press, filter belt press, vacuum filter,** etc. **Trickling filters** are not true filters in any sense, because purification of the sewage is by its **adsorption**, absorption and bio-degradation in the **biological film.** *See* **biodegradable.**

filterability (1) A property of **sludge,** the ease with which water can be extracted from it either mechanically or by **sludge drying beds,** etc. It is related to the type of solids, their adhesion and size, etc., and can be modified by sludge **conditioning.** *See also* **Buchner funnel test, capillary suction time, filter aid, filter leaf test, specific resistance.** (2) The ability of suspended matter to be removed from water in a true **filter.**

filter aid (1) An inert insoluble substance added to a **sludge** to enhance its **filterability.** (2) A chemical added to water to improve the working of a **filter** at a waterworks. It may act as a **coagulant** or **flocculant**, or both.

filter belt press, belt filter p. Two parallel conveyor belts pressed together and moving in the same direction, with conditioned

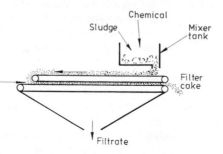

Figure F.3 Filter belt press

sludge between them (*see Figure F.3*). The sludge is slowly de-watered, the water passing through pores in the lower belt. It is an economic alternative to other methods of de-watering sludge, and is continuous in operation. The **sludge cake** after leaving the belt contains 20 to 30% solids. The **chemical conditioner** is often **polyelectrolyte**, 0.2 to 0.3% of the dry weight of sludge.

filter blinding *See* **cloth blinding.**

124

filter cake (1) The solids from a *filter press, vacuum filter,* etc.; occasionally *sludge cake*. (2) In a *fabric filter*, the dust collected on the bag.

filter candle Hollow, candle-shaped ceramic strainers, periodically pre-coated with 3 mm of *diatomite* to make a *diatomite filter* which collects *suspended solids*. They are coated by surrounding them with a water–diatomite suspension and sucking the water through the candle. When, during operation, the *filters* block, they are backwashed (*see backwashing*), which removes the diatomite and the dirt from the water. The candles then must be re-coated. Some types of filter candle are impregnated with a small quantity of metallic silver outside, to kill *bacteria*, and so do not have to be boiled periodically to sterilise them. The pore size is claimed to be only 2 μm. Filter candles are widely used in breweries.

filter fly Many types of small fly breed in *trickling filters*, especially standard-rate filters, and are called *grazing fauna*. In high-rate filters the water flows so fast that it washes most of the *larvae* out into the *humus tank*. Typical species are *Hydrobaenus, Psychoda, Sylvicola fenestralis*. They may create a fly nuisance to people near the sewage works, but can be reduced by flooding the filter for 24 h or by careful use of *disinfection* or an *insecticide*.

filter leaf test A simple way of estimating the *filterability* of a *sludge*, used in the study of *vacuum filters* or to show whether a sludge de-waters easily. The desired vacuum is applied to a perforated disc which is moved through the sludge, covered by the filter medium. The *sludge cake* formed in a given time is measured.

filter loading (1) For *trickling filters* the load can be expressed in several ways: organically as kg of *BOD* per m^3 of filter volume per day; hydraulically as flow of sewage per m^3 of filter; or by surface area, as flow of sewage in m^3 m^{-2} day^{-1}, as in *surface loading*. See *organic loading*. (2) For *sand filters* the loading is expressed as flow of water per unit surface area, in m^3 m^{-2} day^{-1}.

filter press, plate p., pressure filter A collection of parallel, ridged, rectangular, similar, cast-iron recessed plates which can be separated or held together (*see Figure F.4*). The internal surfaces of the plates are covered by textile sheets, the filter medium. Plate presses are used in many industries, including *sewage* and *water treatment*, for de-watering *sludges*. Sewage sludges are invariably conditioned (*see conditioning*) before pressing. After the plates have been joined together, the sludge is forced into the spaces between the cloth sheets. The liquor passes through the cloth as pressure is applied, and drains away through

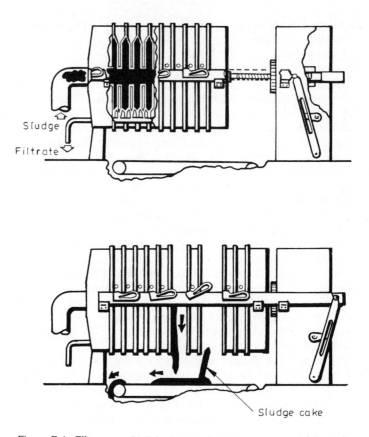

Figure F.4 Filter press (a) Injection of sludge between plates before the sludge is de-watered, (b) sludge cake discharge

holes in the plates. When the filtration cycle is complete, the plates are separated to release the *sludge cake*, of 25 to 30% solids. (With sewage sludge, more than 35% solids is unlikely.) Plate presses are batch-operated. Automatic presses have been developed which have eliminated much of the labour formerly needed for operating them. The press *effluent* is highly polluting and needs treatment.

filter run, service cycle The length of time between one backwash (*see backwashing*) and the next in a *filter* or between *regenerations* in a *carbon adsorption bed*, etc.

filter tube, f. bag A bag in a *fabric filter*.

filtrability *Filterability*.

filtrate Liquid that has passed through a *filter* and so has lost some of its *suspended solids*.

filtration Cleaning of water or air by *filters, air filters,* etc. *Adsorption* on the medium is important in both types.

filtration gallery A *ghanat*.

final clarifier, f. sedimentation tank A *secondary sedimentation tank*.

final cover, secondary c. When a *controlled tip* is completed, it is usually covered with at least 60 cm of topsoil which will allow plants to flourish, sometimes even trees. The top surface should throw off water and so is sloped but only at about 2 to 4%, to prevent scour. Side slopes should be less than 1 vertical to 3 horizontal, preferably 1 in 6.

fine-bubble aeration Aeration of *activated sludge* with porous ceramic *diffusers* at or near the bottom of the *aeration tank*. The pore size of the diffusers is about 50 μm, which results in bubbles between 0.2 and 0.3 mm. The air must be filtered to prevent the diffusers clogging. Greater air pressures are needed for fine-bubble aeration than for *coarse-bubble aeration* because fine-bubble diffusers are deeper in the aeration tank and the pressure loss through the fine holes is high. However, the proportion of oxygen transferred compared with that supplied is three times as much as with coarse bubbles. *See oxygenation efficiency*.

fines Fine solids—e.g. smaller than 0.5 mm.

fine screen In *sewage treatment*, a *screen* with openings not more than 25 mm square or with bars not more than 25 mm apart, usually from 10 to 15 mm. Fine screens are a *preliminary treatment* before grit removal and are usually mechanically cleaned. A *comminutor* may replace a fine screen. *See screenings*.

FINGAL, Fixation In Glass of Active Liquors An early method of *chemical fixation* of long-life and therefore hazardous fission products by casting them into insoluble glass blocks with borax and silica.

fire hydrant A branch pipe, usually in a street, connected to the water *main*, which enables a fire hose to be connected to the main. In low fire-risk areas in UK towns, hydrants are placed every 100 to 150 m; in high-risk areas, sometimes as close as 30 m. The minimum flow from one hydrant should be 1.4 m³/min but 25 m³/min may be needed in a highly populated area; this flow is obtained by grouping several hydrants.

fire point The lowest temperature of a combustible liquid at which enough vapour is formed to burn continuously. It is higher than the *flash point*.

fire side The surface of a boiler tube which is in contact with the flames or *flue gases*, not the *water side*.

fixation of nitrogen Part of the *nitrogen cycle*, the making of *nitrates*. Nitrogen fixation is really unfixing it and making it more available to plants, since the most fixed, least available nitrogen is what is in the air. Some *blue-green algae* are able to use atmospheric nitrogen as their nitrogen source, however, even if the water around them contains appreciable quantities of nitrogen compounds—nitrates or *ammonia. See eutrophication.*

fixed-bridge scraper A rotating *scraper* in a circular *sedimentation tank,* which moves the *sludge* to the *sludge hoppers.* A fixed bridge spans half or all of the tank. The scraper is driven from and attached to a central driving shaft carried by the fixed bridge. *Compare rotating-bridge scraper.*

fixed-film reactor A *biological treatment* process in which a *biological film* adheres to a solid *medium*—e.g. a *trickling filter* or *rotating biological contactor.*

flagellate protozoon, Mastigophora, Flagellata *Protozoa* equipped with *flagella,* unlike the *ciliate protozoa* or the amoeba group of *rhizopods.* They are common in aerobic *biological treatment* and may indicate an *activated sludge* that is in poor condition. *Photosynthetic* flagellates (phytoflagellates) may be considered as either protozoa or algae. Non-photosynthetic ones are known as zooflagellates.

flagellum (plural **flagella**) Latin for a whip, this means a tail in a micro-organism that may help it to move (*see flagellate protozoon*). Various *algae* and *bacteria* also have flagella.

flame photometry A way of measuring very low concentrations of some metals (e.g. lithium, potassium, sodium, calcium, magnesium or strontium), even to micrograms per litre. A spray of the sample is injected into a flame and the colour intensity thereby produced is measured photo-electrically.

flap valve A *non-return valve* with a hinged flap that opens in one direction only.

flash cooling Cooling of *flue gases* in a dry-bottom *evaporative tower.*

flash drying A method of either drying or incinerating (*see incineration*), sewage *sludge,* which is flexible and does not need high investment. The *Atritor flash dryer* is one type. After the sludge has been partly de-watered, it is mixed with previously dried sludge and fed into a stream of gas at 500 to 650°C. In a few seconds the moisture content falls to 10% and the sludge-laden gas is led to a *cyclone* which extracts the fluffy solid.

flash mixer A device to create *turbulent flow* in water and thus to mix chemicals such as *coagulants* quickly and intimately with it. It may be a *weir* with turbulence downstream or a narrow jet through which the water or the admixture passes.

flashover In an *electrostatic precipitator,* sparking between the *electrodes*, a fault that reduces the collection efficiency.

flash point   The lowest temperature at which a substance will burn momentarily when a flame is put to it—often part of the specification of a fuel oil. Compare *fire point*.

flashy stream   A stream that drains a *catchment* of high *runoff*, where the flow rises rapidly after rain and falls again equally fast—e.g. in mountain districts or the large paved areas of cities.

flatworms, Platyhelminthes   A phylum that includes *flukes* and *tapeworms*.

Flavobacterium   *Facultative* bacteria that flourish in *trickling filters*, *activated sludge* and possibly in *anaerobic sludge digestion tanks*.

flax retting wastes   The manufacture of linen from flax may be *anaerobic*, resulting in a coloured, smelly acidic *effluent* with a high *BOD*. With the *aerobic* process, however, the effluent is partly re-used and generally much less polluting.

flies   *See caddis flies, Diptera, dragon fly, filter fly, mosquitoes, stone fly*.

flight scraper   A *scraper* in a *horizontal-flow sedimentation tank*, with horizontal blades (flights) that move the *sludge* along the bottom of the tank towards the *sewage* inlet (sludge outlet). They are pulled along by a chain at each side. Under the inlet to the tank is a deeper part called the *sludge hopper*, into which the

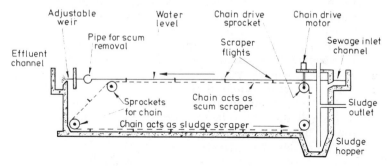

Figure F.5   Flight scraper in rectangular sedimentation tank (Paterson Candy International Ltd.)

scraper sweeps the sludge. Along the surface in the opposite direction the flights return, sweeping the *scum* to the outlet. Each endless chain driving the scraper runs on four sprockets at the surface and bottom of the inlet and outlet ends of the tank. (*See Figure F.5.*)

floating-bed scrubber   A *moving-bed scrubber*.

floating-roof tank   A tank to contain oil or for *anaerobic sludge digestion*, with a hollow steel roof like a pontoon to enable it to float on the liquid or to be raised by the gas pressure. Seals at the rim reduce vapour losses. Vapour lost by 'breathing' is only 10% of the loss suffered from fixed-roof tanks.

floating velocity The least upward flow rate that will hold up a particle in a vertical pipe. Used in *elutriation*, it is determined for spherical particles by *Stokes's law*. For removing *plastics* film in an *air column separator*, a velocity of 1.7 to 2 m/s is needed, and for removing paper at least 2.3 to 2.5 m/s. *See also air classification, rising-current separator, terminal velocity*.

float and sink treatment In *mineral dressing* or laboratory examinations, the separation of a mixture of substances of different densities by suspension in a *dense medium* with a density intermediate between those of the two components of the mixture. This separates the mixture into two fractions—a dense 'sinks' fraction below and a light 'floats' fraction above. Sometimes there is an intermediate product, 'middlings'. The method is used at coal preparation plants, with a sand or galena or magnetite bed suspended in water as the dense medium. Suspensions in air have been used, but are dusty. In the laboratory, *brine*, alcohol, etc., can be used for separating plastics from each other. Other laboratory dense media include bromoform (relative density 2.89), ethylene dibromide (2.17), perchloroethylene (1.6).

floc (Latin *floccus*, a lock of wool) A grouping of solid particles that appears woolly. Unlike granular solids, which settle easily and are dense, a floc breaks up if its shear velocity is exceeded.

flocculant A chemical or physical reagent such as a *polyelectrolyte,* which promotes *flocculation*. It acts as a bridge to bind the suspended particles together. A flocculant may act also as a *coagulant. Compare flocculent*.

flocculation The grouping of solids in water, resulting in *flocs*, often wrongly thought to be the same as *coagulation*. It can be promoted by gentle stirring (*mechanical flocculation*) or by adding chemicals (*flocculants*).

flocculator A stirrer used in *mechanical flocculation*.

flocculator-clarifier, one-pass clarifier A *clarifier*, mainly for drinking water, divided into a flocculating zone and a settling zone. In a simple one-pass clarifier (*Figure F.6*) the two zones are completely separate and the settled solids have no contact with the incoming water, unlike the *solids contact clarifier*. One-pass clarifiers may be rectangular or circular and up to 50 m diameter. In circular clarifiers a central flocculation zone is surrounded by the settling zone. The *surface loading rate* is from 24 to 60 m^3 per day per m of surface area, depending on *raw water* quality.

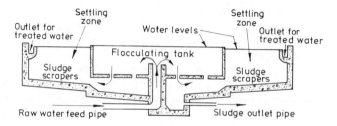

Figure F.6 Flocculator-clarifier, one-pass system, in cross-section, showing the concrete tank. The undermentioned are not shown: the rotating half-bridge and its sludge scrapers in the settling zones; the stirrers in the flocculating tank: the sludge scrapers below the flocculating tank

flocculent Description of particles in a slurry or sludge that are not dense like sand but woolly or gelatinous and have a slow settling speed. *Compare flocculant.*

flocculent settling *See* settling regimes.

flood irrigation Irrigation of crops by flooding the land. In dry climates there may be an excessive loss of water by *evaporation* and therefore *drip irrigation* or *furrow irrigation* may be preferable.

flood routing, streamflow r. Calculations about the passage of floods and their possible storage in *reservoirs* or on flooded low-lying land, so as to reduce peak flows. The greater the area of *catchment* that is flooded, the more will be the reduction of the peak flow downstream of the flooded area.

flora The plants of a region or period.

flotation In *mineral dressing*, as in *sewage treatment*, many flotation processes exist, using fine bubbles of gas or air. In sewage treatment the bubbles lift solids or oil or grease. Light, flocculent particles can often be floated more quickly than they can be settled. Flotation of sewage *sludge* normally uses *dissolved-air flotation. See also dispersed-air flotation, electroflotation, foam fractionation, froth flotation.*

flotation tank, f. cell A tank from which floating matter can be skimmed off.

flow-balancing tank A *balancing tank.*

flue (USA breeching) A usually horizontal or vertical duct that should be gastight and leads furnace gases to a chimney or to *dust arrestors*, etc.

flue-fed incinerator A type of on-site *incinerator*, sometimes used in the USA, in which the refuse chute acts also as a chimney. Consequently, smoke can pollute areas near the points where the refuse enters the chute. In some states they have been made

131

illegal by air pollution laws.

flue gas The hot gas from a furnace of any type, passing along a *flue*. More than 99% of it is *nitrogen, carbon dioxide, oxygen* and water vapour, which are not pollutants, but *gas cleaning plant* may be needed to remove the less than 1% of *dust, grit* and (in the USA) *sulphur dioxide*. Faulty furnace operation may result in the presence of several per cent of *carbon monoxide* as

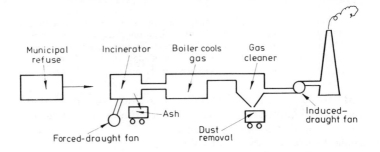

Figure F.7 Essential parts of an incinerator flue gas pollution control system

well as a fraction of 1% of oxides of nitrogen and other polluting gases, but these are rarely if ever extracted. (*See Figure F.7* and below.)

flue gas de-sulphurisation, FGD Removal of *sulphur dioxide* from *flue gases*. Most FGD plants use *sulphur dioxide absorbers* of one sort or another, often with limestone *slurry* as the reagent; consequently, a large volume of gypsum mud has to be disposed of. Other methods for FGD include oxidation to SO_3, sulphur trioxide, using a vanadium catalyst at 400°C. SO_3 is then absorbed in a dilute solution of sulphuric acid. Dry methods of FGD include *adsorption* by copper oxide which has been impregnated on an alumina support. A surface layer of copper sulphate is formed. *Activated carbon* also will absorb SO_2. About half of the new power stations built in the USA in 1977 had to have FGD; the other half used low-sulphur fuel. In Britain FGD is not required by law.

fluid classification The use of *air classification, dense-medium cyclones, float and sink treatment*.

fluidised-bed combustion A bed of mixed inert particles and fuel, through which air or oxygen is blown up to hold them in suspension and thus to increase the burning rate (*see Figure F.8*). It can be regarded as a sort of *suspension firing*, but it has considerable advantages of low *excess air*, high rates of heat

132

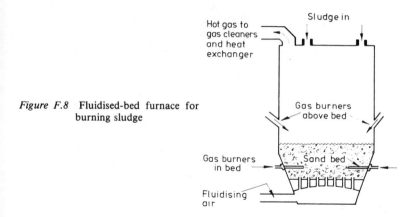

Figure F.8 Fluidised-bed furnace for burning sludge

transfer, the ability to burn low-grade oil or coal and non-slagging combustion. One essential is low-temperature burning, well below the melting point of the ash; otherwise the bed clogs and ceases to work. Even sewage *sludge* can be burnt by feeding it on to fluidised sand or ash. Some ash is carried off in the furnace gases, but can be removed in a *cyclone* separator or in a water spray. By adding limestone grains with a sulphurous coal, 90% of its sulphur can be held in the ash and not released as poisonous *sulphur dioxide* gas.

flukes, Trematoda   Parasitic *flatworms,* such as the schistosomes that cause *schistosomiasis* (bilharzia). Nearly all flukes are *parasites* on or in a vertebrate *host*.

flume   (1) An open channel, usually well above the ground, formerly rectangular and wooden, now also of concrete or steel, and of different cross-sections. It carries water to or within *mineral dressing* plants, *sewage treatment* works, etc., and is often used as a *leat*. (2) Or *measuring flume. See Venturi flume.*

fluorescein   A red solid. Its sodium salt (uranine yellow) dissolves in water to produce a green liquid which fluoresces (emits light) especially in ultra-violet light. It is used as a dye *tracer* and can be detected at very low concentrations.

fluoridation   The UK Ministry of Health has recommended that all drinking water should contain 1 mg/litre of fluoride but many countries have banned fluoride additions. At high levels (50 mg/litre) the bones can be severely damaged. Fluoridation is achieved usually by adding a solution of hydrofluosilicate, sodium silicofluoride or the more soluble sodium fluoride. Children who drink water containing fluoride are less liable to suffer from tooth decay. After teeth have formed, there is no

133

further advantage in drinking water containing fluoride. In toothpaste it can be of some benefit only if the child swallows a little of it. Excess fluoride in water has caused mottled teeth, as in the arid parts of the USA (Texas teeth or Colorado stain), but this does not occur with fluoride contents below 6 mg/litre. *See* Table of Allowable Contaminants in Drinking Water, page 359.

fluorosis Disease caused by too much fluoride in water, air or food, causing mottling of the teeth and then damage to the skeleton. *See* above.

flush cistern *Cisterns* that flush water closets should operate by a *siphon*, delivering 9 litres of water in 5 to 6 s, and be controlled by a float valve (BS 1125). For urinals, automatic flushing should release 4.5 litres per urinal every 25 min.

flushing tank *See sewer flushing*.

flyash Ash that is carried out of a furnace by *flue gases*. Usually the word is applied to the airborne part of pulverised fuel ash (PFA) from the burning of finely crushed coal, the main UK user being the Central Electricity Generating Board. Between 2 and 15% is smaller than 10 μm. The bulk of the rest is smaller than 100 μm. It is used in civil engineering as an additive to cement and as a structural fill in roadbuilding. In *vacuum filters* it has been used as a pre-coat (*see pre-coated filter*) to help in de-watering. Flyash if dry should be moved in sealed containers and the air displaced during the filling of the container should be vented to a *dust arrestor*.

fly tipping Dumping of refuse in an unsuitable place—e.g. on waste land or the drive of an empty house or by the roadside. It is illegal but can be reduced by wise *licensing of sites*.

F/M ratio The *sludge loading rate*.

foamed glass A Swiss development, foamed glass made from *cullet*. It is added to concrete to improve its insulation and has begun to be used for making bricks.

foam fractionation (1) Removal of water-soluble *organic compounds* such as *synthetic detergents* or dyes from *sewage* or water by bubbling air through it and collecting the foam in the exit air stream or skimming it. Solids also may be removed in the foam. (2) *Froth flotation*.

foaming Foam, caused by *synthetic detergents* (syndets) or *polyglycols* in rivers or *digesters* or *activated sludge* plants, can be blown into the air and cause complaints. Syndet foam also reduces the rate of transfer of oxygen from air to the sewage. Spraying with anti-foam chemicals is practised but spraying with final *effluent* is often as effective and cheaper. Foaming becomes noticeable at syndet concentrations as low as 0.5 mg/litre.

fog Tiny drops of water vapour in the air, usually in calm

weather, caused by air cooling below its *dewpoint* and greatly
reducing visibility. Rain dispels fog, as also does warmer
weather. Fog occurs usually during an *anticyclone* and often
under an *inversion*, collecting polluted air.

food chain, trophic c.   A sequence of living organisms that feed
on one another.

food poisoning   Diarrhoea and other complaints, often caused by
Salmonellae other than *S. typhi* and *S. paratyphi*. Food
poisoning is the mildest of the *enteric fevers* and is rarely fatal. It
is spread via the *faeces* and can contaminate food or water.

food-to-micro-organism ratio, F/M ratio   The *sludge loading
rate*.

food processing wastes   *See cannery wastes, dairy and milk
bottling wastes, meat processing wastes.*

food web   A network of *food chains*.

forced aeration of landfill   Blowing air into a *controlled tip*
through pipes that have been buried for the purpose during the
building of the tip. A US investigation found this to be feasible,
that it accelerated biodegradation (*see biodegradable*) and caused
it to be wholly *aerobic*. Thus, no *methane* was produced, or
hydrogen sulphide. Arrangements were made to remove the
waste air and *carbon dioxide,* probably by gravel channels,
ensuring *aerobiosis*.

forced-draught fan   A fan that blows air into a furnace. It
therefore handles only cool air and is subject to much less severe
working conditions than an *induced-draught fan*, but both are
needed in some installations.

force main   A *rising main*.

formazin turbidity unit   A *nephelometric turbidity unit*.

fossil fuels   Coal, oil, natural gas, peat or any other fuel found in
the earth except nuclear fuels.

foul sewer (USA sanitary sewer)   A *sewer* carrying domestic waste
or *trade effluent*. In a *combined system* it also carries
stormwater.

foundries   *See cupola.*

Føyn's process   *Electrolytic treatment of sewage.*

fragmentiser, car shredder, nuggetiser   A powerful *hammer mill*
which can shatter a car body in a few seconds. Tyres, battery,
petrol tank, engine block and transmission are first removed. For
some types the car body first is flattened.

franchise collection   Refuse collection by a private contractor,
uncommon in the UK.

Francisella tularensis   One of the *bacteria* that causes *tularaemia*.

freeboard   The height between the highest water level and the top
of a tank—the height that prevents overflowing.

free-fall velocity, falling speed *See Stokes's law.*

free residual chlorine The unreacted *hypochlorite* ion (OCl⁻) or hypochlorous acid (HOCl) after *breakpoint chlorination*.

free-swimming ciliates *See ciliate protozoa.*

freeze toilet A bucket latrine with the bucket in a deep freeze at $-15°C$. The excreta are frozen, which eliminates the health hazard of the usual bucket latrine, but the capital and operating costs are high. The freeze toilet has been used in Sweden for rural sanitation.

freezing (1) A method proposed for *de-salting* sea-water but not yet commercially used. When water is partially frozen, the salts remain in the liquid fraction and can be washed away from the ice, leaving it purer than the water it came from. (2) Freezing of *sludge* can break it down and reduce it, often to a fraction of its original volume. Freeze-thawing of a difficult, 2% alum sludge concentrates it to 20% solids, but it is expensive.

French degree of hardness *See degree of hardness.*

French drain A sloping trench back-filled with coarse stones, which drains away surface and subsurface water, usually with the help of a *land drain*.

Freons *Haloforms* that are used as *aerosol propellants* and refrigerants.

freshet A clear mountain stream; hence, a 'plug' discharge of *compensation water* from a *reservoir* to help fish to move along a river.

fresh oil spills Oil spills when fresh are the most dangerous to life, because the low-boiling-point liquids such as aviation spirit are the most poisonous. They evaporate or are dissolved quickly and after a day or so in warm weather their toxicity is greatly reduced. An inshore spill remains toxic for longer, because inshore there is less dilution.

fresh sewage *Sewage* with some *dissolved oxygen*. It does not smell unpleasant. Compare *septic sewage*.

fresh water Water with low *total dissolved solids. ASTM 31* defines it as water with less than 1000 mg/litre of dissolved matter; other authorities consider 500 mg/litre as the limit. *See agricultural water, drinking water standards.*

freshwater mussels Water *mains* have been known to suffer from *molluscs* such as *Dreissena polymorpha* in temperate climates or *Limnoperna fortunei* in warm climates. They have been flushed out by heavy *chlorination* (50 mg/litre), during which the main cannot be used.

freshwater shrimps, Amphipoda A *biological indicator* of clean water, requiring a fairly high *dissolved oxygen* level—e.g. *Gammarus pulex.*

froth flotation *Flotation* of finely divided, even *colloidal*, solids helped by adding a suitable soap, oil or other *surfactant*. When air is blown in, the froth contains some of the solids, but not others.

Froude number A dimensionless number that describes *open-channel flow*. It is given by $v/(dg)^{0.5}$ in which v = velocity, d = depth of liquid and g = acceleration due to gravity.

FTU Formazin turbidity unit. *See nephelometric turbidity unit.*

Fucus vesiculosus Bladder wrack.

fuel efficiency The proportion of the *calorific value* of a fuel that is converted into useful energy. It is only about 35% in the most modern purely electrical generating stations, but combined with heat production, as in district heating plants, the efficiency can exceed 70% at the boiler.

fuel reprocessing wastes A *nuclear reactor waste.*

fume In the USA and according to BS 3405, fume is airborne material smaller than 1 μm and may be as small as one ten-millionth of a micron. But the *Alkali Inspectorate* uses the definition 'less than 10 microns', pointing out that the real distinction between *dust* and fume is that fume disperses as a gas. The Inspectorate therefore allows a lower concentration of fume than of dust. Fume must not exceed 0.115 g/m³, while dust may not exceed 0.46 g/m³, although this is reduced to 0.23 g/m³ for large volumes of gas.

fume hood A *dust-extraction hood*, used for extracting chemical fumes.

fume incineration Burning of poisonous or polluting fumes so as to make them safe. Burnable fumes can be lean or rich. Lean fumes contain enough *oxygen* for them to burn if a flame is applied to them. Rich fumes come from activities such as paint drying, solvent recovery or the rubber industry and need added air so as to have enough oxygen to burn.

fumigation (1) A rapid increase in ground-level air pollution. (2) Exposure of germs to poisonous gas or smoke, so as to kill them.

fungi (singular **fungus**) (or **moulds**) Tiny *aerobic, heterotrophic, protists* containing no chlorophyll. They can tolerate drier and more acid conditions than most *bacteria* and also are often many-celled. They live in the earth, fresh water and sea-water. Often they grow so large that they can be seen with the naked eye (mushrooms). Many grow as filaments and may be seen in polluted rivers or *trickling filters* or *activated sludge*. The optimum *pH* for most types is between 5 and 6 but they can exist with water at pH values between 2 and 9. Because fungi are wholly aerobic, they can, in animals, exist only on the skin or in the bloodstream or lungs. Consequently, there are relatively few

137

fungi that cause disease (apart from athlete's foot) in humans, although many cause disease in plants—e.g. potato blight, Dutch elm disease. Fungi include yeasts, which are unicellular.

fungicide   A *pesticide* that kills *fungi*.

furrow irrigation   A type of *irrigation*, widely used for row crops, with small flooded furrows between the rows of crops. This method has only one-fifth to one-half the top surface of *flood irrigation* and therefore loses less water by evaporation. Treatment of *settled sewage* by furrow irrigation is often known as *land filtration*.

Fusarium aquaeductum   A true fungus that grows on *trickling filters* and in rivers. It may give an orange-pink colour to the surface of the filter.

FWPCA   The former US Federal Water Pollution Control Administration, taken over by the *EPA*.

G

gage height   US term for *stage* (water level).

galvanic corrosion, electrolytic c.   *Corrosion* of metal in which *anodic* and *cathodic* areas are distinguishable—e.g. two dissimilar metals in contact or one metal in two different environments. The cathode does not lose metal; the *anode* may.

galvanic series, electro-chemical s.   A list of metals that are increasingly *anodic* in a particular medium—e.g., for sea-water:

(*cathodic*, less corroded, end)
silver
copper
brass
cast-iron
mild steel
aluminium
(anodic, more corroded, end)

increasingly anodic

galvanised steel, g. iron   Steel or iron coated with a thin layer of zinc to protect it from *corrosion*. The zinc may be attacked by acid or strongly alkaline water or by dissolved *chloride*. Galvanised metal has no scrap value, unlike pure zinc or pure iron or steel.

game fish   The families of salmon and trout need a *dissolved oxygen* level of at least 4 to 5 mg/litre, unlike coarse fish, which can exist with less.

gamma radiation   Electro-magnetic radiation resembling X-rays but of shorter wavelength. That from cobalt-60 has been used for disinfecting both water and sewage. Because of its high penetrating power, it can be a hazard to health even when delivered some distance from the body.

Gammarus pulex   A type of *freshwater shrimp* which needs high levels of *dissolved oxygen*.

Gammexane   *Benzene hexachloride.*

ganat   A *ghanat.*

gangue   In an ore, the rock that contains so little metal as to be worthless. Most of it is extracted in *mineral dressing*.

garbage   Kitchen waste, such as potato peelings, fish heads. It decomposes more quickly than other municipal waste. In the USA it can mean other refuse.

garbage farm   A disposal method for house refuse, tried in the USA, in which the refuse is ploughed into the topsoil of a working farm. The main doubt is whether the paper that forms the bulk of the refuse will degrade quickly enough neither to hinder plant growth nor to blow away.

garbage grinder, kitchen g., home g., disposal unit, waste disposal unit, pulping unit   A machine with an electrical input of about 250 W which shreds food wastes so that they can be washed down the sink. It increases the *water demand* by about 5 to 10 litres per person per day.

garnet   A group of silicate minerals with a specific gravity around 4.2, much denser than quartz (s.g. 2.65). It is used in *deep-bed filters* in the 1 mm size as a layer 70 to 80 mm thick between the fine sand and the gravel around the *underdrains*, to prevent them mixing. Garnet is too dense to fluidise during *backwashing* and therefore ensures that the gravel below it is not fluidised.

gas-cleaning plant   *Dust arrestors, sulphur dioxide absorbers, wet scrubbers.*

gas/cloth ratio   *See air/cloth ratio.*

gas-lift pump   An air-lift pump used for circulating the sewage in a digestion tank. Instead of air, it uses sludge gas from the digester to avoid forming an explosive mixture in the tank.

gas–liquid chromatography, GLC   *Chromatography* in which the sample is vaporised and carried along in an inert gas (*nitrogen* or helium) through a column packed with an inert solid (e.g. crushed firebrick) and an inert oil (e.g. silicone oil). The components of the sample move through the column at various rates. Their spacing, measured at the exit from the column, indicates their identity. GLC can be used quantitatively or qualitatively in examining for organic pollutants, particularly *pesticides*.

gastro-enteritis   Inflammation of the stomach and intestine, sometimes caused by *enteric fever* or *food poisoning*.

gate valve   A *sluice valve*.

gathering ground   A *catchment area*.

gauge well   A *stilling well*.

generation time, doubling t.   The time taken for one cell to become two, or for two to become four, etc. For *bacteria*, generation times vary from a few days to 20 min, according to the conditions—especially *pH*, adequacy of *nutrients*, temperature and type of bacteria.

genus (plural **genera**)   A division in *taxonomy*, which groups closely related species. Names of *species* include two (occasionally more) words, of which the first is always the genus. Thus, in *Escherichia coli*, the genus is *Escherichia*.

Geotrichum candidum   A *fungus* that may be found in *trickling filters, sewage fungus*, etc.

ghanat, ganat, kanat, filtration gallery   A long, ancient tunnel or network of tunnels dug to collect water in arid countries, usually in alluvial ground, probably first used 2500 years ago in Persia. The tunnels are dug like an *adit* at a slight upward slope into hillsides to obtain *groundwater*.

Giardia lamblia   A *flagellate protozoon* responsible for giardiasis, a *waterborne disease* that can cause severe diarrhoea. Both humans and animals can act as *host* to it.

Gibraltar fever   *Brucellosis*.

gladioli   Gladiolus leaves indicate the presence of airborne fluorides at a very low percentage. At 0.4 $\mu g/m^3$ in the air the leaves turn yellowish brown.

glasphalte, glassphalt   Asphalt road surfacing in which the sand, or part of it, has been replaced by crushed glass.

glass   *See cullet*.

GLC   *Gas–liquid chromatography*.

go-devil, pig, sewer pill   A wooden ball with spikes projecting about 25 mm from its surface, sent down a *sewer* or other pipe to clear blockages.

grab sample   For fluid, a single, rather than a *composite* sample. Solids also may be grab-sampled.

gradient (US **grade**)   A slope. *See also hydraulic gradient*.

grain size   *See particle-size distribution*.

grains per cubic foot   A unit sometimes used for measuring the amount of solids suspended in *smoke* or air: 7000 grains = 1 lb, or 15 432 grains = 1 kg, or 15.43 grains = 1 g, and 1 m^3 = 35.31 ft^3. Consequently,

$$1 \text{ g/m}^3 \quad = \quad \frac{15.43}{35.31} = 0.437 \text{ grain/ft}^3 \text{ (see page 357)}$$

or

$$1 \text{ grain/ft}^3 = 2.28 \text{ g/m}^3$$

Gram stain An immensely practical staining technique that distinguishes groups of *bacteria*. A heat-fixed smear of bacteria is stained with the dye crystal violet, and then with dilute iodine solution. Finally it is washed with alcohol or acetone. Bacteria that keep a deep blue stain are Gram-positive. Those that finally lose colour are Gram-negative (Christian Gram, 1884).

granular activated carbon *Activated carbon* coarser than 0.1 mm diameter (powdered carbon is finer). It has been used in *carbon adsorption beds* in the USA since 1961 and has the very high surface area of 500 to 1400 m^2/g, depending on the grain size, the larger figure applying to the smaller size. Some commercial active carbon grains are made as large as 1.6 mm. The best carbons are made from anthracite, which is hard-wearing and dense.

granular carbon filter A *carbon adsorption bed*.

granular gas filter A fixed or moving mass of closely packed granules of chemical that can absorb a noxious gas. For *incinerators* the absorbent for *sulphur dioxide* may be alkalised alumina. A preliminary *dust arrestor* should prevent blockage by flyash. Compared with *wet scrubbers*, granular filters have the advantage that they do not pollute water.

graphitisation The *corrosion* of cast-iron by a decay that leaves only a weak network of graphite flakes.

grass plots, overland-flow land treatment Treatment of *raw sewage* or sewage effluents involving grassed areas sloping at 1:15 to 1:100. Clay soils should be well drained to avoid waterlogging and to encourage the growth of grass. In cold weather, channel distribution is better than sprays, which freeze easily. Feed channels and collection channels should be brushed out and the soil may have to be rested regularly for tillage (ploughing, re-grading, seeding). Typical results for removal of suspended solids are 60 to 90% with a *BOD* reduction of 50 to 95%, some reduction in *bacteria, nitrogen* and *phosphorous,* and an increase in *dissolved oxygen*. The amount of liquid fed to the plots varies with the soil, weather and other conditions from 10 litre m^{-2} day^{-1} with raw sewage to 100 litre m^{-2} day^{-1} for *secondary effluent.*

grate (1) In an *incinerator* a grate supports the refuse and allows ash to fall through to the ashpit. It also admits *underfire air* through openings, and its section nearest the entry point dries the incoming wet refuse. It moves the refuse from the drying area through the combustion area to the burnout area of the grate and

141

into the ashpit, with agitation to ensure full burning. Many types of incinerator grate exist: stationary, rocking, travelling, rotating and reciprocating. Furnaces with *suspension firing* have no grate, nor do fluidised beds. (2) Or *grille* or *screen*. The parallel bars in the lower part of the casing of most *hammer mills*, through which the crushed refuse passes when it has been made small enough. The grate may be adjustable to increase or reduce the bar spacing, thus increasing or reducing the mill's throughput. For *primary shredding* the grate bar spacing is usually 6 to 12 cm, sometimes 20 cm.

gravel-bed filter, granular-bed f. A *dust arrestor*, usually for hot gas, with a bed of grains sometimes of 6 mm size, which can collect fine dust at temperatures up to 400°C. The bed, carried on springs, is periodically vibrated to release the dust.

gravimetric dust sampling Determining the weight of the *respirable dust* in unit volume of air rather than the number of particles in it, a way of measuring the danger from dust breathed in, since the harm to the breather's health increases with the mass of the respirable dust, not with the number of particles.

gravitational water (1) Soil water that moves downwards in the *zone of aeration*. (2) *Irrigation* water in canals or rivers, which is not pumped.

gravity current A *density current*.

gravity filter A *filter* that is subject only to the *head of water* over it, and is always a downflow filter. *Compare pressure filter* (2).

gravity main A *main* through which the water flows because of difference in level, without the need for pumping.

gravity sand filter A downflow filter, either a *rapid gravity sand filter* or a *slow sand filter*.

gravity separation The use, in *mineral dressing* of a *mechanical sorting plant*, of *air classification, float and sink treatment, dense media, jigs, rising-current separators*, and so on. Many of these methods involve wet separation. In *sewage treatment* also gravity is the basis of *sedimentation*.

gravity thickening *See sludge thickening*.

grazing fauna Insects and their larvae, *nematodes* and oligochaetes, and *rotifera* which graze on the *biological film* in a *trickling filter*. They restrict excessive growth of the film and reduce *ponding. See sloughing off*.

greases and fats Greases or fats from *meat-processing wastes, woollen industry wastes* or the leather or soap industries, or even from kitchens, can coat the walls of *sedimentation tanks* and putrefy there, as well as increasing the amount of *scum*. Both scum and grease coatings are unpleasant and smelly. Consequently, they should be removed where possible by *flotation* in *preliminary treatment* or at the polluting works before its *effluent* reaches the *sewer*. Domestic *sewage* contains 50 to 150 mg/litre of grease. *See ether-extractable material.*

grease trap A small *skimming* tank with a *detention period* of 10 to 30 min, large enough to hold and cool the grease delivered in one day from any *sullage* source, and installed near it. It should be regularly emptied. The inlet is below the surface and the outlet is at the bottom, separated from the inlet by a baffle that holds back the grease.

green algae, Chlorophyta A large *phylum* of *algae* that includes *Chlorella* and *Chlamydomonas*. They may have *flagella*, and can be unicellular, multicellular or *filamentous organisms*.

greenhouse effect The fact that the heat from the sun enters a greenhouse more easily than it can leave it. On the atmosphere as a whole, water vapour acts similarly, because clouds can blanket the earth at night, preventing frost. *Carbon dioxide* also has a greenhouse effect. It has therefore been claimed that the CO_2 in the air, if it were to rise much above its present level of 0.03%, because of high consumption of fuels, could warm the earth by a greenhouse effect. This may be true, but the *aerosols* also emitted by industry might have an opposite effect.

green water Water goes green in ponds in summer because of green *algae* which develop fast in sunlight. Their existence, however, may encourage other growths, including some that eat them.

grey water *Sullage. Compare black water.*

grit In air pollution, particles larger than 76 μm. In *sewage*, grit is much larger but includes sand, gravel, ashes, metal, glass, earth from vegetable washing, etc. It settles more easily than the *organic* solids in sewage and usually increases in amount after rain. In desert areas the grit content may be high, because of wind-blown sand. In cold climates gritting of roads in icy weather increases the grit content of *stormwater*. The volume of grit removed from domestic sewage is normally 15 to 30 cm^3 per person per day.

grit chamber (1) General name for any grit removal device. (2) Or *detritus tank*. An old-fashioned tank for removing grit from *sewage*, usually with a flow velocity below 0.3 m/s, which consequently allows *organic* solids to settle with the grit and putrefy.

grit channel A long narrow tank that extracts grit from *sewage*, as in the *constant-velocity grit channel*.

grit removal *See above; see also aerated grit chamber, Detritor, Pista grit trap.*

grit washer *See classifier.*

143

grizzly A bar *screen* used in *mineral dressing* and usually made from rails, with at least 10 cm gap between them. The bars are sloped to enable the material to slide down them.

gross heating value *See calorific value.*

ground-level pollution The weight of a pollutant (usually in μg) per unit volume (m^3) in the air between the ground and 1.5 or 2 m above it.

groundwater Water contained in the soil or rocks below the *water table*—i.e. in the *zone of saturation*. It is the immediate source of well water. If too much is drawn off and the water table (*piezometric surface*) is lowered, the ground surface may sink disastrously, as in Venice, but much depends on the type of ground. About 5.75 million m^3/day were obtained from wells as *raw water* in England and Wales in 1974, compared with 9.9 million m^3 from rivers and lakes.

groundwood pulp, mechanical p. Wood pulp for making newsprint, obtained by the crushing of wood. It is brittle, with short fibres, but uses few chemicals, and is the cheapest and environmentally least harmful way of making paper from wood.

growth phases of bacteria The main growth phases in the *population dynamics* of microbes in sewage treatment are: *lag phase, logarithmic growth* phase, *declining growth* phase, and *endogenous phase of growth*, all of which are seen in a *sigmoid growth curve*.

Guinea worm, *Dracunculus medinensis* The cause of *dracontiasis*; a *nematode* worm that lies beneath the skin, producing arthritis of the joints when it matures in the rainy season. This may interrupt the planting in a village community. Mature *larvae* washed from the infected skin may then infect *Cyclops* which in drinking water passes on the dracontiasis. The disease, common in west Africa, could be reduced by *disinfection* of the water.

gulley, street inlet, gutter inlet A grated opening at the edge of a street, for *road drainage*.

gulley emptying The task of extracting accumulated mud from the *gulley* pots located every 100 m or so in the gutters beside a city street. A gulley emptier (gulley sucker) is a tanker vehicle with a powerful vacuum pump to suck out the mud. It has two tanks, one to contain the mud, the other to hold clean water that re-seals the gulley. Gulley emptiers may also be used for emptying *cesspools* and removing *night soil*.

gutter inlet North American term for a *gulley*.

Gyractor A *pellet reactor*.

H

habitat The natural environment or location of a plant or animal.

half-life In a *radio-isotope*, the time taken for half of its atoms to disintegrate to form another isotope. *Isotopes* with half-lives of years or tens of years may be unsafe, because no way is known whereby *radioactive decay* can be accelerated. Alternatively, if their level of radioactivity is low, they may be safe. A pollutant, though not radioactive, also may have a half-life.

haloforms, halomethanes Compounds of *halogens* with carbon and hydrogen, in which the halogen replaces one or more of the hydrogen atoms in methane, CH_4. Thus, chloroform (trichloromethane) is $CHCl_3$. Bromoform (tribromomethane) is $CHBr_3$. CCl_4 is carbon tetrachloride. The Freons are also haloforms; one of them is dichlorodifluoromethane, CCl_2F_2. They are impurities sometimes found in drinking water.

halogens *Chlorine, bromine*, fluorine and *iodine*, a family of elements that combine with other elements in similar ways, and can sometimes be interchanged, as with the *haloforms*.

halophile An organism that lives in salt water or salty soil.

hammer mill The commonest type of *mill* for the dry shredding of municipal refuse. Many types exist, with or without *grate*, with single or double rotors, with horizontal or vertical shaft, having swing hammers or rigid ones, apart from distinctions of size or output (*Figure H.1*). They all run very much faster than *drum*

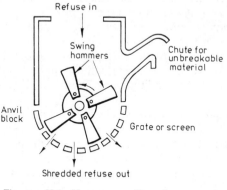

Figure H.1 Hammer mill, diagrammatic section

pulverisers and consume more electricity. The hammers can weigh over 50 kg and need regular attention so that the inevitable wear can be taken up. The rotors turn at 900 to 3000 rev/min and the tip speed, according to US makers, is 30 to 70 m/s. The power installed should be not less than 8 kW per tonne of hourly throughput, preferably 16 kW, to enable the motor to cope with the surges of power demand caused by massive objects entering the mill. *See ballistic separator.*

hard detergent A *synthetic detergent* that does not easily biodegrade and so may cause *foaming* in rivers at concentrations as low as 0.3 mg/litre. Modern soft detergents are about 90% *biodegradable* in *biological treatment*, and so have greatly reduced foaming.

hardness The hardness of water is the extent to which it prevents soap lathering. It has little effect on *synthetic detergents*; hence their popularity in hard-water districts. It is caused usually by *calcium* or *magnesium* salts, occasionally also by those of iron or aluminium, which react with soap to form the typical *scum* seen around baths or washbasins. *Carbonate hardness* also causes the *scale* that is precipitated when water is boiled and the bicarbonates of calcium and magnesium are converted to insoluble carbonates. Hardness is therefore a nuisance both to homes and to industries. It may be expressed in mg/litre or in *milligram equivalents* of *calcium carbonate*, $CaCO_3$, per litre of water, even though salts of metals other than calcium may be causing the hardness. In the UK 'soft' water has less than 50 mg/litre of $CaCO_3$; 'hard' water has more than 150 mg/litre. In very soft water *heavy metals* may be dissolved from pipes, e.g. plumbo-solvency. In several countries it has been found that soft water is associated with heart disease but the reason is not known. *See also alkalinity, degree of hardness, non-carbonate hardness, water softening*; *see also* Table of Allowable Contaminants in Drinking Water, page 359.

Hardy Cross method A method of successive approximations due to Professor Hardy Cross (1936), which can be applied to many complex networks, including pipe flows, bending moments in building frames, etc.

harvesting of algae *See algal harvesting.*

hazard diamond A usually coloured, square label set with its diagonals vertical and horizontal, on the right-hand side of the *Hazchem* sign, on which is shown the particular danger from the chemical in the tanker, whether corrosive, spontaneously flammable, etc.

Hazchem code A code of the signs displayed on lorries, or road or rail tankers, used in this transport hazard information system, devised largely by the UK Chemical Industries Association. For the use of the emergency services, the code gives information on suitable fire-fighting media—in particular, whether the substance if spilt may safely be washed into drains or

146

Figure H.2 Hazchem sign seen on the the rear of tanker lorries and the sides of rail tanker wagons. The background colour of the sign is orange except for the diamond-shaped hazard warning label which has a white background. 1789 is the United Nations number for HCl. The lorry driver carries a Hazchem card explaining how to deal with a spillage. Advice about spillage or other accidents may be obtained from the Road Haulage Association, the National Association of Waste Disposal Contractors, the Chemical Industries Association or their members. (Reproduced by courtesy of the Department of the Environment.)

watercourses. Each substance has therefore an indication 'dilute' or 'contain' to prevent unnecessary pollution of rivers.

haze   A faint mist, slightly reducing the clarity of the air, caused by *airborne* dust or liquid particles about 0.2 to 1 μm in size.

Hazen unit   A number that defines the *colour of water*. One unit is the intensity of colour produced by a solution of 1 mg/litre of platinum as chloroplatinic acid with 2 mg/litre of cobaltous chloride. Drinking water should have fewer than 5 Hazen units of colour.

Hazen–Williams formula   A formula widely used in the USA to evaluate *pipe flow* in water distribution systems:

$$v = 0.849 \; CR^{0.63} s^{0.54}$$

where v = velocity (m/s), C = Hazen–Williams roughness coefficient, R = *hydraulic mean depth* (m) and s = *hydraulic gradient*.

HDPE   *High-density polyethylene* (polythene).

head loss, pressure loss   Reduction in *head of water* in a flowing pipe or channel, usually because of the friction of flow or change in a direction of flow. The higher the speed of flow, the higher is the head loss.

147

head of water, pressure of w.   The energy in water that enables it to do mechanical work. The total head is the sum of three components: (1) elevation head—for example, the height above sea level; (2) pressure head, measured by a *piezometer tube*, is the height of the static column of water that can be carried by the pressure at that point; (3) velocity head, the energy in water due to its movement (its kinetic energy) is equal to $v^2/2g$, where v is the mean velocity of the water and g is the acceleration due to gravity. This kinetic energy is measured by a *Pitot tube* with the opening facing upstream.

heat drying of sludge   *Sludge cake* can be dried at about 360°C but with the production of offensive smells. To destroy the smells a temperature of at least 750°C is needed. *See band dryer, flash drying.*

heat-recovery wheel, thermal w.   A large wheel that slowly turns through hot outgoing air while its other half is turning through cool incoming air. In this way much of the heat from the outgoing air is transferred to the incoming air, as in the *rotary regenerative air heater*.

heat treatment of sludge   Two processes of heat treatment (not drying) of *sludge* exist. In the process developed originally by Porteous the sludge is heated with live steam at 180°C and high pressure, 10 to 15 atm, for 30 min or so. The smell can be disposed of by discharging the gases under the boiler grate as combustion air for the boiler. The liquor resulting from the filter presses is also foul, with a high *BOD* and must be returned for *biological treatment*. For the other process, *see wet air oxidation.*

heavy-medium cyclone   A *dense-medium cyclone.*

heavy-medium separation   *Dense-medium treatment.*

heavy metals   Elements commonly used in industry and known as 'heavy metals' are generally toxic to animals and to *aerobic* and *anaerobic* processes, but not all are dense, nor are all entirely metallic. Including *arsenic, cadmium, chromium, copper, lead, mercury, nickel, selenium* and *zinc*, they are allowed only in small concentrations in any waste discharged to a *sewer* and even smaller to a river or lake. Many of their compounds may be undetected in river or lake water because of their *adsorption* on solids suspended in the water, which concentrates them in the bottom mud, as at *Minamata Bay. Ion exchange* can enable a works to re-use its water and thus to reduce the amount of these pollutants sent out. Occasionally, also, some of the metal may be recovered. Heavy metals are sometimes defined as those with a density relative to water of more than 5. The truly dense metals, with relative density around 20, include gold (19.3), iridium (22.4), osmium (the densest: 22.48), platinum (21.5), uranium

(18.7) and tungsten (19.1), but these are all so rare and valuable that they are unlikely to be found as pollutants.

heavy water Deuterium oxide, D_2O; but *see also tritium oxide.*

helminths Any *parasitic worms; see* below.

helminthic diseases Diseases caused by *tapeworms, flukes* or *nematode* worms. Flukes cause *schistosomiasis* and *schistosome dermatitis.* Nematodes cause *ascariasis, Guinea worm, hookworm disease, onchocerciasis, pinworm* and *whipworm* diseases. Probably half the world's population are affected by a helminthic *parasite.*

hemi-celluloses Soluble *carbohydrates*, with properties like those of *cellulose*, and often found with it—e.g. in wood. Being soluble, they dissolve in the water used to make paper from wood pulp and often reach half the *organic* content of the *effluent* from a pulp mill. Hemi-celluloses can be partially fermented to alcohol.

Hemiptera *Water bugs.*

hepatitis *See infectious hepatitis.*

herbicide A chemical, especially any selective weedkiller, that kills vegetables.

herbivore A plant eater.

hetero-aerobic lagoon A *facultative lagoon.*

heterotroph, heterotrophic organism Any organism, including *bacteria* and *fungi*, that needs *organic* matter as an energy source, unlike *autotrophs*, which do not. Heterotrophic *bacteria* in *sewage treatment* biodegrade the organic pollutants and thus help the process.

hexa-chrome, hexavalent chromium *Chromium* in the form of chromate, $-CrO_4$, and dichromate, $-Cr_2O_7$, is more toxic to plant and animal life than trivalent chromium (chromium as the cation). *Metal-plating wastes* include this poison. Effluents containing it should be treated by chemical reduction to trivalent chromium in acid conditions. The pH is then raised to precipitate out chromium hydroxide. Alternatively hexa-chrome may be taken out by ion exchange. *See* Table of Allowable Contaminants in Drinking Water, page 359.

high chimney policy *See tall-stack policy.*

high-density baling, high-pressure b. Reducing the volume and increasing the density of a material by crushing it into wired or strapped or loose blocks with a hydraulic press and thus making it easier to handle. So far as municipal refuse is concerned, it means cubes of about 1 m³ weighing about 1 t. The bales occupy 30% less space on the tip than refuse from a compacted container. Densities above $1200 \ kg/m^3$ have been obtained, using pressures up to 340 atm. *See balefill, springback.*

149

high-density polythene, HDPE   One of the *thermoplastics*, of which good-quality and light dust bins are made.

high-efficiency cyclone   *Cyclone* types of *dust arrestor* that can reliably remove particles as small as 4 μm, though with a high pressure drop, up to 10 cm water gauge.

high-energy scrubber   *See low-energy scrubber.*

high-rate activated sludge process   The operation of *activated sludge* plants at *organic loads* above 1 kg of *BOD$_5$* per kg of *mixed liquor* solids per day. High loadings in activated sludge plants may result in *bulking*.

high-rate aerobic lagoon   A shallow (20 to 40 cm) *waste stabilisation pond*, in which the light can reach the bottom, enabling *algae* to grow fast. It is used to treat *settled sewage* or the *effluent* from *anaerobic ponds*, and *detention periods* of only 1 to 3 days are needed. The main problem is separating the algae from the effluent flowing out of the lagoon. It usually has no *surface aerators*.

high-rate anaerobic digestion   *Anaerobic sludge digestion* at about 30 to 35°C with a short *detention period* (less than 20 days) and a high *organic loading*, greater than 1.6 kg of *volatile solids* applied per day per m^3 of digester capacity—3 to 4 kg m^{-3} day^{-1} being typical. The digester must be fully mixed and closely controlled, with regular laboratory analyses. Much success has been claimed in the USA, but sludge digestion can be difficult to control and the fast reaction gives the operator little time to manoeuvre. For example, a typical high-rate digester is designed to operate at 15 days' detention for an inlet sludge concentration of 7% solids. If the inlet solids drop to 5%, the detention time will drop to less than 7 days, which is critical for successful operation of the digester, even with excellent process control. The high-rate digester is followed by a secondary digester or by some other tank to separate out the *digested sludge*. Digested sludge is sometimes re-cycled from the secondary digester to the primary digester to increase the active microbial population in the primary digester. *See methane fermentation.*

high-rate trickling filter, roughing filter   (1) A *trickling filter* with both a high *hydraulic loading* (1 to 4 m^3/day per m^3 of filter) and a high *organic loading* (1 to 5 kg BOD$_5$ per day per m^3 of filter); often built as a tower containing *plastics filter medium*, with the *sewage* distributed, as usual, at the top of the tower. A minimum *wetting rate* of 30 m^3/day per m^2 of filter plan area is normally needed, and *effluent* must be re-circulated if necessary to maintain this flow. The tower is followed by a *humus tank*. *Ponding* of the filter is avoided because of the large spaces in the plastics media. Typical *BOD* removals are 45% at an organic

loading of 4 kg BOD_5 per day per m^3 of filter, varying up to 85%
at 1 kg BOD_5 per day per m^3 of filter. It is used as a roughing
treatment for industrial effluents of high BOD before discharge
to the *sewer* or before further *biological treatment*. *See extended
filtration.* (2) A conventional, shallow (1 to 2 m deep) trickling
filter packed with stone, in which effluent is re-cycled from the
humus tanks. A high wetting rate of 10 to 30 m^3 per m^2 of
trickling filter per day helps to slough off the slime and stop
ponding. It is claimed that *BOD* loadings up to 2 kg of BOD_5 per
m^3 of filter per day can be achieved without ponding.

hindered settling   *See settling regimes.*

hog, hogger   A *mill* or *shredding* machine.

hog louse   *Asellus aquaticus.*

holophyte   An organism that obtains food by *photosynthesis.*

holoplankton   Organisms that spend their whole lives as *plankton*,
unlike *meroplankton.*

holozoic organisms   Organisms that prey on others, whether plant
or animal, and thereby affect their population. *Compare
saprobe.*

home grinder   A *garbage grinder.*

homoiostasis, homeostasis   The maintenance of an *ecosystem*
against external stress.

homoiotherm, homeotherm, idiotherm   Warm-blooded birds and
animals. *Compare poikilotherm.*

hookworm   A *nematode* with hooks in its mouth, which is a
parasite in man.

**hookworm disease, miner's anaemia, ankylostomiasis,
necatoriasis**   A *waterborne disease* that in 1963 was believed to
affect more than one-third of the world's population. It is caused
by parasitic *nematode* worms such as *Ankylostoma duodenale* or
Necator americanus. The infection is known either as
ankylostomiasis or necatoriasis, depending on the worm present
in the human. The worms enter the body through the skin of bare
feet and attach themselves with the hooks in the mouth to the
sufferer's duodenum. The worms are passed in the *faeces* of the
sufferer. The disease can be reduced by adequate *sewage*
hygiene.

hopper-type clarifier   A sample *sludge-blanket clarifier* with the
inlet at the bottom of the *sludge hopper.* The coagulated water
flows up and through the *sludge blanket* (*see Figure H.3*). The
surface loading is 40 to 100 m^3/day per m^2 of surface area of
tank, depending on the quality of the *raw water.*

horizontal-flow sand filter   A sand filter in which the flow is radial
(Simater filter) or from one side to the other (Bohna filter).

horizontal-flow sedimentation tank, rectangular s. t.   *Sedimen-*

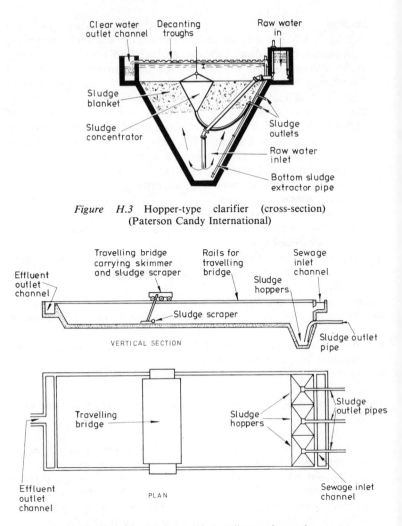

Figure H.3 Hopper-type clarifier (cross-section) (Paterson Candy International)

Figure H.4 Horizontal-flow sedimentation tank

tation tanks (*Figure H.4*) in which the *sewage* flows in at one end and out at the other. The *sludge* is usually scraped to *sludge hoppers* at the inlet end of the tank and is removed once or twice daily. The length/breadth ratio is normally about 4/1 and the greatest length is not usually above 100 m. The depth is from 2.5 to 3.0 m and the floor slopes down towards the inlet end at about 1:100. The tanks are used for *primary sedimentation* of sewage or for *de-silting* raw water. In the UK *travelling-bridge scrapers*

152

are more common but in the USA *flight scrapers* are used.

horizontal-shaft hammer mill   The oldest type of *hammer mill*, in existence at least since 1875. It has a screen of perforated steel plate or parallel steel bars below it, known as the *grate*. The shaft carries swing hammers whose maximum swing diameter is called the tip circle. The difference between the tip circle and the internal diameter of the casing partly decides the maximum size of the output from the mill, but the spacing of the grate bars also is important.

host   An organism that is lived on by a *parasite*. In many diseases humans are hosts to a parasitic worm (*see helminthic diseases*).

humus   (1) Remains of plants and animals that have been broken up (biodegraded) by microbes in soil—a valuable part of topsoil for plants. (2) *Humus sludge*.

humus sludge   *Biological film* that has sloughed off a *trickling filter* and settled out in the *humus tank. See* above.

humus tank   A *secondary sedimentation tank* that purifies the *effluent* from *trickling filters* by separating the *humus sludge* from it.

hydrated lime   *Calcium hydroxide*.

hydraulic gradient (USA **h. grade line**)   For fluid flow through a *pressure pipe*, a line drawn to show the levels to which the fluid would rise in open tubes connected to the pipeline—i.e. the *piezometric surface*. In *open-channel flow* the hydraulic gradient coincides with the water surface. (See pages 210 and 239).

hydraulic jump, standing wave   In *open-channel flow*, the change from *supercritical flow* to *subcritical flow*—i.e. a sudden increase in depth from less than to greater than the *critical depth. See standing-wave flume*.

hydraulic loading   (1) The flow rate per day through a treatment tank or filter, per m^3 of capacity of the unit. (2) US term for *wetting rate*. (3) *Surface loading rate*.

hydraulic mean depth, h. radius   The cross-section of the water in a channel, stream, etc., divided by its *wetted perimeter*. For a pipe flowing full the hydraulic mean depth is one-quarter of the diameter.

hydraulic residence time, h. detention t.   *See detention period*.

hydraulics   The study for practical purposes of fluid flows and pressures, including the forces exerted by and on water in motion.

hydraulic yield   The *yield of a water source*.

hydrazine, N_2H_4   A chemical added to boiler *feed water* to de-aerate it and thus to reduce the hazard of boiler *corrosion*. It decomposes in the boiler to form *nitrogen* and water.

hydro-, hydr-   A prefix indicating a connection with water.

Hydrobaenus A genus of *Diptera* midges that are common *grazing fauna* in *trickling filters*. *See filter fly.*

hydrocyclone A *wet cyclone.*

hydrodynamics The theory of fluid motion.

hydro-extraction A method of *concentrating viruses* by placing a sample of water containing the virus in a cellulose dialysing bag immersed in polyethylene glycol. This fluid absorbs water from the bag, leaving half the viruses inside it and achieving a concentration of about 100:1. Other molecules of high molecular weight may also be concentrated with the viruses.

hydrofracture *Encapsulation* of less hazardous *nuclear reactor wastes* by mixing them with Portland cement, *flyash*, attapulgite, illite, delta-gluconolactone and tributyl phosphate, to make a concrete that sets after it has been injected through a *disposal well* into a suitable geological formation. The illite attracts caesium and the flyash attracts strontium, while the delta-gluconolactone retards the setting until injection has been completed, at a depth of 200 to 300 m.

hydrogasification High-temperature conversion of *garbage* or manure with added hydrogen to produce *methane* or oil fuel—a subject of research by US Bureau of Mines since the early 1970s.

hydrogen ion concentration *See pH.*

hydrogen peroxide, H_2O_2 A liquid that may be used to disinfect water. It is sold as a 70% solution, and diluted to 50% before it is added to the water. A strong oxidising agent, it reduces smell from *hydrogen sulphide* and is much less corrosive than *chlorine*. However, it leaves no residual *disinfection* and is expensive. It has also been used to re-aerate *sewage*.

hydrogen sulphide, sulphuretted hydrogen, H_2S A poisonous gas produced in *anaerobic sludge digestion, sewers* and wherever *sewage* putrefies. Its smell of rotten eggs can easily be detected at one part per million in air but not at high concentrations. It is toxic to man and at 1000 p.p.m. kills immediately. The gas progressively ionises in water as the *pH* rises:

$$H_2S \rightarrow HS^- + H^+$$
$$HS^- \rightarrow S^{2-} + H^+$$

Only the undissociated H_2S comes out of solution to produce the smell; hence, a low pH favours removal and flushing out by *aeration*. However, if water containing H_2S is aerated, some of it may be oxidised to *elemental sulphur*, giving a milky look to the water. H_2S production in sewers is particularly important in *sulphide corrosion*. If a well water smells of H_2S, this does not necessarily mean it is polluted, since *groundwater* is often de-oxygenated. *See de-oxygenation; see also* Table of Allowable

Contaminants in Drinking Water, page 359.

hydrogeology All of *hydrology*, with emphasis on geology, chemistry and *groundwater* exploration.

hydrograph A graph that shows the flow rate (discharge) of a river or stream. The area under the graph, the product of flow rate and time, is the volume of water delivered in that time. Long-period hydrographs are essential for planning either a hydro-electric or a water supply scheme. Short-period hydrographs (*unit hydrographs*) show the pattern of *runoff* from a given intensity (number of mm or inches) of rainfall and are useful in the design of flood prevention schemes.

hydrological cycle, water c. (*Figure H.5*) Water evaporates from a sea or lake and the resulting vapour rises to form a body of damp air over the water. Further heating of this damp air causes

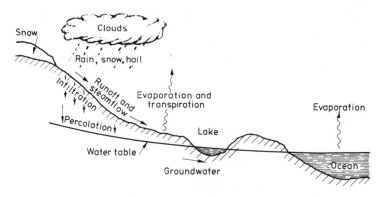

Figure H.5 Hydrological cycle

it to rise through the colder air above it, high into the atmosphere, until it is cooled by the increasing height. The water vapour then condenses to form clouds in which the water droplets enlarge. When they become too heavy to remain airborne, they fall from the cloud as rain, snow or hail. This is the water cycle. *Evaporation* takes place from rain and when wind blows over water or land. Water also evaporates by *transpiration* from vegetation and as a result of human activity—for example, in exhaust steam. When water falls over land, some is held by vegetation and never reaches the ground surface. Some reaches the ground on roads, roofs, etc., and flows straight into drains, streams and rivers for a quick return to the sea or a lake. Other water will flow into the ground rather than over it and will percolate down to the *water table*. This *groundwater* may re-appear on the surface at springs or wells or move horizontally to a river or sea. Such movement may be very slow, and water held deep in the ground may have been there for many thousands of years. Rain, snow or hail can also fall over permafrost and remain for a very long time as ice or snow. Glaciers and polar ice contain 35% of the world's fresh water, compared with 0.03% in the atmosphere and 0.003% in the rivers.

hydrology   The study of the *hydrological cycle* and of water.

hydrolysis (verb **hydrolyse**)   The formation of compounds by combination with water. For example, *urea* ($(NH_2)_2CO$) is readily hydrolysed in *sewage* to ammonium carbonate. *Polysaccharides* may be hydrolysed to their component sugars. Metal ions form complex *ions* with water. *Bacteria* can often hydrolyse insoluble or long-chain *organic compounds* to simpler soluble ones.

hydrolytic tank   A two-storey *septic tank* used in the UK in 1903 by Travis, from which the *Imhoff tank* was developed.

hydrometer   A thin-stemmed instrument that floats vertically in fluid to indicate its *specific gravity*.

hydrometry   The measurement of water, particularly its flow rate.

hydrophilic   Description of something that is easily wetted.

hydrophobic   Description of something not easily wetted.

hydrophobic chalk, Nautex H   An oil-attracting powder, of which 3000 t were used by the French for sinking oil slicks on the sea off Brittany at the time of the *Torrey Canyon* disaster. A similar material, stearated limestone dust, was used in the Firth of Tay in 1968. Such sinking methods cannot be used over fishing grounds, because of the danger that the sunken oil will foul fishing nets and taint any fish caught.

hydrophobic colloid   A colloidal suspension that has a low affinity for water—i.e. it settles readily.

Hydropsyche   A genus of *caddis fly* that is more tolerant of *organic* pollution than other caddis flies.

hydrosphere   The water of the Earth's crust, including *groundwater*, surface water, the oceans, etc.

hydrostatic head, hydrostatic pressure   The pressure in water caused by its difference in level between one point and another. A *head of water* is expressed as its height in metres. Pressure is force per unit area.

hydrostatic valve   *See telescopic valve*.

hygrometer   An instrument that measures the air's water vapour content or its *relative humidity* or both.

Hymenolepis nana, dwarf tapeworm   A *parasite* on man which is passed in the *faeces* and transmitted without an intermediate *host* via soiled clothes, food, *sewage*, contaminated water, etc.

hyper-filtration *Ultra-filtration*.

hypo *Sodium thiosulphate*.

hypochlorination The use of calcium or sodium *hypochlorite* instead of chlorine gas for *chlorination*. This change-over was made in New York City in 1965 because liquid chlorine, previously brought in 1-ton containers, was thought to be too dangerous.

hypochlorite Any salt of *hypochlorous acid*, usually NaOCl or Ca(OCl)$_2$. *See chlorination*.

hypochlorous acid, HOCl One of the most powerful *chlorine* disinfectants for water, which dissociates to OCl$^-$ and H$^+$ at *alkaline* pH. At *pH* 7.5, 50% is dissociated; and at pH 5, none. Undissociated HOCl is said to be 70 times as bactericidal as OCl$^-$; therefore, during *disinfection* with chlorine, low pH is desirable. At pH 7, 99% kill of *Escherichia coli* is achieved in 20 min contact time at 0.04 mg/litre of *free residual chlorine* or in 5 min at 0.12 mg/litre of free residual chlorine.

Hypogastrura viatica A species of *springtail*, an insect that can help *trickling filter* operation by feeding on the *algae, bacteria* or *fungi* that clog it during *ponding*. *Filter flies* also help.

hypolimnetic Concerned with the *hypolimnion*.

hypolimnion The coldest, lowest layer of water in a stratified lake or *reservoir* or sea, bounded above by the *thermocline*. Its water is stagnant and has little or no *dissolved oxygen* but a high *carbon dioxide* content and a fairly uniform temperature of 4 to 6°C. *See thermal stratification*.

I

ICE Institution of Civil Engineers, Great George Street, London SW1, the UK's foremost civil engineering body, with members in all the civil engineering fields, including environmental control.

ICOLD The International Commission on Large Dams, a body that is particularly interested in dam safety. Its UK host is the Institution of Civil Engineers. *See ICE*.

ilmenite A *dense mineral* of relative density 4.8, used like *garnet* in *deep-bed filters* to prevent disruption of the gravel surrounding the *underdrains*.

imago The adult stage of an *insect*, following after either the *pupa* or the *nymph*, depending on the species.

Imhoff cone A conically tapered, open-topped glass vessel of 1 litre capacity, with the point of the cone down, graduated up to about 50 ml from the bottom. It measures the *settleable solids* in *sewage* quickly and cheaply. Domestic *raw sewage* from western communities produces 3 to 12 ml of solids per litre after 45 min of settlement in an Imhoff cone.

Imhoff tank A type of *septic tank*. It is a two-storey *sedimentation tank* with a *sludge* compartment directly below the *sedimentation* compartment, connected with it by a slot through which the solids continuously slide in (*see Figure I.1*). Gases

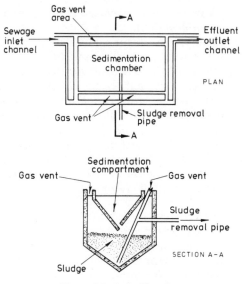

Figure I.1 Imhoff tank

from the sludge compartment are directed away from the falling solids so as not to hinder their descent, and the sludge can be removed by a pipe from the sludge compartment. In warm weather it therefore functions better than the *septic tank*, in which the digesting sludge, yielding gas, may hinder settling or lift the sludge to the surface.

Immedium sand filter An *upward-flow sand filter* about 1.5 m deep, in which the sand rests on a grid, the coarse grains settling naturally at the bottom and the finer ones at the top.

impact block, liner Replaceable steel or iron wear plates inside a *mill*.

impact breaker, crusher, impactor A machine sometimes used with municipal refuse for breaking bulky wastes such as sofas or crates. The feed opening should be at least 1.3 by 2 m.

impact deceleration A *signature method* of distinguishing steel, aluminium, glass and wood by the impact of the material on a tool carrying an accelerometer. The individuality of each of these

158

materials is seen in its 'impact signature'—the shape of the graph of deceleration after impact against time. The accelerometer's mini-computer recognises each impact signature and indicates the appropriate chute into which the material should be directed.

impact statement *See environmental impact statement.*

impeller, rotor The rotating wheel carrying blades that move the water or air in a *centrifugal pump* or *axial-flow pump* or fan.

impeller collector A *fan collector.*

impermeability factor, i. coefficient, runoff coefficient (USA coefficient of imperviousness) A factor that indicates how much of the rain that falls on a surface will run off it. This fraction, the amount of rain that runs off divided by the amount that falls, enables the surface *runoff* to be calculated. There is more runoff after rain and less after dry weather; consequently, two figures are stated for each material:

watertight roofs	0.70 to 0.95
asphalt	0.85 to 0.90
concrete flags	0.5 to 0.85
macadam roads	0.25 to 0.60
gravel road	0.15 to 0.20
level grass	0.10 to 0.20

impingement scrubber, plate scrubber A *wet scrubber* consisting of a tower with horizontal perforated plates across it, perpendicular to the gas flow. Opposite each hole is a wetted plate which the gas strikes, surrendering its pollutant to the liquid. The number of plates is decided by the pressure drop that can be tolerated and the amount of pollutant to be removed. The pressure drop at a gas velocity of 2.5 m/s would be from 5 to 8 cm (water gauge) per stage, collecting particles mainly of 10μm or larger. Some makers claim that particles of 1 μm size can be caught.

impinger Any device for collecting a pollutant by projecting the gas containing it against a wetted surface that catches it.

impounding Bringing water into a *reservoir* for storage, an activity that, in the UK, is forbidden unless the dam has received a certificate from one of the civil engineers on the panel of dam engineers approved by the *Department of the Environment.*

impounding reservoir, i. lake A *reservoir* in which *raw water* may be stored for periods ranging from a few days to several months or even longer, as opposed to the *service reservoir.* Reservoirs filled by pumping from a nearby river usually contain only two or three weeks of normal demand, but even this allows *abstraction* to be stopped when the river is contaminated by floods or other dirt. Reservoirs provide a significant improvement in water quality, because they allow *suspended matter* to settle and *bacteria* to die. However, if the water contains *nitrate* and phosphate, *algal blooms* can occur in summer, giving the water an unpleasant taste and smell. *River-regulating reservoirs* and *direct-supply reservoirs* are also impounding reservoirs. *See raw water storage.*

inactivation of viruses Inactivation is the word used for the destruction of *viruses,* since they are not living *microbes* and so cannot be killed. Certain *bacteria* destroy them. *See break-point chlorination, Coxsackie virus.*

incineration Burning of waste solids, liquids or gases. For solids it produces a hygienic ash, reducing them to about one-fifth of their original volume and a half of their original mass. In the UK in 1977 it cost about £14 per tonne and was the most expensive of refuse treatments. Poisonous gases or fumes that result from incineration are best removed by *wet scrubbers.* Sewage *sludge cake* at 50 to 55% moisture can be burnt without added fuel in efficient burners such as those that use *fluidised-bed combustion*, but they must reach a temperature above 760°C to avoid smell. *See* below, *fume incineration* and *waste-derived fuel.*

incineration with energy recovery Europe has some 200 refuse *incineration* schemes that can profitably use the heat obtained from refuse, but only four of them are in the UK. The energy can be recovered either as electricity or for district heating as steam or hot water. At 1977 prices the investment needed for a UK incineration plant with heat recovery was £250 000 per tonne of refuse burnt per hour. Without heat recovery the cost was £200 000. *See also energy recovery from solid waste.*

incinerator A furnace for burning wastes or poisons. Incinerators must be inoffensive—i.e. they must not produce *grit, dust, smoke,* smell, etc. The earlier batch types have been superseded by expensive, large, automatic continuous *grates* that demand less operating labour, although *trade-waste incinerators* at hospitals and elsewhere will continue to be batch-operated. Furnace temperatures should fall between about 930 and 1050°C. Below 750°C there are smells and smoke. Above 1200°C the ash melts excessively, causing blockages that eat into the *refractory* lining. Continuous incinerators work throughout the week without the stoppages for emptying that are needed by the old batch type. Many types exist, most of them having 'downhill' grates. In some incinerators the refuse is turned over at steps in the grate so as to complete the burning. The allowable dust emission varies from 0.91 g of dust per m³ air for an 880 kW unit to 0.23 g/m³ for a 14 600 kW unit. *See chute-fed* *incinerator, flash drying, flue-fed incinerator, fluidised-bed combustion, multiple-hearth furnace, on-site incinerator, residuals, rotary kiln, shaft furnace, two-stage incinerator, vortex incinerator,* and below.

incinerator corrosion *Corrosion* of the metal in *incinerators* is of two types—high-temperature, mainly from *chlorides*, and low-temperature. For example, at Edmonton, near London, the steam temperature is only 460°C, so as to reduce high-temperature corrosion. Because of low-temperature corrosion below the *acid dewpoint, flue gas* temperatures are kept above 230°C. *See excess air, water wall.*

inclined conveyor separator, cinder s. An endless belt conveyor, sloping at about 45° to the horizontal, has been used in the UK for separating the *garbage*, carried to the top, from cinders, which roll down to the bottom. It is a type of *inertial separator.*

inclined-tube settling tank, settling module, tube clarifier, tube settler Water or *sewage* to be cleaned flows upwards through settling tubes rather than through a conventional deep *sedimentation tank*, thus reducing the distance that the solids travel before they reach the base of the tubes. Often the tubes are steep, at 60° to the horizontal, and the *sludge* that settles in them should slide down inside them, but this does not always happen.

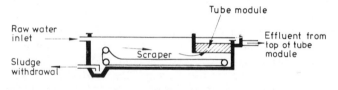

Figure I.2 Inclined-tube settling module in horizontal-flow sedimentation tank

Other types, with more gently sloping tubes (*see Figure I.2*), need periodic **backwashing** to clear the sludge from the tubes. Overloaded sedimentation tanks in water or sewage works can be uprated some 10 to 100% by placing inclined tubes near the outlet weir. In some waters solids accumulate around the tube tops and eventually block them. To remove this accumulation, air diffusers may be installed below the tubes to give an occasional backwash of air. Alternatively, a water jet may be installed above the tubes to clear the deposit. Inclined tubes have also been used as oil separators by reversing the flow so that the water flows down them, leaving oil floating on the surface of the water. *See also lamella separator.*

incombustibles Materials that do not burn. In refuse these are sometimes also called the inorganics—metals, stone, glass, etc. It is important to extract them from refuse that is being made into *waste-derived fuel*, because they add weight and ash without adding heating value, thus lowering its usefulness. They can reach 40% by weight of UK refuse.

INCPEN The Industry Committee for Packaging and the Environment, formed in 1974 by some of the largest food and drink manufacturers in the UK, with a view to reducing litter and over-packaging and their unfavourable effects on the environment.

incremental loading *Step aeration* of an *aeration tank*.

incubation Holding a sample, microbial *culture*, etc., for a suitable time at a temperature, appropriate to the investigation, at which microbes are likely to multiply. Typical temperatures are 55°C for *thermophils*, 44°C for *Escherichia coli*, 37°C (human body temperature) for the *coliform group* of bacteria and 20°C for incubating samples in the *BOD* test.

indicator organisms *Biological indicators of pollution*.

indirect catchment *See catchwater*.

indirect re-cycling The re-use of a material for making something different from the original object (*see re-cycling of solid waste*). For municipal refuse the phrase has even been used to mean *incineration with energy recovery*. Ordinarily it includes recovery of iron and steel by *magnetic separators,* sometimes followed by *de-tinning* of tinplate.

indirect re-use of water The *abstraction* of water by a water undertaking from a river downstream of a *sewage* effluent *outfall*. This is usually called 'uncontrolled' indirect *re-use*. 'Controlled' indirect re-use is, e.g., the *artificial recharge* of *aquifers* using properly treated waste water, often after *advanced wastewater treatment*. Rivers in populated areas, such as the Thames, may pass through many water treatment and sewage treatment works before reaching the sea.

Indore process *Composting* in hot climates (originally Indore, India) using layers of *night soil* and vegetable waste alternately. It is kept *aerobic* by regularly turning for two or three months and then is left to mature.

induced-draught fan A fan that sucks *flue gases* out of a furnace and pushes them up the chimney. It is therefore placed between the boiler and the chimney. Since hot flue gases pass through it, it is liable to be corroded and its service conditions generally are more severe than those of a *forced-draught fan*. Fans made of stainless steel can work at temperatures up to 600°C, provided that the bearings are water-cooled.

industrial effluent *See trade effluent*.

Industrial Pollution Inspectorate The body in Scotland that corresponds to the *Alkali Inspectorate* in England, except that it is concerned also with water pollution and refuse disposal.

industrial river   A river whose flow in dry weather consists mainly of *effluents* from domestic or industrial waste-water treatment plants. The Irwell and the Mersey in Lancashire became industrial early in the nineteenth century. Such a river may have little *dissolved oxygen* and be without fishes.

inertial separator   (1) A separator, used in *mechanical sorting plants,* whose operating principle is based on the differences in relative densities of materials being separated (*see Figure I.3*).

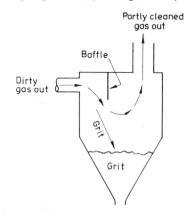

Figure   I.3 Inertial   separator   (an expansion chamber greatly improved by the addition of a baffle)

Ballistic separators, cyclones, inclined conveyor separators and *secator conveyor separators* are four common types. (2) A *dust arrestor*, such as a cyclone, *fan collector* or *expansion chamber*, that uses the inertia of a particle to extract it from fast-flowing *flue gas*, air, etc.

inert refuse   *Clinker*, rock demolition refuse or any other substance that neither pollutes nor is *biodegradable.* It can be used to cover a *controlled tip* or to fill a wet pit. Plastics in *landfill*, though inert, are not regarded as useful fill, because they are too compressible and elastic.

infectious hepatitis   This disease, known to be transmitted by a *virus*, can be waterborne. The worst recent *epidemic,* of 20 000 clinical cases and perhaps another 80 000 less serious, occurred in Delhi in 1955–56 with a number of deaths. The drinking water was *turbid*. The *turbidity* particles are known to have helped to harbour the virus, although the *chlorination* was otherwise effective, because there was no parallel increase in other bacterial *waterborne disease* during this outbreak.

infiltration   (1) The passage of water through the soil surface—i.e. the entry of rain, snow, etc., into the soil by drainage, as opposed to *runoff* and **evapo-transpiration.** (2) Water that enters *sewers* from the ground because of leaks resulting from their poor condition, loose joints, etc.

infiltration capacity   The maximum rate at which soil can absorb rainfall in a given condition. *Infiltration* decreases as the rain continues, because some of the soil particles swell and block the pores; also, some of the pores fill with capillary water. Infiltration tends to increase with porosity.

infiltration gallery   A tunnel or well dug near a river to withdraw water from it. The water is partly filtered as it flows through the soil between the river and the gallery.

influent   Description of fluid that flows in; usually water, *sewage*, etc., entering a treatment system. *See* below.

influent stream   A stream type, common in an arid climate, that loses water into the ground, thus recharging the *groundwater*, unlike the *effluent streams* that are usual in wetter countries.

infraction   Breaking the law, a term used by the *Alkali Inspectorate* to describe the operation of a works that does not use the *best practicable means*. The first official action after warnings have been tried is to send an 'infraction letter'. If letters have no effect, a prosecution is started. About 8% of recent infractions have been followed by prosecutions by Alkali Inspectors.

infra-red spectroscopy   Observation of the intensity of the infra-red light that is emitted or absorbed by various materials. In solid waste treatment it is a *signature method*, which monitors the characteristic absorption spectrum of the substance to be separated.

infusorial earth   *Diatomite.*

inhibitory toxicity   In a *bio-assay* ASTM 31 defines this as any reduction by a pollutant of the rate of reproduction of *diatoms*, within 7 days.

in-house treatment   *See pre-treatment.*

injection aerator   An *aerator* that injects air directly into a *main*, thus avoiding the loss of pressure that occurs with *spray aerators* or *cascade aerators*, when the water is released from the pressure of the water main. A disadvantage of injection aeration is that excess air may bubble out at the consumer's tap. *See also air injection.*

injection well   A *disposal well* or a well for *artificial recharge.*

INKA process   A method of Swedish origin for operating *aeration tanks* in the *activated sludge* process. *Coarse-bubble aeration* is obtained from pipes with 2.5 mm diameter holes on the underside, immersed only about 0.75 m deep instead of on the tank floor. The pipes are to one side of the tank, to produce spiral flow (*see Figure I.4*). Running costs are low, because of

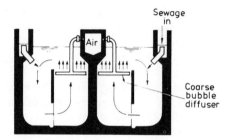

Figure I.4 Cross-section of Inka aeration tank (From 'Disposal of Sewage and other Waterborne Wastes' by K. Imhoff, W. J. Muller and D. K. B. Thistlethwayte, Butterworths, 1956, 1971)

the shallow immersion, requiring only low-pressure air, but the oxygen transfer efficiency also is low, because of the coarse bubbles, and, hence, more air has to blown in than with *fine-bubble aeration*.

inlet time   Of *stormwater,* the time required for *runoff* to enter a *sewerage* system. *See time of concentration.*

in-line   Description of the processing of a liquid within a pipe system that contains it.

inoculation   *See seeding.*

inorganic matter   Matter that does not contain carbon, except as carbonates or bicarbonates. *Carbon dioxide* and *carbon monoxide* also are regarded as inorganic.

inorganics in refuse   *See incombustibles.*

Insecta, insects   The largest class of *arthropods.* Mainly land creatures, they breathe through an external tube, have antennae and six legs, and are often also winged. They are important *grazing fauna* in *trickling filters* and are *biological indicators of pollution.* They include beetles, *springtails, caddis flies, Diptera, dragon flies, mayflies, stone flies, water bugs.* Some are *vectors of disease.*

insecticide   A *pesticide* that kills insects.

insoluble   Description of a substance of low *solubility.*

inspection chamber   *See manhole.*

Inspector of Nuisances   A very early name for *Environmental Health Officer.*

INSTAB   Information Service on Toxicity and Biodegradability, a service of the *Water Research Centre.*

instability   Of a *sewage*, its tendency to putrefaction (*anaerobic* decomposition), measurable by the *methylene blue stability test.*

Institute of Solid Wastes Management   Formerly the Institute of Public Cleansing, the main UK professional association of specialists in public cleansing, municipal waste collection, treatment and disposal, at 28 Portland Place, London W1.

Institute of Water Pollution Control The main UK institute for the management and operation of *sewage treatment* works at London Road, Maidstone, Kent.

Institution of Public Health Engineers The only UK institution concerned with all aspects of *environmental control engineering*, at 13 Grosvenor Place, London SW1.

interaction of toxicants Two or more compounds in water may increase or reduce each other's toxic effect. For instance, calcium *ions* may counteract the toxicity of the *heavy metal* zinc. Toxic effects are generally increased, for fish, by low *dissolved oxygen*, because more water has to be taken in through the gills; consequently, more poison also is brought in. *See antagonism, synergism.*

intercepting sewer, interceptor A main *sewer*, one that receives its flow from *laterals*. It delivers either to another main sewer or to a treatment plant or an *outfall*. Intercepting sewers were originally so called because they intercepted polluting drains flowing to a river.

intercepting trap, interceptor, disconnecting trap (1) A water seal sometimes installed between a house drain and a *sewer* it flows into, to separate the air between the two. It is usually placed at a *manhole*, where it can easily be inspected. (2) A *grease trap*.

interception Short-term retention of rainfall by leaves, branches, grassland, etc., which may later add to *runoff, percolation* or *evapo-transpiration*.

interface turbidity monitoring Use of a *turbidimeter* to monitor water samples taken from the interface in a *filter* between, e.g., coarse anthracite and fine sand. This shows the operator whether the anthracite removes as much solids as it should, and the *coagulant* dose can be altered accordingly.

interflow Water in the ground which, after meeting a relatively impermeable layer, flows more or less horizontally to an outlet at a spring, stream, etc.

intermittent downward filtration, intermittent filtration *Land filtration*.

International Commission on Large Dams *See ICOLD.*

inter-tidal zone The area of sea beaches between high and low tides. This area was the most highly polluted by oil spills and by the *dispersants* that followed them after the Torrey Canyon disaster.

inversion, temperature i., i. of the lapse rate Up to about 12 km above sea level, the air usually becomes cooler at higher levels, but occasionally it warms up with increase in height. This is an inversion of the *lapse rate*. Warm air above cool ground is an inversion. Inversions often occur in temperate climates between ground level and 100 m above it on calm clear nights (nocturnal inversion), because the ground cools more quickly than the air above it. During an *anticyclone* an inversion may rise higher than usual and persist for several days, causing dense *fog* or *smog* in winter. In any case polluted air concentrates under an inversion. *See also subsidence inversion.*

invert The bottom of a pipe, open channel, sewer, etc.

invertebrates Animals with no backbone. They are much smaller and more numerous than those with a backbone and are important in *sewage treatment*. Many of them live in water—in, e.g., *trickling filters* or *activated sludge*.

inverted siphon A U-shaped pipe that flows full and passes *sewage* or water under a railway, road, river or other obstacle. *See siphon.*

inverted weir A *scum board*.

involute cyclone A *cyclone* with the usual tangential inlet.

iodine, I A *halogen* with the advantage over *chlorine* in *disinfection* that it does not react with *ammonia, phenols* and other *organic compounds*. The potency of iodine for killing *cysts* and of hypoiodous acid (HIO) for *inactivation of viruses* is thus not hindered by these nitrogenous pollutants, as chlorine is. Iodine is, however, scarce and expensive compared with chlorine, so is unlikely to be used in municipal water supply.

iodine number The number of mg of *iodine* adsorbed by a known mass of *activated carbon* from a known *solution* of iodine. The amount adsorbed is proportional to the internal surface area of the carbon and includes pores as small as 1 *nanon* across. If the iodine number is reduced by 50%, it is time to change the carbon. *See also adsorption, phenol adsorption test.*

ion An atom or group possessing an electrical charge, important in *electrolysis*. Most salts of metals, when they dissolve in water, partly dissociate to yield a metallic *cation* with a positive charge and an acid-radical *anion* with a negative charge. Ions exist also in gases—e.g. within *electrostatic precipitators*.

ion exchange A treatment of *raw water* in which one *ion* in the water is exchanged for another that is less harmful, by adding a chemical; commonly used in *water softening* and to make *de-mineralised water*. Ion exchange is ordinarily reversible.

ion-exchange membrane *Ion-exchange resins* formed into sheets; first achieved about 1950. Ion-exchange membranes made *electro-dialysis* possible.

ion-exchange resins Synthetic resin beads made for *ion exchange*. *Ions* which are held by electrostatic forces to electrically charged groups on the surface of the bead interchange with similarly charged ions in solution in the water being treated. Beads with SO_3^- or COO^- ions permanently attached will accept and exchange cations such as Na^+, H^+, Ca^{2+}. These are cation-exchange resins. Anion-exchange resins have amine groups (e.g. NH_3^+) permanently attached to the bead and will accept and exchange anions, such as Cl^-, SO_4^{2-}, OH^-. Resins are regenerated for re-use by passing a concentrated solution of the original ion through them. Different resins have even been developed to extract uranium preferentially from low-grade ore. One of the commonest uses of ion exchange is in making *de-mineralised water*. Here the exchange of the unwanted ions for a hydrogen ion or a hydroxyl ion (OH^-), as the case may be, enables the high quality of a distilled water to be achieved at low cost. The H^+ or OH^- ions are changed into water by combination with each other. Cation-exchange resins are used first and then anion-exchange resins, but *mixed-bed ion exchange* is also used. The capacity of the resins, before *regeneration* is needed, varies from 300 to 400 m^3 of *raw water* per m^3 of resin. During regeneration the plant has to shut down. *Continuous ion exchange* can be achieved only by using several plants. *See also water softening, zeolite.*

ion-selective electrodes *Electrodes* that may be used to measure the concentration of *ions* in water, such as ammonium, *nitrate*, sulphide, *chloride, potassium*, etc. The presence of other elements may interfere with their accuracy and this may limit their use.

ion-selective membrane An *ion-exchange membrane* that holds either *cations* or *anions* but not both, used in *electro-dialysis*.

iron bacteria A diffuse group of *bacteria* that form natural colonies heavily encrusted with ferric oxides. They are all *chemosynthetic*, although some are *heterotrophs*, such as *Leptothrix* and *Sphaerotilus*, and these may grow in the absence of iron. Others are probably *autotrophs*, requiring *ferrous salts* which they oxidise to ferric oxide. *Thiobacillus ferro-oxidans* is both a sulphur-oxidising and an iron-oxidising bacterium. Iron bacteria may create rusty slime that smells, stains and blocks *sand filters*. Since iron bacteria are always likely to flourish in a water with a high iron content, very low levels are recommended. *See* Table of Allowable Contaminants in Drinking Water, page 359.

iron in water The *solubility* of ferrous and ferric compounds in water is very low at neutral *pH*. If oxygen is present, *ferrous salts* readily oxidise to the less soluble ferric compounds. In *anaerobic* conditions ferrous compounds, however, predominate.

Dissolved iron is not poisonous, but even in tiny concentrations of 0.2 mg/litre can stain clothes in washing machines or spin dryers. It is therefore best to bring the iron content down to 0.05 mg/litre, which can be achieved by *aeration* and raising the *pH* above 6.5. The iron is thereby oxidised and precipitated as ferric oxide which can be removed by *sedimentation* and *filtration*. When iron compounds undergo *precipitation*, they produce a heavy *floc* that settles quickly but has the disadvantage that, if any passes into the distribution system, it may stimulate the growth of *iron bacteria* in the pipes. *Groundwater* with pH below 8.0 may contain up to 10 mg/litre of ferrous iron, but on taking up oxygen, the iron slowly precipitates as ferric hydroxide, making the water *turbid* and eventually brown. Occasionally, ferrous compounds are present in concentrations up to the exceptional level of 20 mg/litre, for which the water should be aerated at a pH above 8.0, followed by sedimentation to allow the ferric hydroxide to settle. The water can then be filtered. For small amounts, of the order of 1 mg/litre, the ferrous iron is oxidised merely by a small injection of air or by trickling over coke or pumice before filtration. Highly acid waters from springs or mines may contain up to 6000 mg/litre Fe. Since much water pipe is of steel or cast-iron, it is important that the *acidity* and, hence, the *corrosiveness of water* should be low before it enters the distribution system. *See also ferruginous discharges*.

iron pyrite   Ferrous sulphide (FeS_2). *See acid mine drainage*.

irrigated filter   A *filter* for air or other gas, with liquid, usually water, flowing over its collecting surface to improve its collecting efficiency; not a *trickling filter*.

irrigation   Artificial watering of farm land by pipe or channel. Total dissolved solids in irrigation water should be less than 1000 mg/litre; there should be less than 0.5 mg/litre of boron. *Heavy metals* may *ion-exchange* into the soil, rendering it infertile. Sewage disposal by irrigation is possible where land is plentiful, cheap and well drained, the sewage contains no poisons and the climate is warm. Usually at least one hectare per 30 persons is needed, so irrigation is extravagant with land. *See centre-pivot irrigation, drip irrigation, flood irrigation, furrow irrigation, land treatment, grass plots, sodium absorption ratio, spray irrigation, subsurface irrigation*.

iso-   This prefix often indicates 'equal' or 'same'. Thus, an isotherm or isothermal line is a line joining points at equal temperature. An isoconcentration line is one joining points with the same concentration of *suspended solids*.

isochlor   A line of equal concentration of *chloride* in a *groundwater* or an *aquifer*, used for mapping contamination by chlorides (*saline intrusion*).

iso-electric point, iso-electric pH   The *pH* value required to produce a zero charge on a *colloidal* particle, *bacterium, alga,* etc. *See electrophoretic mobility.*

isohyetal line, isohyet   A line that connects points with equal rainfall.

isokinetic flow   *See sampling train.*

isolating valve   A stop valve, to isolate an area of a pipe system, a tank, etc.

isothermal   At constant temperature.

isotope   Elements can have several varieties of isotopes, all with the same chemical properties except for their different atomic masses; for example, hydrogen has mass 1, deuterium has mass 2, tritium has mass 3, but all are isotopes of hydrogen. Tin has ten isotopes. If these disintegrate with the emission of alpha or beta or gamma radiation, they are *radioactive isotopes.*

J

Jackson turbidity unit, JTU, Jackson candle unit, JCU   A US measure of *turbidity* in water. It can be defined in two ways: (1) The turbidity resulting from 1 mg/litre of fuller's earth suspended in water. (2) The depth of the column of water that just obscures the image of a burning standard candle viewed vertically through the sample (ASTM 31). The *EPA* requires 1 JTU as a monthly average in drinking water but some US states require only 5 JTU.

jar test   A simple test to determine the correct chemical dose for a water or *sewage*, most commonly used for estimating the *pH* and *coagulant* requirement for a *raw water*. Various amounts and combinations of chemicals are added to several samples of water of the same volume, placed in jars and stirred slowly.

Jet aeration   An *activated sludge* process using *coarse-bubble aeration* introduced at about a quarter of the depth of the *aeration tank* through perforated pipes.

jet ejector scrubber   An *ejector Venturi scrubber.*

jigs   *Mineral dressing* baths that pulsate to separate coal or other mineral from stone and could be used in a *mechanical sorting plant* where the wetness does not matter. The light materials flow over the top. The heavies (sinks) flow out below the water surface. A disadvantage of jigs is their high water consumption and pollution of the water, but they can be operated in such a way that no water is rejected to the river or *sewer.*

JTU A *Jackson turbidity unit.*
juvenile water Water that has never been *meteoric water.* It comes from deep rock that originally was molten, not sedimentary.

K

kanat A *ghanat.*
katabatic wind A wind that blows downhill along a slope cooled by radiation, frequently at night. It may be the cause of an *inversion.*
katabolic metabolism *Catabolic metabolism.*
kathode, kation *See cathode, cation.*
kelp, *Laminaria digitata* A *brown alga,* a seaweed. This *alga,* sometimes used as a *biological index of pollution,* grows near the low-water mark.
Kessener brush An early method of *surface aeration* in the *activated sludge* process, devised by Dr H. H. Kessener of Holland in 1925, with rapidly rotating stainless steel brushes on a horizontal shaft. The brushes are partly submerged in the *mixed liquor,* swirl it around and thus aerate it. It has been superseded by the *TNO rotor.*
kier liquor The waste liquid from boiling cotton, flax, straw, or other material with a high *cellulose* content, in a strong *alkaline* liquid. Kiering removes dirt, grease, colour, etc., thereby allowing the cellulose to be used for papermaking. Cotton yarn also is kiered before dyeing. Kiering in papermaking produces a waste liquor with *BOD_5* of 30 000 to 40 000 mg/litre and *pH* of 12, and in cotton yarn kiering a BOD_5 of 4000 to 8000 mg/litre.
kieselguhr *Diatomite.*
kinematic viscosity *See viscosity.*
kingdom A major division of *taxonomy.*
kitchen waste disposal unit A *garbage grinder.*
Kjeldahl technique A standard method for measuring the unoxidised *organic nitrogen* in water. It involves digestion in sulphuric acid with a *catalyst* to produce ammonium sulphate from the *organic* nitrogen. The concentration of *ammonia* can then be measured by distilling it off at *pH* 10 followed by *nesslerisation.* Total Kjeldahl nitrogen is the *ammoniacal nitrogen* plus the unoxidised organic *nitrogen.*
Kniebühler tank A *Dortmund tank.*
knife discharge A long blade parallel to the face of a *vacuum filter* drum which scrapes the *filter cake* from its face. In some applications the cake may have to be detached by blowback (internal air pressure).

Kolkwitz–Marsson saprobic system   The *saprobic classification* of rivers devised by Kolkwitz and Marsson in 1911.

konimeter   A dust-sampling instrument formerly used in mines, which caught the dust particles on a greased glass slide. They were later put under a *microscope*, counted and measured.

kraft or alkaline sulphite process   A papermaking process that produces the strongest (*kraft* = German for strength) brown paper. Wood chips are cooked in caustic soda, sodium sulphite and other chemicals for several hours under pressure at 176°C and then released. The spent liquor is de-watered down to about 65% dry solids and then incinerated to reduce pollution of water. *See paper and pulp manufacturing wastes.*

Kraus process   An *activated sludge* process designed for treating *sewages* that are deficient in *nitrogen*, an essential ingredient for bacterial growth. The *supernatant liquid* from *anaerobic sludge digesters* is aerated with *digested sludge* and some *return sludge* in a special *aeration tank* for 24 h, which converts the *ammonia* to *nitrate*. This nitrified liquor is then mixed with the rest of the return sludge and sent to the aeration tank for conventional activated sludge treatment.

Kubel test   A test for small amounts of *organics* in water, which involves boiling the water with *potassium permanganate* for 10 min. The value is about three times as high as that obtained in 4 h at 27°C in the *permanganate value* test.

L

labelling, tagging   The use of *tracers*.

laboratory water   According to BS 3978, laboratory water in Britain should have a *pH* between 5 and 7.5; electrical *conductivity* below 10 μS/cm at 20°C; and maximum concentrations of 0.1 mg/litre *ammonia*, 0.5 mg/litre *chloride* and 1 mg/litre *sulphate*.

lactose broth   A reagent used in *most probable number* testing for the *coliform group* of bacteria, which consists of lactose sugar in a *nutrient* broth.

lacustrine   Concerned with lakes.

lagoon   (1) *See aerated lagoon, waste stabilisation pond.* (2) A pond where dirty water is allowed to settle and clarify before discharge to a river—in particular, one for *sludge thickening*, or *de-watering of sludge* or its temporary storage, whether from waterworks, *sewage treatment* plants or other industry. *Underdrains* are not usual. *See liquid waste disposal on tips, sludge lagoon.*

lag phase An early *growth phase of bacteria* during which the microbes acclimatise to their new environment. The rate of growth is very low.

Lake Tahoe, California, USA Known to ecologists because the barrenness of the surrounding land has prevented *eutrophication*. They aim to keep the exceptionally clean water of the lake in its original state. An *activated sludge* plant in South Tahoe treats 11 000 m^3/day and its *effluent* is polished by chemical *coagulation, filtration, air stripping* and *activated carbon* treatment to adsorb *trace organics*, followed by *chlorination*. The filtration is high-rate with *multi-media filters*, the carbon is given *regeneration*, and *lime* coagulant is recovered. In 1971 the average effluent quality was: BOD_5 0.7 mg/litre, *COD* 9 mg/litre, *suspended solids* 0, *phosphorus* 0.06 mg/litre, *methylene-blue-active substances* 0.15 mg/litre.

lamella separator, tilted-plate s. A modern settling device like the *inclined-tube settling tank* but with parallel, steeply inclined plates, about 50 mm apart, instead of the tubes.

laminar flow, streamline f., viscous f. Motion of fluid at low speeds, in parallel paths and without eddies, unlike *turbulent flow*. In pipes laminar flow normally occurs at *Reynolds numbers* below about 2000.

Laminaria digitata   Kelp.

lamphole A narrow shaft built of pipes over a hole through the centre of the crown of a *sewer*. It enables a lamp to be lowered on a cord into the centre of the sewer, which then can be inspected by a man looking along from the next manhole.

land disposal of wastes *See controlled tipping, sludge disposal, toxic waste disposal.*

land drain, field d., agricultural d. A buried pipe with loose joints, dug into wet farmland to keep it dry enough for cultivation. The drain slopes down either to a ditch or to a bigger drain pipe. It is buried about 1 to 1.5 m deep in peaty soils and about 0.9 m deep in silts. *See French drain, mole drain.*

landfill, land reclamation One of the most constructive functions of *controlled tipping* is to raise low-lying ground and thus to make it available for playgrounds or, after years of *compaction*, for housing. Between 1947 and 1967 in the UK nearly 11 000 ha of land were reclaimed by controlled tipping and more than 75% of the area was used for farming or amenity purposes.

land filtration (1) *Land treatment* of any type. (2) Or *intermittent downward filtration*. A type of rapid-infiltration land treatment which consists of horizontal beds of sand over a sandy subsoil, which are periodically flooded with *settled sewage*. The liquid may percolate into the groundwater or be removed through land

173

drains 100 mm or more in diameter, laid in a bed of gravel, 1.5 to 4.5 m below ground and 9 m apart. As the sewage flows through the soil, it undergoes filtration as well as *biological treatment*. The effluent may be very good, with a *BOD$_5$* less than 5 mg/litre. In the USA slow-rate land treatment implies an application rate of 1 to 10 cm per week of settled sewage, but rapid-infiltration land treatment can be used on very porous soils with applications of up to 300 cm per week of settled sewage, according to the *EPA* design manual for land treatment of municipal waste waters. The method was established in Britain by 1900 and may now be a tertiary treatment. *See absorption pit, grass plot.*

land treatment Any method of treating sewage or sewage effluent by irrigation. Land treatment was used in Silesia in A.D. 1543 and even earlier in ancient Athens.

Langelier saturation index (symbol *I*) An indication of the *non-corrosiveness* of a water, measured by the oversaturation or undersaturation of *calcium carbonate* in it. It is defined as the difference between the *pH* of the water under consideration and its pH when in equilibrium with calcium carbonate ($CaCO_3$) (the saturated pH or pH$_s$). Thus $I = $ pH $-$ pH$_s$. A negative value of *I* implies undersaturated water that will dissolve $CaCO_3$ and will not deposit a protective film of $CaCO_3$ on the pipe to stop corrosion. A positive value of *I* implies that $CaCO_3$ will be deposited on the pipe. The pH$_s$ value can be calculated by measuring the *alkalinity*, calcium ion concentration, *total dissolved solids* and temperature of the water. The *chalk test* is not essential. The calculations are described in *Standard Methods for the Examination of Water and Waste Water*.

lapse In meteorology the change of temperature, usually cooling, of air as its height increases.

lapse rate The rate at which air temperature changes with increasing height. *See adiabatic lapse rate, environmental lapse rate.*

larva (plural larvae) An immature stage of an invertebrate insect or animal, developed from an egg, often shaped like a maggot.

larvicide A *pesticide* that kills *larvae*.

LAS *Linear alkyl sulphonate.*

lateral, l. sewer, collecting sewer A branch *sewer* buried in the street, collecting *sewage* from the house drains, and delivering into an *intercepting sewer*.

launder A *flume*, usually within a *mineral dressing* or other plant, especially an outlet from a tank or peripheral weir.

laundry wastes The waste water from a large laundry has a high *turbidity,* a pH of 8 to 10 and a *BOD$_5$* of 400 to 1000 mg/litre. Water consumption is about 35 litres per kg of clothes.

LC$_{50}$, LD$_{50}$ The *median tolerance limit*—i.e. the 50% lethal concentration or 50% lethal dose.

LDPE *Low-density polyethylene.*

leachate, percolate Water that leaks from tipped refuse. It can enter streams and possibly *groundwater* unless the tip is on clay or other impervious ground. At a *controlled tip* water pollution can be avoided by the heavy expense of leading all leachate into a *sewer*. It costs much less to lay out the tip with any surface water diverted around it and to slope the final surface so that it throws off the rain. The *final cover* may be some material such as compacted clay which will not allow water through. Leachate is nearly always highly polluting. In the early years after completion of a tip, its leachate has high *BOD*, contains plant *nutrients* and is both coloured and *turbid*. With their early high porosity, tips at first have an *aerobic* population of microbes which gradually changes to an *anaerobic* one as the tip subsides and becomes more airtight and less porous, and the *oxygen* is used up. In such conditions iron and manganese compounds became soluble as Mn^{2+} and Fe^{2+}, the manganous and *ferrous salts*. The Institution of Chemical Engineers stated in 1971 that if the bulk *permeability* of the soil under a tip was such that the leachate would take 250 years to reach an *aquifer*, the tip was safe. Other authorities say that tips should be underlain by at least 15 m of impermeable ground such as clay, which may mean the same thing.

leachate re-circulation, l. re-cycling Pumping *leachate* back into the *cell* of the *controlled tip* from which it came. It has been found to stabilise the tip more quickly, especially when the *pH* was controlled to be less acid than it normally would be at the start of a *landfill*. It also reduced the final concentration of most pollutants and the total pollution entering the environment. The moisture content is held at about 70% of the dry weight, preferably by adding sewage *effluent*, to reduce expense. The main extra cost is in the pumping, pipes and pumphouse.

leaching field, tile f. US description of the buried *land drains* used in *subsurface irrigation* or *subsurface filtration*, which provide *self-purification* in the ground for the *effluent* from *septic tanks*.

lead, Pb Lead is a poison whose compounds accumulate in the body, but it is rarely found in natural waters, except in the discharges from lead mines. Modern methods can detect it at 0.02 mg/litre, but see the Table of Allowable Contaminants in Drinking Water, page 359. If lead pipes are used, the water should be checked for *plumbo-solvency*. Lead compounds accumulate in oysters and other shellfish and they may also be leached by a water supply from pipes made of plastics stabilised

175

by lead compounds. From waste water, lead can be removed by chemical precipitation with *lime*. *See* below.

lead dioxide, l. peroxide, PbO$_2$ Lead dioxide has been used (BS 1747: part 4) to compare the absorption of sulphur compounds in air between one month and another at a particular site. An area smeared with lead dioxide was left exposed in a louvred box for a month and the amount of sulphate found by analysis.

lead in petrol Tetra-ethyl lead used since 1923 and tetra-methyl lead used since 1959 to raise the quality of petrol at low cost. They are the main sources of lead in city air. Unleaded and low-lead gasolines are used in the USA because lead hinders the working of the *catalyst* in modern car exhaust systems which restricts *hydrocarbon* and *carbon monoxide* emissions. Lead compounds are emitted from exhaust pipes as solids smaller than 1 μm and form *respirable dust*.

lead trap A device, in the exhaust pipe of an internal combustion engine, that catches about 40% of the lead particles that would otherwise be emitted. Its useful life is about the same as that of the pipe.

leaf test The *filter leaf test*.

leaping-weir overflow A *stormwater overflow* in which the *dry-weather flow* drops through a hole in the *invert* of the *sewer*. *Stormwater* leaps across the opening to flow elsewhere.

leat An open channel or *catchwater* cut along a contour of a hillside, but with a slight fall, to bring water to drive a water wheel, into a *reservoir*, etc.

leather industry wastes *See tannery wastes.*

leech A class of *annelids* which are quite tolerant of mild *organic* pollution.

lentic Concerned with still waters.

Leopold filter bottom Rectangular perforated burnt clay blocks used in the USA as *underdrains* for *filters*.

Leptomitus lacteus A dominant member of *sewage fungus* in Continental rivers, a true aquatic *fungus*.

Leptospira icterohaemorrhagiae **Bacteria** belonging to the *spirochaetes*, carried to man by rats, which excrete them in their urine, causing *leptospirosis*.

Leptospirosis, spirochaetosis, spirochaetal jaundice Diseases caused by coil-shaped, motile *bacteria, Leptospira icterohaemorrhagiae*. They enter the blood through cuts or scratches or the mucous membrane by infection from domestic or wild animals, rats or mice. Infection also may be waterborne from streams polluted by the urine of animal carriers. *Weil's disease* is a severe form of leptospirosis.

Leptothrix A filamentous *iron bacterium* found in organically polluted water, which may be capable of growth in the absence of iron.

lethal dose *See median tolerance limit.*

Leucothrix Filamentous *heterotrophic* bacteria associated with *bulking* sludge.

Leuctra A species of *stone fly* found in well-aerated water with no *organic* pollution, although it may tolerate pollution by *heavy metals*.

levee An earth bank.

licensing of sites Under the UK *Control of Pollution Act 1974*, all waste disposal sites have to be licensed by the local *waste disposal authority*, which is free to attach any conditions to the licence. The purpose of licensing is to ensure that refuse is treated and disposed of without harm to the environment or to public health. Realistic waste disposal officers are aware that over-restrictive conditions may lead to *fly tipping*, which can be dangerous and is illegal.

lichen A dual organism of blue-green or green *algae* and a *fungus* which are in *symbiosis*, living attached to rocks or trees. It is an early coloniser of bare rocks.

LIDAR, light detection and ranging An 'optical radar' system used for detecting and measuring smoke *plumes* that are invisible to the naked eye. Even in the dirty air of the UK it can detect a smoke plume at 10 km distance by its laser technique.

lift In *controlled tipping*, the vertical height of the layer of refuse in each *cell*, usually 2 to 3 m, completed in one day. A tip may have many lifts. In California 100 or so lifts have been used in a canyon, with a total tipping depth of 220 m or more.

light-obscuration instrument A *smoke-density meter*.

light oils, light refined petroleum products, white products Motor spirit and similar liquids, which are by far the most poisonous of all petroleum cargoes, being in part water-soluble. They are especially dangerous to marine life when spilt close inshore where fish breed, and unfortunately much of this traffic is coastal. However, they evaporate easily.

lime Strictly calcium oxide (CaO; quicklime), but often meaning *calcium hydroxide* ($Ca(OH)_2$; slaked lime). *See* below.

lime–soda softening of water During *lime softening* or *excess lime softening*, soda ash (Na_2CO_3) may be added to remove *non-carbonate hardness* by forming precipitates of *calcium carbonate* and magnesium hydroxide. The magnesium hydroxide is precipitated only in conjunction with excess-lime softening.

lime softening A common method of *water softening. Lime* is added to the water before *clarification.* It reacts with the *calcium* *bicarbonate* dissolved in the water to form a precipitate of *calcium carbonate*. After clarification the water contains 20 to 40 mg/litre of dissolved calcium carbonate, some of which may slowly deposit on the walls of the pipes in the distribution system. *See excess-lime softening.*

limit of resolution, resolving power The size of the smallest object that can be seen through a *microscope*, telescope, etc. In the *optical microscope* it is about 0.25 μm or 250 nm; in the *electron microscope* about 0.1 nm. Some microbes are more difficult to see than others.

limnobiotic Description of freshwater life.

limnology The study of freshwater rivers, lakes or ponds; compare *potamology*.

Limnoperna fortunei A *freshwater mussel* that infects water *mains* at Hong Kong, breeding for 10 months of the year, unlike *Dreissena polymorpha* in colder countries, which breeds only for 2 to 3 months.

limnoplankton Freshwater *plankton*.

Lindemann–Newell plant A German–American *fragmentiser* which can break up 100 cars per hour.

linear alkyl sulphonate, LAS A *biodegradable* (soft) *synthetic detergent* used since 1965. It is a linear *alkyl benzene sulphonate* which has a straight carbon chain attached to the benzene ring. This straight chain is readily attacked by *bacteria*. LAS is the basis of most household detergents.

linear induction separator An *eddy-current separator*.

liners *Impact blocks*.

lipids A collective name for fats, oils and waxes.

liquid chlorine *Chlorine* does not liquefy above − 34.6°C unless it is compressed in steel cylinders, but the term may be applied to sodium *hypochlorite* (NaOCl), which is always sold in solutions containing a maximum of 15% chlorine. Chlorine gas, though cheap when sold liquid under pressure in cylinders, is a dangerous poison. To eliminate the hazard or reduce it, waterworks may buy NaOCl or make their own chlorine by electrolysing *brine*. *See hypochlorination*.

liquids cyclone A *wet cyclone*.

liquid waste disposal on tips Certain liquid wastes may not be discharged into a *sewer* and must be disposed of otherwise, on a *lagoon* or a *controlled tip* or elsewhere. Before lagooning is planned, the drainage of the site must be investigated. Where the drainage is vertically downwards into an *aquifer*, lagooning is undesirable, since the aquifer will be contaminated by the *leachate*. If all the surface water drains into a stream, lagooning may be permissible, provided that no harm is done to the water

178

or to its users downstream. Much sewage *sludge* is harmlessly disposed of on controlled tips. *See toxic waste.*

lithium salts   Lithium is rare in water or waste water and its salts are readily soluble. Therefore it is a useful chemical *tracer*. Lithium chloride or sulphate is often used. Lithium can be measured by *flame photometry.*

lithostatic pressure   *Overburden pressure.*

lithotroph   Literally 'stone-eating'; *bacteria* such as *Thiobacillus* are lithotrophs, because they eat inorganic materials like mineral sulphides, including even *hydrogen sulphide* gas.

Litonotus   This *predator*, a free-swimming *ciliate protozoon*, is found in *activated sludge* and among *plankton.*

litter fence   A movable, wire-mesh fence set up on the lee side of a *controlled tip* so as to catch windborne paper, plastics film, etc. There may be several, one behind another. The litter caught on the fences must be periodically removed or a strong wind will blow the fence down.

littoral zone   (1) The *inter-tidal zone* of a beach. (2) The shallow part of a lake, where rooted vegetation still grows.

live capacity   The volume of a *reservoir* that can be used for storing *raw water*, and from which water can be withdrawn. It is the total volume of the reservoir minus the *dead storage.*

livestock wastes   *See cowshed wastes, piggery wastes, poultry house wastes.*

Lloyd–Davies method or **rational method for the design of sewers**   A way of estimating *runoff* from a known number of mm of rainfall in a known area by the simple formula: runoff (m^3/min) = 0.167 ARC, in which A = catchment area (ha), R = rainfall (mm/h), C = *impermeability factor.*

load-on-top system   *See oil tanker washings.*

logarithmic growth   A *growth phase of bacteria* (or other life) that involves very rapid increase in the numbers of a population. For *bacteria* it means that there is no limitation of food. If a bacterium reproduces itself every 30 min and its successors also do so, the growth will be logarithmic and after a day there will be 2.8 × 10^{14} or 280 million million which have grown from one bacterium.

London smog   Coal smoke contained in thick yellowish *fog*, causing coughing, smarting of the eyes and blackening inside the nose. In the smog from 5 to 9 December 1952 there was, as usual with fog, an *anticyclone*, and the air temperature was from 0° to minus 6°C. Nearly 4000 people in London died from respiratory illnesses. Domestic burning of coal was the main cause of the pollution. As a result eventually the *Clean Air Act 1956* banned the burning of smoky coal in smokeless zones and smog is no longer expected in London.

looping The behaviour of a chimney *plume* in light, eddying winds, which occasionally bring *air pollutants* to ground level.

Los Angeles smog *Photochemical smog.*

lotic Concerned with flowing water.

low-density baling, low-pressure b. Compression of refuse, usually into containers from which it is eventually released. The compression achieved is only about 4 to 1. *High-density baling* compresses the refuse much more but needs much stronger machines.

low-density polythene, LDPE *low-density polyethylene* One of the *thermoplastics*, of which cheap but light *dust bins* are made, which can become brittle in cold weather.

low-energy scrubber, low-pressure-drop s. A *wet scrubber*, such as a *packed tower*, that collects the coarser dust—e.g. from 2 to 5 μm in size. Operating at 15 cm (*water gauge*) resistance to the gas flow, it could collect nearly 100% of the dust larger than 5 μm and 90% of the 2-μm size. If the gas flow velocity and the resistance to it (pressure drop) were to be doubled or tripled, much of the fine 1-μm size would be caught and the system might become a high-energy or high-pressure scrubber—more efficient as a *dust arrestor* but also more expensive in power consumption. *See Venturi scrubber.*

low-rate filter A standard-rate *trickling filter*.

Loxophyllum A *ciliate protozoon* found in *activated sludge* where *nitrification* is well established.

LSI The *Langelier saturation index.*

lugworm, *Arenicola* A beach worm used in tests by an oil pollution research unit for measuring the toxicity of Kuwait crude oil and of a beach *dispersant*. It is convenient because one wormcast on the sand corresponds to one buried worm. Waterlogging of sand or silt reduces the penetration of the pollutant and thus its toxicity.

lumbricid, Lumbricidae Large worms that include the common earthworm. Some members graze in *trickling filters*.

Lumbriculidae Small thin worms which may be numerous in *trickling filters* as *grazing fauna*.

luxury uptake of phosphorus Removal of *phosphorus* from *sewage* by incorporation into *activated sludge*, possibly by microbes but in excess of their needs for growth, particularly at high *oxygen* levels and a *pH* of 7 to 8. The phosphorus returns into solution if the sludge is stored anoxically and the pH is slightly lowered.

lysimeter An enclosed volume of soil and plants arranged in a tank so that they can be weighed periodically or continuously.

The lysimeter is in the open so that the rainfall, *runoff* and *groundwater* flow can be measured and the *evapo-transpiration* determined.

lysis (verb **lyse**)   Disintegration; the word describes the death and break-up of a bacterial *cell*, making its *protoplasm* available as food for its neighbours. *See **autolysis, cryptic growth, endogenous respiration**.*

M

M   Symbol for mega-, one million times or 10^6.

m   Symbol for milli-, one-thousandth or 10^{-3}.

MacConkey's broth   A bacterial *medium* that is *selective* for *coliforms*, used in *bacteriological examination* of water and waste water. It consists of salts, peptone, lactose, sodium chloride and a *pH* indicator. Coliform *bacteria* ferment this broth at 37°C to produce a gas and an acid which is shown by the solution changing from purple to yellow, with a bubble of gas collected in a small Durham tube. At 44°C *Escherichia coli* is the dominant coliform that produces acid plus gas in MacConkey's broth.

macerator   (1) A *screenings disintegrator*. (2) A *garbage grinder*.

McGowan strength   A measure of the strength of *sewage*, estimated from the expression $4.5N + 6.5P$, in which N is the *ammoniacal nitrogen* plus *organic nitrogen* and P is the *permanganate value* using N/8 *potassium permanganate*. The factor 6.5 varies in accordance with the industrial waste content in the sewage.

macrobiota   The larger organisms in soil, including plant roots, larger insects and earthworms. *Compare **microbiota, mesobiota**.*

macroscopic   Description of what can be seen with the naked eye.

MAFF   The UK Ministry of Agriculture, Fisheries and Food.

magnesium, Mg   Although usual in hard waters, magnesium is almost always present in water in a much smaller concentration than *calcium*. WHO states that its maximum should be reduced if sulphate is present, since magnesium sulphate, $MgSO_4$, is the purgative, Epsom salt. At 250 mg/litre of SO_4 the desirable maximum of Mg is 30 mg/litre. *See also **water softening** and* Table of Allowable Contaminants in Drinking Water, page 359.

magnetic pulley   A *magnetic separator* in the form of a return pulley at the end of the belt conveyor that carries municipal refuse from the *trommel*. Less expensive than the *overband separator*, it is also less efficient and removes more rags and paper. It can also damage the belt by dragging nails or other sharp metal pieces between it and the belt.

magnetic separators Devices to extract iron and steel from material carried past them on a conveyor belt, an important part of many *mineral dressing* plants. They are usually operated electromagnetically but sometimes by permanent magnets. There are four main types: the *overband separator*, the *magnetic pulley*, the *rotary screen magnet* and the *multipole overband separator*. The last helps to reduce the amount of rags and paper extracted with the iron and steel. Magnetic separation is often preceded by *air classification* and *shredding*.

magnification in a food chain *Biomagnification*.

main An important pipe, often one carrying water. If buried, it should have at least 0.9 m of earth cover as frost protection. *See ring main, secondary main, service main, trunk main.*

make-up water The proportion of the *feed water* that is needed to make up the losses of water in a boiler system.

malaria A tropical disease caused by *parasitic, spore-forming protozoa* and transmitted by the *anopheline mosquito*. To transmit the *parasite* to humans the *mosquito* must bite a person at a particular stage in the life cycle of the protozoon. The protozoa develop in the mosquito, entering the human from its salivary glands.

Malta fever *Brucellosis*.

Mammoth rotor A *brush aerator* 1 m in diameter with a horizontal drive shaft, similar to the *TNO rotor*. It dissolves more oxygen per unit length of rotor than the TNO and has been used in *oxidation ditches* serving populations from 10 000 to 250 000. Mammoth rotors can be used in *aeration* channels 2 to 4 m deep. A single rotor may be from 4.5 to 9 m long. At maximum immersion the gross power required is about 5 kW per m length of rotor. *See oxygenation efficiency.*

manganese, Mn Manganous ions (Mn^{2+}) are soluble in water and exist in *anaerobic* conditions, but with *aeration* of the water they form dark brown precipitates of manganese dioxide which have an unpleasant taste as well as colour. Large quantities are toxic. Manganese is rarer than *iron in water* but may be present in water from moorland or wells or *impounding reservoirs*. Manganese compounds can be removed from water in the same way as *iron in water*, in a *filter* or by intense *aeration* at a pH of 6 to 7 followed by filtration through pyrolusite, MnO_2. *Chlorination* can replace the intense aeration. *See* Table of Allowable Contaminants in Drinking Water, page 359.

manhole (1) An access opening into any tank, boiler, etc., large enough for a person to pass through. (2) Or inspection chamber. A vertical shaft giving access to a *sewer* or to an underground duct, so arranged that a man can enter from the surface. In small sewers, less than 1 m diameter, manholes should be provided at every change of direction but not further apart than 90 m on straight lengths, because 45 m is the greatest length that can be rodded. On large sewers, through which a man can walk, the maximum spacing can be 120 m, so that a sewerman need never be further than 60 m from a manhole. *See backdrop, rodding.*

Manning formula   A formula used widely in the USA to evaluate *open-channel flow:*

$$v = \frac{1.486 \, R \, G^{0.5}}{n}$$

where v = average velocity, R = *hydaulic mean depth*, G = *hydraulic gradient* and n = Manning's coefficient of roughness.

manometer   A pressure-measuring instrument consisting of a U-tube containing a liquid or system of liquids. When one leg of the U is under higher gas pressure on its surface than the other, the liquid in it is correspondingly lower. The difference in liquid level corresponds to the pressure difference between the liquid surfaces.

marble test   The *chalk test.*

marine pollution   The sea is polluted in three main ways: (1) By raw or imperfectly treated *sewage* both from the land and from ships and by rubbish from ships. Ships are now able to have *activated sludge* plants for which the activated sludge may be supplied dry. (2) By *trade effluents*, often discharged untreated. *See Minamata Bay.* (3) By oil or oil-derived materials such as tar from petroleum refineries and similar works on the coast, as well as from ships, since 60% of the world's vast oil output is carried over water. Nutrients from pollution of types (1) and (2) can cause coastal areas to become *eutrophic*, especially if there is little flushing action by coastal currents. The chief immediate victims of oil pollution are diving sea birds. Their plumage becomes coated with oil and spoilt. They cannot avoid swallowing oil when they try to preen themselves clean and it may poison them. In the North Sea and North Atlantic alone some 150 000 to 450 000 sea birds are estimated to die every year because of oil pollution. Phytoplankton, on which all sea life depends, may be prevented from growth by an oil slick.

mass balance   Matter is neither created nor destroyed; therefore in any treatment process, physical, chemical or biological, whatever goes in must come out.

mass curve   In *hydrology*, a graph of the cumulative volumes of *runoff* or other flows, plotted as ordinates against time. The curve may be used to estimate the storage capacity needed by a *reservoir* to supply a given flow.

Mastigophora The *flagellate protozoa*.

MATC *Maximum acceptable toxicant concentration*.

maturation pond An aerobic *waste stabilisation pond*, usually following a *facultative pond*, typically 1.5 to 3 m deep. Two maturation ponds in series, each with a *detention period* of 7 days, are needed after facultative ponds to reduce the final effluent BOD_5 to 25 mg/litre. *Algae* are fewer in maturation ponds than in facultative ponds and even fish may be present in the last maturation pond.

maturing (1) Or ripening. Of a *trickling filter*, the process of acquiring a balanced population of *bacteria*, etc., to purify the *sewage* passing through it. A new trickling filter may take 6 months to mature in a cold winter but only a few weeks in warm weather. Early summer is therefore the best time to start a trickling filter. The term is also used of an *activated sludge* process that is starting, but mature sludge can be introduced from another *aeration tank* very easily. (2) Of compost, *curing*.

maximum acceptable toxicant concentration, MATC The highest observed concentration in air or water of a pollutant which has no effect on reproduction, growth, behaviour, etc., of a plant or animal. For air it has been superseded by *threshold limit values* so far as humans are concerned.

mayflies, Ephemeroptera Water insects that are on the wing in May and are eaten by trout. Their presence, except for *Baetis rhodani*, normally shows that the water is unpolluted. They are thus a *biological index of pollution*.

MBAS, MBS *Methylene-blue-active substances*.

MBF A *moving-bed sand filter*.

MCL, maximum contaminant level The highest permissible concentration of a substance in a drinking water, as laid down by the US *EPA*. Such levels have been published also by the *World Health Organization* and other authorities.

mean cell residence time, MCRT, solids retention time, sludge age, cell age, θ_c The total sludge in a biological treatment process, divided by the daily waste (excess) sludge. Thus, for an *activated sludge*:

$$\theta_c \text{ (days)} = \frac{\text{aeration tank volume (m}^3\text{)} \times MLSS \text{ (mg/litre)}}{\text{waste sludge (m}^3\text{/day)} \times SS \text{ in waste sludge (mg/litre)}}$$

In activated sludge θ_c is a measure of the rate at which *cells* are wasted from the process and it has been related to the growth rate of *bacteria, sludge loading rate,* etc. In conventional plants the sludge loading rate is about equal to

$$\frac{3}{2\theta_c} + 0.1$$

θ_c may be used for design or for process control of the activated sludge process. Thus, for 95% removal of **BOD** in activated sludge plants, θ_c should be from 5 to 10 days, while for nitrification it should vary from 10 days at 20°C to 20 days at 10°C. *Extended aeration* operates at θ_c of 20 to 30 days. For *anaerobic sludge digestion* without re-cycling of *digested sludge*, θ_c is normally considered equal to the *detention period*.

meat processing wastes Liquid wastes derived from slaughtering, cooking meat or rendering bones and offal into lard or grease. Although blood is often recovered as a rich source of *protein*, slaughtering wastes remain reddish brown. Their BOD_5 varies around 2000 to 5000 mg/litre. To reduce the amount of solids, the waste should be screened before entry to the sewer. Meat cooking wastes are usually less .polluting but still have a high grease content.

mechanical aeration The use of *surface aerators* or *turbine aerators* in *activated sludge* treatment, *aerated lagoons*, etc.

mechanical collector This vague term for *inertial separation* of any type is usually reserved for *cyclones,* especially in the USA.

mechanical composting, accelerated c. The use of enclosed chambers for *composting,* as opposed to open-air *windrowing*. Stationary multi-storey structures or rotating drums divided into compartments may be used, with, in some processes, microbial *seeding*. Digestion time is short (2 to 6 days) but a long maturing time is needed (1 to 3 months). Total composting time is thus about the same as for windrows, but the investment in plant is ten times as high. *See vertical cell digestion.*

mechanical de-watering of sludge *De-watering of sludge* using mechanical extraction of the water by *centrifuge, filter belt press, filter press, tube press, vacuum filter, sludge thickening*, etc.

mechanical filtration This may mean one of several things: *mechanical de-watering of sludge, micro-straining,* use of a *pressure filter* (2) or *screening.*

mechanical flocculation A low-cost method of *flocculation*, capable of improving the clarity of *raw water* or *raw sewage*, as well as of *effluents*, by the gentle movement of a *picket-fence stirrer* or of paddles at speeds of the order of 70 mm/s for 15 min, followed by 30 min of quiescence. It is often used after *coagulation.*

mechanical sorting plant, mechanical separation p., automatic

separation p., mechanised reclamation p. (or in USA **resource recovery facility**) Equipment which separates the iron, steel, glass, non-ferrous metals, *cellulose, plastics,* etc., in municipal waste, with a view to selling them or forming a marketable *waste-derived fuel* (WDF) from the cellulose, etc. The separation processes are based on particle size, using *screening*; on *specific gravity,* using vibrating tables, *float and sink treatment, dense media, air classification*; also on optical properties. Chemical separation methods are unusual except for *incineration. Eddy-current separators, magnetic separators, mills, textile removal equipment, trommels* and other plant are used. For making WDF dry processes are preferred to wet ones. About two-thirds of the treatment costs may be recovered from the sale of materials and one-third in fees from waste disposal contractors for disposal of their wastes. Mechanical sorting reduces the volume of the material that goes to the tip. If only a WDF is produced, the reduction by weight is not more than about 50%. If, also, all metals and glass are reclaimed, a further 20% by weight is saved from tipping but 30% still must be tipped. *See inertial separator, particle-size distribution, rising-current separator.*

media Plural of *medium.*

median tolerance limit, TL_m, or 50% lethal dose or LD_{50} The concentration of a pollutant in a water under investigation, at which 50% of the fish can survive. The test usually lasts 48 or 96 h. TL_m may vary according to *pH,* temperature, *dissolved oxygen,* chemicals present in the water and species of fish used. The maximum concentration for a discharge to a river is normally determined by using an *application factor. See interaction of toxicants; compare maximum acceptable toxicant concentration.*

Medina worm *Guinea worm.*

medium (plural **media**) (1) Or culture medium. In the laboratory, *bacteria* or other microbes grow in a favourable environment—a medium—which may be liquid, using water (aqueous), or solid, using *agar-agar.* Solid media allow the microbes no movement but local growth clusters or colonies can exist. Media contain all the necessary foods and are held at the right temperature, *pH, salinity* and *dissolved oxygen* content, and are autoclaved or tyndallized beforehand. *Non-selective media* allow the growth of all types of bacteria. *Selective media* encourage one group either by the lack of an ingredient needed by the others or by the presence of one that harms all the others. Thus, if *organic compounds* are left out, the medium is selective for *autotrophs.* If bile salts are put in, the medium is selective for intestinal bacteria. *See autoclave, culture, slope, tyndallisation.* (2) Or packing. The clinker or stones or special *plastics filter medium* that fills the space in a *trickling filter* and provides surfaces on which the *microbes* can live and form *slime.* (3) Or porous septum. In a true *filter*, the strainer that collects solids from the fluid being filtered. It may be sand or other mineral grains in a *deep-bed filter*, or a *membrane*, but is often merely a filter cloth.

megalitre One million litres, 1000 m³, 1000 t of water.

membrane A skin. A membrane filter is consequently one of skin thickness, not a *deep-bed filter*, but the term usually means a *polymer* membrane as used in a *membrane process. Compare membrane filtration; see semi-permeable membrane.*

membrane electret A *membrane* filter with a permanent internal electrical charge, a concept now under development so as to achieve a membrane for *ultra-filtration* that will not block.

membrane filtration A method of direct counting of coliform or other *bacteria* in water, which is faster than *most probable number* techniques. A measured diluted volume is filtered through a *membrane* of pore size small enough (0.22 μm) to hold back bacteria. The membrane is then incubated face upwards in a *selective medium* at a suitable temperature and after 18 h the colonies that have developed are counted. As in the MPN method, these counts are only an estimate. Counting is done under a hand lens and suspect colonies can be transferred to liquid or solid *medium* for confirmation of their properties. Membrane filters are also used for sampling *algae* or for concentrating bacteria from a water in which they are known to be scarce—e.g. a pure drinking water, where the expected count is less than about 30 bacteria per ml. Another use of the method is for sterilising certain culture *media* that break down when heated. If the culture medium is passed through a filter of pore size 0.22 μm, all the bacteria will be held on it.

membrane processes *Electro-dialysis, osmosis, reverse osmosis* and *ultra-filtration* are processes that use *membranes* to hold back dissolved material or fine solids. Their cost depends on the power for operation or the cost of the membranes, which, because of their short life, are expensive. Membranes are being intensively studied and many methods of forming them exist—e.g. inside or outside a porous cylinder, as a flat sheet of *cellulose acetate* interleaved with a packing and wound into a spiral, etc. *See de-salination.*

meq/litre *Milligram equivalent per litre.*

mercaptans Organic compounds containing a sulphur–hydrogen group—i.e. a sulphur and a hydrogen atom attached to the carbon chain. They often have an unpleasant smell.

mercury Many tonnes/yr of mercury were used in US farming in the 1970s, mainly as *fungicide*. It can be highly toxic and is used also in drugs and paper products and is found in *plastics industry wastes*. *See Minamata Bay Japan,* and Table of Allowable Contaminants in Drinking Water, page 359.

meringue de-zincification *De-zincification* that shows a white crust.

meroplankton Sea-bed organisms that spend the egg and larval stages of their lives as *plankton*, unlike *holoplankton*.

mesh sizes The mesh numbers used in the past to describe sieve openings indicated the number of holes (or wires) per lineal inch; thus, a higher number had smaller openings. The diameter of the wires affects the hole size. Sieves therefore are now defined by the size of opening and the wire diameter in mm, under BS 481.

mesobiota Medium-sized soil organisms including *nematodes, oligochaetes*, small insect *larvae* and small *arthropods*. *Compare macrobiota, microbiota.*

mesophile Any *bacterium* that thrives at temperatures between 20 and 45°C, unlike *thermophiles* and *psychrophiles*.

mesophilic digestion *Anaerobic sludge digestion* at about 30 to 35°C. *Compare thermophilic digestion.*

mesosaprobic river A river with medium organic pollution, intermediate between *polysaprobic* and *oligosaprobic* rivers. The sub-division 'α-mesosaprobic' contains less oxygen than 50% saturation but the surface of the bottom mud is not *anaerobic*, although *macrobiota* are limited. β-mesosaprobic waters always have *dissolved oxygen* above 50% and a greater diversity of plants and animals. *Caddis flies, may flies* and *freshwater shrimps* are rare in α-mesosaprobic rivers, but reasonable numbers exist in β-mesosaprobic ones. *Asellus aquaticus* is common in α-mesosaprobic rivers.

mesotrophic Having a moderate content of *nutrients*—i.e. between *eutrophic* and *oligotrophic*.

metabolism The changes, *biochemical*, chemical and physical, in living matter needed for life to continue, involving the building of complex from simple substances, mostly with *enzymes*. Metabolism is either *anabolic* or *catabolic*.

metabolite A substance taking part in *metabolism*, an intermediate product.

metalimnion The *thermocline*.

metal-plating wastes Liquid wastes from metal plating are not voluminous but often contain poisons such as the salts of *copper, chromium, lead, nickel,* tin, *cyanide compounds* and de-greasing agents. Cyanides strip the oxide layer from the metal before electro-plating. Baths of salts of *hexa-chrome* are used for chrome plating and the metals remain in the wash water from the rinse tanks. The pollution load can be reduced substantially if the plated metal is allowed to drip above the plating tank before it is rinsed. This reduces the drag-out of plating solution into the rinse tank.

metals *See arsenic, barium, boron, caesium-137, cadmium, calcium, cast-iron, chromium, copper, heavy metals, hexa-chrome, iron bacteria, lead, lithium, magnesium, manganese, mercury, micro-nutrient, nickel, potassium, selenium, silver, sodium-restricted diet, stainless steel, steel, strontium, zinc,* and Table of Allowable Contaminants in Drinking Water, page 359.

metal salvage Metal often has to be extracted from refuse, so as to protect a *shredding* plant or a furnace from blockage. Iron and steel are the easiest to separate with *magnetic separators*. If tinplate cans are burnt, their removal from the *clinker* will improve its market value and they can also be sold as scrap. If removed from the refuse before *incineration*, cans can sometimes be sold for *de-tinning*. Scrap metals have to be baled and marketed separately, except that iron and steel can stay together. *Aluminium,* brass, bronze, *copper, lead* and *zinc* should each be separated.

meteoric water Water from the atmosphere, which has been rain, dew, hail, sleet, snow, etc.

meteorology The study of the weather. Weather greatly affects pollutant concentrations in *inversions* and their dispersal by winds or rain. High air pollution is usually impossible in high wind.

methaemoglobinaemia *Cyanosis.*

methane, CH_4 A gas with about half the density of air, explosive at concentrations between 5 and 15% in air, with very little, if any, smell. It is formed by the *anaerobic* degradation of *organic* materials.

methane fermentation The action of *methanogenic* bacteria in releasing *methane* and *carbon dioxide* during *anaerobic* decomposition of *organic* material, as in *anaerobic sludge digestion.* The total production of gas ($CH_4 + CO_2$) is about 1 m^3 of gas per kg solids (mainly carbon) destroyed, of which up to 70% is methane. *High-rate anaerobic digestion* reduces the volume of gas and possibly the percentage of methane. *See sludge gas.*

methane recovery from landfill *Methane* recovery from *landfill* becomes economic only at sites for cities of several million people, as at Los Angeles, California.

methanogenic bacteria, methane formers Strict *anaerobes* that, in *anaerobic sludge digestion,* generate *methane* mainly from

simple organic acids or alcohols made by the *acid formers.*
Carbon dioxide also is formed by some methane formers but
others may consume it. They cannot function at a *pH* below 6.2
and prefer a pH from 7.0 to 7.5. Their slow growth rate
necessitates long *detention periods* compared with *aerobic*
treatments.

methemoglobinemia US spelling of methaemoglobinaemia,
cyanosis.

methylene-blue-active substances, MBAS, MBS Anionic
synthetic detergents that react with methylene blue to produce an
intense blue colour. Other *organics* or inorganics may interfere
with this reaction; therefore the test should be used only to
examine drinking water.

methylene blue stability test An estimation of the *stability* of a
sewage *effluent*, based on the time taken to decolorise methylene
blue solution. This is related to the time for the sample to become
anaerobic. If the blue colour persists for more than 5 days, the
sample is considered to be stable.

methyl mercury, H_3CHgCH_3 or $(CH_3)_2Hg$ A highly poisonous
mercury compound with a tendency to accumulate up the *food
chain.* It poisoned many people at *Minamata Bay, Japan.*

methyl orange alkalinity The *total alkalinity.*

mg/l, mg/litre Milligrams per litre, an SI measurement that has
superseded parts per million (p.p.m.) but usually means the
same.

mho A siemens, a unit of *electrical conductance*, the reciprocal of
the resistance in ohms.

micro- Prefix that indicates one-millionth, or 10^{-6}, for short
written with the Greek letter μ.

micro-aerophile An organism that prefers low to high levels of
oxygen.

microbe A micro-organism. *See microbiota.*

microbial film A *biological film.*

microbiology The sciences of microbes, including *bacteriology,
cytology, enzymology, mycology, virology.* The main living
beings studied are, in order of complexity: animals, plants and
protists (the least complex). Animals in this sense are tiny
invertebrates, including *rotifers* and *crustaceans.* Plants include
mosses and ferns as well as seed plants. However, no individual
plants or animals are visible except under the microscope.

microbiota, microbes, micro-organisms *Algae, bacteria, fungi,*
and other *microscopic* organisms found in the soil, the air, the
water, the body, as well as in *trickling filters, activated sludge*
and other treatment processes.

Micrococcus An important genus of *bacteria* in *activated sludge,* also an *acid former*. They are *facultative anaerobes*.

microcurie, μ**Ci**   One-millionth of a curie, 10^{-6} curie, one million picocuries, a unit of radioactivity.

microfauna   *Microscopic* animal life.

microfiltration   *Ultra-filtration*.

microflora   *Microscopic* plant life.

microgram, μ**g**   One-thousandth of a gram;
   1 μg = 0.001 milligram = 10^{-6} g.

micrometre, micron, μ**m**   A measure of length, 10^{-6} metre, one-millionth of a metre, 1000 nanons, a thousandth of a millimetre.

micromho, microsiemens   A millionth of a *siemens* or of a *mho*, a unit of *electrical conductance*.

micron   *See micrometre*.

micro-nutrient   A food required in minute quantities that can be measured in parts per thousand million; it may be toxic in excess. *See boron*.

micro-organisms   *Microbiota*.

micro-pollutants   Substances that pollute or poison even in minute concentrations—an ill-defined group that can include *viruses* or *bacteria*. *Mercury, cadmium, lead* or *organo-chlorine compounds* could be regarded as micro-pollutants.

micro-Ringelmann chart   A photographic reduction of the *Ringelmann chart*, made to a size that enables it to be used by a single observer holding it at arm's length.

microscope   Normally an *optical microscope*, but *see also electron microscope*.

microscopic   Description of things too small to be seen except under the microscope.

microscreen   A *micro-strainer*.

microsiemens   *See micromho*.

micro-strainer, micro-screen   A drum covered with very fine

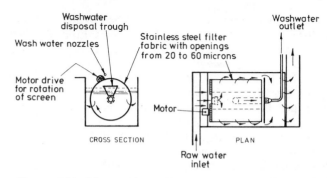

Figure   M.1   Micro-strainer (Water Treatment Handbook', Degremont-Laing)

191

stainless steel wire mesh which rotates on a horizontal shaft and is partly submerged in the water being screened (*see Figure M.1*). A jet of filtered water from inside the drum washes off the solids caught outside. Drums are from 1 to 3 m diameter and 0.3 to 4.5 m long, with openings from 15 to 65 μm across. Regular inspection of the mesh is needed and the *biological film* on it has to be removed continually by water jets unless it is inhibited by *chlorination* or by *ultra-violet radiation*. Micro-strainers demand less land than *sand filters* and are cheaper to operate. The feed may need good *flocculation*. A sewage *effluent* of 10:10 quality would be achieved with a 15 μm fabric and 15:15 with a 35 μm fabric, starting from an effluent of Royal Commission standard (30:20—i.e. 30 mg/litre *suspended solids* and 20 mg/litre *BOD_5*). If a *raw water* is relatively pure (e.g. upland catchment water), micro-straining followed by disinfection may be the only treatment the water needs to make it drinkable. *Hydraulic loadings* on micro-strainers are in the range 250 to 1500 m³/day per m² of total surface area for *sewage* effluents and 700 to 2300 m³/day per m² for raw water. The finer mesh needs the lower loading.

midges at trickling filters *See filter flies.*

mill, shredder, shredding machine, pulveriser, size-reduction machine Most *shredding* machines are probably *hammer mills* running at high speed without added water. *Drum pulverisers* turn slowly and so need less maintenance and less electricity to drive them but do need added water. A third general type more used in Holland than the UK is the *rasp mill*. Some of the mills used are: *alligator shears, cage disintegrator, chipper, disc mill, fragmentiser, pulveriser, shredder.*

milled refuse See *shredding.*

milligram, mg One-thousandth of a gram; 1 mg = 1000 micrograms = 10^{-3} gram.

milligram equivalent per litre, milli-equivalent per litre, meq/litre A way of expressing the concentrations of substances in water instead of using mg/litre. The concentration in mg/litre is divided by the *equivalent weight*. A solution with a strength of 1 meq/litre has one-thousandth of the strength of a *normal solution*. *See also equivalent per million.*

millimole per litre, mmol/litre The concentration in mg/litre divided by the gram-molecular weight of the substance. A solution with a strength of 1 millimole/litre has one-thousandth of the strength of a *molar solution*.

millirem One-thousandth of a *rem*. *EPA* in 1976 allowed a maximum annual dose of 4 millirems in drinking water, a figure that was disputed by atomic power stations as being much too low.

millisiemens, mS A unit of *electrical conductance,* one-thousandth of a *siemens.*

Minamata Bay, Japan Where thousands of people became seriously ill and more than 100 people died from taking in compounds of *methyl mercury* concentrated in the fish. These compounds settled in the mud of the bay and came from the waste discharges of a factory of the Chisso Corporation making acetaldehyde with *catalysts* containing *mercury.* The mercury content of the estuary mud near the factory outlet channel exceeded 2000 p.p.m., but dumping continued until May 1968. Mental deficiency among children born between 1955 and 1959 in the most heavily contaminated part of the town reached the very high level of 29%, measured in 1962. By 1975 the company had paid more than 80 million dollars in compensation to victims of the disease. Plans in 1975 were to fill in parts of the bay. The safe level for methyl mercury in fish to be eaten is now thought in the USA and Japan to be 0.4 p.p.m. but even the minute total of 30 mg in a 50 kg adult caused serious illness and 200 mg caused death. *See bottom sediments, osteomalacia.*

mineral dressing, m. beneficiation, m. preparation, m. separation The raising of the metal concentration in an ore and the reduction of its waste rock (gangue) content. Coal is not ore but is treated very similarly. Mines and their associated preparation plants produce more solid waste than any other industry and certainly much more than the tonnage of domestic refuse. *See dense medium, flotation, mechanical sorting plant.*

mineralisation In *sewage treatment* mineralisation implies the reduction of the proportion of organic material, by its conversion to *carbon dioxide.* Since all water contains minerals, sewage treatment increases the ratio of mineral to organic substances in the water, but not the absolute quantity.

mineralised sludge Fully *digested sludge.* Before digestion, *sludge* is some 70% *organic* matter and 30% mineral, if we exclude the water. Digestion can transform two-thirds of the organics into gases and water, so that in fully digested sludge the organic proportion is reduced to 45% and the mineral is increased to 55%.

minestone *Colliery spoil.*

mineral water Any water containing inorganic salts and a very low organic content. Strictly speaking, all *groundwater* is mineral water but the term is usually reserved for bottled drinking water or other waters thought to have health-giving properties.

miner's anaemia *Hookworm disease.*

193

mine water Water drained from mines may contain the mineral being mined and a high *total solids* content, but usually little *organic* pollution. The most common pollution from mine water is caused by *acid mine drainage* or by *ferruginous discharges*. *Alkaline* mine water also exists.

miniature smoke chart A modified, small *Ringelmann chart* designed to be read at about 1.5 m from the eye. Like the *micro-Ringelmann chart*, it is easier to use than the Ringelmann chart.

minimum acceptable flow The *prescribed flow*.

mist Drops of water smaller than fog (2 μm) hanging in the air and resulting in a visibility below 2 km but usually more than 1 km.

mist eliminator, de-mister An *entrainment separator*.

miticide, acaricide A *pesticide* that kills mites on animals or humans.

mixed-bed ion exchange A mixture of *anion-* and *cation-*exchange materials in a treatment bed. High purities can be achieved with confidence—for example, a *conductivity* of only 0.5 μS/cm and a silica content of 0.05 mg/litre. Even better results can be achieved but with less certainty. Mixed beds are standard practice for de-mineralising feedwater for high-pressure boilers.

mixed fission products The waste from the re-processing of nuclear fuel, one of the *nuclear reactor wastes*.

mixed-flow pump, diagonal-flow p. A rotodynamic pump intermediate between a *radial-flow pump* and an *axial-flow pump*. *Volute* casings are suitable for dirty water but are less efficient than the alternative, the *diffuser* casing. They may be used for moderate heads, up to 30 m. *See turbine pump*.

mixed liquor In the *activated sludge* process, the mixed activated sludge and *settled sewage* in the *aeration tank. See MLSS, MLVSS*.

mixed-media filter A *multi-media filter*.

mixed sludge A mixture of *primary sludge* and *secondary sludge*. Mixed sludge from a *primary sedimentation tank* is usually about 3 to 4% solids.

mixing of lake water *See de-stratification of lakes and reservoirs*.

Ml, megalitre One million litres, 1000 m³, about 1000 t of water.

MLSS, mixed-liquor suspended solids The dry *suspended solids* in *mixed liquor* (mg/litre). Conventional *activated sludge* plants operate at 2000 to 3500 mg/litre but *oxygen-activated sludge* plants may operate at 6000 to 8000 mg/litre. *See also return activated sludge*.

MLVSS, mixed-liquor volatile suspended solids The dry volatile *suspended solids* in mg/litre of *mixed liquor*. MLVSS is often considered to be 70% of the *MLSS* and is used in the USA in

194

preference to MLSS.

modified aeration A high-rate *activated sludge* treatment with the same flow sheet as conventional aeration but with much shorter aeration times, below 3 h, consequently with lower-quality *effluent*. The *MLSS* is about 500 mg/litre. The *sludge* often does not settle well.

modular units The use of several adjoining small tanks instead of one large one. Although modular units may be more expensive to build, the controller obtains operating versatility which otherwise he does not have. In particular, *activated sludge* plants built on a modular basis can be operated to simulate *plug flow*.

Mogden formula The calculation method for the charges for treating industrial *effluent* at Mogden *sewage* works near London is the sum of three costs for (1) the flow rate; (2) the *biological treatment* based on the *McGowan strength* of the waste; (3) the treatment of the *sludge*, based on the *suspended solids* content of the waste.

Mohlman index The *sludge volume index*.

moisture-holding capacity The *field capacity* of a soil.

molar solution A solution containing 1 g molecular weight (*mole*) of the *solute* per litre of solution. *Compare normal solution.*

mole In chemistry, the molecular weight of a substance in grams, formerly the gram-molecule.

mole drain A *land drain* about 80 mm diameter through stiff clay, made without a drain pipe. The hole should be formed when the clay is moist and plastic. It stays open for many years and drains the clay effectively if it is driven at the right slope. A tractor of at least 50 h.p. is needed to pull the mole plough.

moler earth *Diatomite.*

mollusc, Mollusca A *phylum* of invertebrate animals that includes snails, marine and *freshwater mussels*, limpets, clams, oysters and other *shellfish*.

molluscicide A *pesticide* that kills molluscs.

monitoring well, observation w. A water well sunk solely for the purpose of watching the quality of the water (not for *abstraction*) and to determine how far a particular contamination of a *groundwater* has reached. Where a *disposal well* is in use, a monitoring well or a series of them is an essential part of the installation.

monomer A simple molecule from which a *polymer* can be chemically built up.

mosquito Of the many mosquitoes (40 British types alone) some are *vectors of disease*, but only the *anopheline mosquito* carries *malaria. Filariasis* is carried mainly by the culicine mosquito (*Culex fatigans*) and sometimes by the anopheline. *Yellow fever* and *dengue* are carried by *Aëdes aegypti*. Mosquito *larvae* develop on water, and an increase of water area may increase the mosquito population. Malaria epidemics have occasionally been associated with man-made lakes, *irrigation* schemes or *waste stabilisation ponds*.

most probable number, MPN, multiple-tube method   A method of estimating a count of the number of microbes in unit volume of water, based on statistics. The positive and negative results obtained when testing multiple tubes of geometrically increasing dilution are recorded. The MPN method gives an estimate of the number of coliforms or *Escherichia coli* in water or *sewage*. Fivefold or tenfold dilutions of the sample are made successively, so that some of the results are negative. Samples from three successive dilutions are innoculated into *MacConkey's broth* and incubated for 48 h at 37°C for the total coliform count or (for *E. coli*) at 44°C. A positive test is shown by a change of colour from purple to yellow, with evolution of gas, collected in a small tube. The full technique and the statistical tables are published in *Bacteriological Examination of Water Supplies*, HMSO. The result is a *presumptive coliform count*. *See also membrane filtration*.

moving-bed sand filter, sand recirculation f.   A *sand filter* used for cleaning coagulated *raw water* or sewage *effluent*, in which the sand moves continuously down through the *filter*. At the bottom of the filter the sand–sludge mixture is removed to a washing device and then re-fed into the top of the filter (*see Figure M.2*). Because of the continuous cleaning, the sand can be very fine-grained, which ensures good filtration of the water.

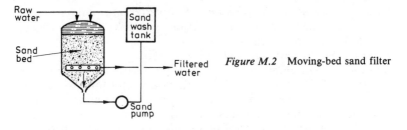

Figure M.2   Moving-bed sand filter

Various designs exist, some with an upward flow of water *countercurrent* to the movement of the sand. The filters may be conical, circular or rectangular. *Hydraulic loadings* are similar to those in the *rapid gravity sand filter*. *Compare bi-flow filter*.

moving-bed scrubber, floating-bed s.   A *wet scrubber* containing spheres of plastics, glass, marble, etc., which are fluidised by dirty gas rising up through them (*see Figure M.3*). The wash

196

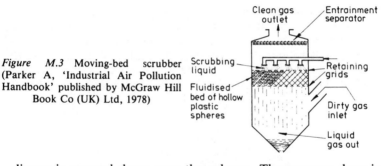

Figure M.3 Moving-bed scrubber (Parker A, 'Industrial Air Pollution Handbook' published by McGraw Hill Book Co (UK) Ltd, 1978)

liquor is sprayed down over the spheres. The pressure drop is directly proportional to the flow rate of the liquor. Particles of 0.1 μm size can be collected, as well as viscous liquids. If the spheres are of hollow plastics, they may need to be held down by an upper constraining plate. There may be more than one bed. These scrubbers have been claimed to have such high effective throughputs of liquor and gas that the absorption of *sulphur dioxide* by the caustic soda scrubbing liquor is quintupled by comparison with a stationary *packed tower*. *See turbulent-contact absorber.*

MPN *Most probable number.*

multicellular collector, multicyclone A nest of *axial-inlet cyclones*.

multi-media filter, multi-layer f., mixed-media f. A modern downflow *rapid-gravity sand filter* that uses *coarse-to-fine filtration* with three or more media, often anthracite at the top (the coarsest), sand in the middle and *garnet* (the finest) at the bottom. Polystyrene may be added as an extra layer on top and magnetite as an even finer, denser layer at the bottom. Other materials have been used and they should be chosen to suit the water or *effluent* to be filtered. The grain size varies from 1 to 2 mm at the top to less than 0.5 mm at the bottom. Despite the much greater total area of the grains compared with rapid gravity sand filters or *dual-media filters*, the *head loss* is the same for the same flow rate, but flow rates can be 2.5 times as high with the advantage of cleaner water. Consequently, several US waterworks that have replaced their rapid-sand filters by multi-media filters in conjunction with *inclined-tube settling tanks* in their old *sedimentation tanks* have been able to double their throughput and produce cleaner water without enlarging the works.

multiple-flue stack A chimney with several flues in it, usual at modern power stations. The outer structure is called a windshield.

multiple-hearth furnace A continuous *incinerator* some 8 m diameter and 10 m high, capable of burning sewage *sludge* or solid waste, consisting of a hollow, central, vertical, air-cooled shaft that rotates slowly, raking the sludge gradually from the coolest hearth at the top to the hottest at the bottom (*Figure M.4*). There may be as many as eight horizontal hearths directly

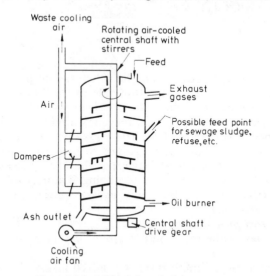

Figure M.4 Multiple-hearth furnace

above one another, and oil firing may be introduced near the middle hearth. If used for burning solid refuse, this must first go through a *mill.* Solid waste and sewage sludge have been burnt simultaneously, the solid waste providing the heat to burn the sludge. The sludge is generally 60 to 75% water, although some designs allow sludges with 90% water to be burnt.

multiple-tube fermentation *See most probable number.*

multiple-tube press A *tube press.*

multiple or **multi-purpose use of water** The aim of modern water development is to satisfy as many as possible of the community's needs for water, whether for amenity, recreation, fishing, drinking, industry, *irrigation*, navigation, electrical generation or *effluent* disposal. Some of these needs oppose others and compromises must be made. *See river-regulating reservoir.*

multipole overband separator An *overband separator* with a number of magnets which alternately release and pick up the magnetic material from a belt, thus shaking out the non-magnetic material and producing cleaner iron and steel.

multi-stage flash distillation, vacuum separation A *de-salination* process for distilling drinking water from sea-water or *brackish* water. The vapour from each stage condenses on tubes cooled by

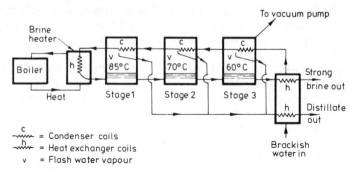

Figure M.5 Multi-stage flash distillation.
Pressure in stage 1 is lower than in brine heater and some of its water therefore flashes into vapour. Correspondingly stages 2 and 3 have even lower pressures. Stage 3, connected to the vacuum pump, is at the lowest pressure

the sea-water flowing into earlier stages (*see Figure M.5*). One of the largest plants, in Hong Kong, has an output of 180 000 m³/day.

multi-stage shredding *Shredding* of refuse may take place in as many as three stages, with *primary shredding* to a nominal maximum size of some 15 cm, *secondary shredding* to about 3 cm and the unusual stage of tertiary shredding to about 1 mm nominal maximum size. *Two-stage shredding* is commoner. The second shredding always involves less wear on the machine and a smaller expenditure of energy than the first. The maximum size of shredded refuse may be 25 cm for *landfill* (in one stage) or 3 cm (for *composting*) in two or more stages.

municipal refuse *See solid waste.*

mussels *See freshwater mussels, Mollusca.*

mutant In biology, an individual with different characteristics from the rest of its species, which may be transmitted to its descendants. *Pesticides* kill all the pests except those mutants that can resist them, and often the pests develop a strain that is immune to the pesticide.

Mycobacterium The members of this genus are mostly actinomycetes and harmless to humanity, but two species cause disease. *M. tuberculosis* causes tuberculosis but the disease is believed to be only rarely waterborne. *M. leprae* causes leprosy, which is not waterborne.

mycology The science of *fungi.*

199

N

Nais A true segmented worm, an *annelid* belonging to the family Naidides, typical of water with mild to serious organic pollution.

nanon, nanometre A unit of length, 10^{-9} metre, 10^{-3} micron, formerly the millimicron. It supersedes the ångström, and is equal to 10 A.

National Environmental Policy Act 1969, Public Law 91-190, NEPA Probably the most far-reaching of the US federal laws on the environment. It established the *Council on Environmental Quality*, a national policy for the environment, also a requirement in all major federal projects for a detailed, comprehensive *environmental impact statement* (EIS). Its omissions in detail about the EIS have been rectified by many court decisions of federal judges, who have not hesitated to interpret the law in spirit rather than in letter.

natural draught The height of a chimney and the warmth of the air in it cause the air to be sucked up from the fire below. The higher the chimney and the warmer the air, the greater is the draught. An electrically driven fan, however, is much cheaper than a high chimney. *Forced-draught fans* and *induced-draught fans* are now used.

natural purification *See self-purification.*

Navicula A freshwater *diatom* abundant in organically polluted water, which is tolerant of a wide range of *pH* and may occur in salt water. It absorbs and concentrates pollutants such as *Dieldrin*, posing a threat to fish that live off it.

Neanthes A genus of *polychaetes*, some species of which may be used in *bio-assays* of tidal mud and estuaries.

Necator americanus *See hookworm disease.*

necton, nekton Tiny organisms that swim strongly, unlike *plankton* and *neuston*.

nematicide A *pesticide* that kills *nematodes*.

nematode, Nematoda A large class of small, unsegmented round worms, of which a few are parasitic and mentioned under *helminthic diseases*. Most are free-living creatures of the water or mud, found also in *trickling filters*, and possibly *activated sludges*, and harmless except that, like other small beings, they can harbour *viruses* and protect them from *chlorination*.

NEPA The US *National Environmental Policy Act 1969*.

nephelometer A type of *turbidimeter* that measures *turbidity* through the amount of scattering of the light that falls on turbid water.

nephelometric turbidity unit, NTU, formazin turbidity unit, FTU A unit of *turbidity* defined by the turbidity of a given solution of formazin. One NTU equals one *Jackson turbidity unit* approximately.

neritic zone   The sea bed between low water and the edge of the continental shelf.

nesslerisation   A measurement of the *ammonia* content of water by the formation of a yellow colour when it is added to Nessler's reagent—a solution of potassium mercuric iodide with potassium hydroxide.

neuston   The water life at the air—water interface, either above it (e.g. the water skaters) or below it (e.g. the *mosquito* larvae). *Compare benthon, nekton, plankton.*

neutralisation   In *chemical treatment*, elimination of the *acidity* or *alkalinity* of a water, bringing its *pH* value near to 7.0.

nickel   A *heavy metal* that may be present in *metal-plating wastes*. It is poisonous to many crops.

night soil   The contents of *cesspools,* etc., so called because they are often removed by night. In hot climates they can be treated by *facultative lagoons* or the *Indore process.*

nitrate removal   *See de-nitrification.*

nitrates, -NO₃   Substances whose chemical formula ends in $-NO_3$ are derived in domestic *sewage* from the oxidation of *ammonia*. Too much nitrate in water causes *cyanosis* in infants. In the USA nitrates have occasionally been added to polluted streams to stop hydrogen sulphide being given off. *See eutrophication; see also* Table of Allowable Contaminants in Drinking Water, page 359.

nitrification   Conversion of the *ammoniacal nitrogen* in *sewage* to *nitrites* and eventually *nitrates*, reducing the *oxygen demand* in the *receiving water*, since the oxidation of *ammonia* to nitrite requires 4 to 5 kg *oxygen* per kg ammonia oxidised. Nitrification can be fairly complete in standard *trickling filters* but not in *high-rate trickling filters*. In *activated sludge* low *sludge loading rates* are needed for nitrification. The process is very temperature-dependent. Usually in winter in the UK, little nitrification occurs in conventional trickling filters. To nitrify an activated sludge the maximum sludge loading rates are: at 10°C 0.15 kg kg⁻¹ day⁻¹; at 15°C 0.23 kg kg⁻¹ day⁻¹; and at 25°C 0.38 kg kg⁻¹ day⁻¹. The *pH* should be between 7.5 and 8.5 or nitrification will be slow. It can also be prevented or delayed by small amounts of *trade effluents*, including *heavy metals, cyanide compounds, herbicides, insecticides* or *allyl thiourea*. *See also* below and *de-nitrification, nitrogenous BOD, nitrifying.*

nitrification—de-nitrification   *Nitrification* followed by *de-nitrification*, resulting in bubbling away of *nitrogen* gas. The two processes may be separate or contained in one *activated sludge* plant which consists of *anoxic* zones with 2 h detention alternating with aerated zones of 6 h detention, and no intermediate *sedimentation*. The first zone is anoxic and the last zone aerated.

nitrified effluent *Effluent* that has enjoyed full *nitrification* contains up to 50 mg/litre of *nitrogen* in the form of *nitrate*.

nitrifying bacteria, ammonia oxidisers *Bacteria* common throughout nature in earth and water, which achieve *nitrification* by *chemosynthesis*. In streams, lakes and estuaries *Nitrosomonas* and *Nitrobacter* are commonest, while in the sea *Nitrosocystis oceanus* often oxidises *ammonia* to *nitrite*. They are *autotrophs* and use *carbon dioxide* as their carbon source, while the oxidation of ammonia gives them their energy. Nitrifying bacteria have a slow initial rate of growth, so in most *sewages* their *oxygen demand* is negligible for the first week or ten days. Consequently, its effect does not appear in the BOD_5 test.

nitrifying filter A *trickling filter* may be used to nitrify an effluent and reduce its *BOD* after *activated sludge* treatment or a *high-rate trickling filter*. The reactor may be an ordinary trickling filter or one with *submerged-bed aeration*. A further *sedimentation* may be needed to settle the *humus* sloughed off the filter.

nitrilo-triacetic acid, NTA A compound whose sodium salt has been proposed as a *builder* that could replace *sodium tripolyphosphate* in *synthetic detergents* and thus reduce the *phosphorus* content of *sewage*. Disadvantages are that it may chelate (*see chelating*) toxic metals and hold them in *solution*, and may partially biodegrade to form nitrosamines that cause cancer.

nitrites, -NO$_2$ Compounds whose formula ends in -NO$_2$ are an interim stage of *nitrification* and readily oxidised to *nitrates*, -NO$_3$. In *sewage* they hardly ever exceed a *nitrogen* content of 1 mg/litre. In *raw water* underground or on the surface, nitrites rarely exceed 0.1 mg/litre. *Ammonia* is converted to nitrites by *bacteria* of the genera *Nitrosococcus* and *Nitrosomonas*.

Nitrobacter A nitrifying bacterium that converts *nitrite* to *nitrate*. It is a chemosynthetic *autotroph*.

nitrogen, N Nitrogen exists in *sewage* in four main forms and its *biological treatment* cannot proceed without some at least of them: *organic nitrogen*; *ammonia* and its salts, such as $(NH_4)_2CO_3$; *nitrite* nitrogen, usually only in small amounts; and the final product of *nitrification, nitrate* nitrogen. In any combined form (not dissolved N_2) its presence is undesirable. Sewage *effluent* may contain 40 to 50 mg/litre of nitrogen compounds. Some *algae* consume atmospheric nitrogen. *See eutrophication.*

nitrogen cycle The breakdown and formation of *nitrogen* compounds (*see Figure N.1*). The bulk of the nitrogen on earth is probably in the atmosphere, which is 80% nitrogen gas. Only a

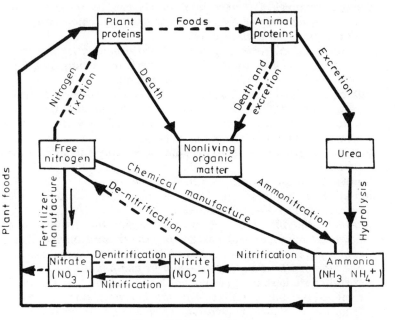

Figure N.1 Nitrogen cycle

few organisms can absorb it as a *nutrient*; one of these is *blue-green algae*. Others must take it up in combined form. Making *nitrates* from nitrogen gas is called *fixation of nitrogen* and it may be industrial or natural. Like the *hydrological cycle*, it is a circulation from earth to atmosphere and back, but with many biochemical changes on the way.

nitrogen dioxide, n. peroxide, NO_2 A poisonous, reddish-brown gas produced at high temperatures in small concentrations in generating station boilers and internal combustion engines from the burning of the *nitrogen* in the air. It has been found to harm people's health even in tiny concentrations (over 6 months) of 0.11 to 0.16 mg/m^3. Mixed in air that also contains *sulphur dioxide*, which is usual, it has a much more serious effect on plants than it or SO_2 alone would have. It is an important part of *photochemical smog*.

nitrogenous BOD, NOD The *oxygen demand* exerted in the *BOD* test by the oxidation of *ammonia* to *nitrite* or *nitrate*. It normally does not start until after 10 days of *incubation* of the sample and therefore cannot be included in the BOD_5 value. *Nitrification* can be suppressed by adding allyl thiourea in the BOD test.

nitrogen oxides *See nitrogen dioxide*; see also *NO_x.*

nitrogen removal *Nitrogen* compounds can be removed from waters by *algal harvesting, ammonia stripping, chlorination, denitrification, ion exchange*, etc.

Nitrosococcus, Nitrosocystis oceanus, Nitrosomonas, nitrosofying bacteria. These are Nitrifying bacteria that oxidise *ammonia* to *nitrite.* They are *chemosynthetic* and *autotrophic.*

Nitzschia A genus of *diatoms* in fresh water and the sea, which may be abundant in ponds, ditches or occasionally in polluted water. It tolerates a wide range of *pH*, 4.2 to 9.0, as well as *copper, phenol* and *chromium* salts.

NOD, nitrogenous oxygen demand US term for *nitrogenous BOD.*

nominal retention period A *detention period.*

non-alkaline hardness A *hardness* of water, usually called permanent or *non-carbonate hardness.*

non-biodegradable Description of something that is not broken down by microbes and has an *oxygen demand* only if it is a chemical reducing agent. It has no *biochemical oxygen demand. See biomagnification, organo-chlorine compounds.*

non-carbonate hardness, permanent h. *Hardness* that cannot be removed by boiling, unlike *carbonate hardness.* It results mainly from the *sulphates* or *chlorides* of *calcium* or *magnesium. See water softening.*

non-community water supply As defined by the *EPA*, a drinking water supply that, the year round, serves fewer than 15 service connections or 25 permanent residents. Sampling conditions are less strict than for community supplies.

non-ferrous metals Metals other than iron and steel.

non-filterable residue One of the goals set by the American Water Works Association is that non-filterable residue (the solids that remain in drinking water after filtration) shall not exceed 0.1 mg/litre. This would ensure low *turbidity.*

non-ionic detergent *See synthetic detergent.*

non-return valve, clack v., check v. A pipe fitting that allows flow in one direction only. It can work by a flap that swings in one direction or by a spring-loaded ball valve, etc.

non-selective medium A culture *medium* that encourages the growth of many types of microbe.

non-uniform flow *Open-channel flow* in which the water surface is not parallel to the *invert* in a channel of uniform cross-section. It can occur at a *hydraulic jump* or with gradually varying flow. *See uniform flow.*

normal solution A *solution* containing the *equivalent weight* in grams (the gram-equivalent) of *solute* per litre of solution. For monovalent *ions* it is the same as a *molar solution*.

NO$_x$ Any mixture of *nitrogen dioxide* (NO_2), nitric oxide (NO) and nitrous oxide (laughing gas: N_2O). N_2O, unlike the other two, is not a pollutant but is often present in air in small concentrations. NO is slowly oxidised to NO_2 in the presence of *oxygen*, but the ultra-violet radiation of bright sunlight dissociates NO_2 to NO and oxygen. The interaction between NO and hydrocarbons in bright sunlight leads to a build-up of *oxidants*, forming *photochemical smog*. In furnaces, if temperatures stay above 1100°C for long, NO$_x$ can be produced, since 79% of the air passing through is *nitrogen*. Under the US Air Quality Act 1970 the allowable NO$_x$ emission from a furnace with an input of 250 million Btu/h (73.3 MW) or more is, per million Btu (expressed as nitrogen dioxide, NO_2), 90 g for a gas-fired unit; 135 g for an oil-fired unit; and 310 g for a coal-fired unit. *See air quality standards.*

NSSC Neutral sulphite semi-chemical pulp, a papermaking process.

NTA *Nitrilo-triacetic acid.*

NTU *Nephelometric turbidity unit.*

nuclear reactor wastes, radioactive wastes *Cooling water* and fuel re-processing water are the two main liquid wastes from nuclear reactors. Cooling water even at its worst is a minor problem and generally now flows through *closed recirculation systems* that do not pollute. The fuel-reprocessing waste is much more difficult to dispose of. Decay products of uranium or plutonium have to be removed from the reactor fuel to allow it to continue to do its work. These 'mixed fission products' contain, apart from unused fuel and decay products, metal from the fuel element, aluminium or *stainless steel* or zirconium. The solid waste products that cause the worst problems are *strontium-90* and *caesium-137*. These are long-lasting, high-energy *radio-isotopes*. After 50 years they still produce 0.1 Ci/litre. Considerable heat is generated for all this time, which means that, apart from secure storage for centuries, the waste water also needs to be cooled. This is why waste disposal controls the rate of development of nuclear power.

nuggetiser A *fragmentiser.*

nuisance In law, a nuisance is one of many types of pollution by *dust* or soot, noise, smells, smoky chimneys, litter, water pollution, offensive tips, etc., which may be a matter for prosecution by the local authority, sometimes also by an individual. Most nuisances, like those mentioned above, are common law nuisances. Statutory nuisances are those that have been created by law (statute). All are inspected by *environmental health officers.*

nutrient The foods for microbial and plant life, mainly compounds of *nitrogen* and *phosphorus* but also of *potassium, magnesium,* iron, *calcium,* cobalt, *copper,* sulphur, *zinc* and others. Although *oxygen* and water are essential to life, they are not commonly called nutrients. *See eutrophication, micronutrient, nitrogen removal, phosphorus removal, substrate.*

nutrient cycles The *regeneration* and cycling of *nutrients,* particularly *carbon, nitrogen, phosphorus, sulphur. Primary production* releases inorganic nutrients from mineral sources, which are then available for other life forms. *See nitrogen cycle.*

nutrient enrichment *See* **eutrophication,** and above.

nymph An immature adult of some insects such as *stone flies* and *mayflies,* which develop into the *imago* (adult) without a pupal stage (*see pupa*).

O

obligate aerobe A *microbe* that can live only in an environment containing *oxygen,* unlike a *facultative anaerobe.*

obligate anaerobe A *microbe* that can live only in the absence of free oxygen, unlike a *facultative anaerobe.*

obligate parasite A type of life that can be only a *parasite*—e.g. *viruses* and the *Plasmodium* that causes *malaria.*

observation well A *monitoring well.*

ocean dumping Under the Oslo Convention of February 1972, 12 west European countries agreed to ban certain types of dumping in the North Sea and Atlantic between Greenland and Europe, as far south as Gibraltar. Any dumping of severely *toxic wastes* allowed is to take place in sea more than 2 km deep and at least 185 km (150 sea miles) from shore. An absolute ban was placed on *mercury, cadmium* and their compounds, *organo-chlorine compounds* that are pesticides, other poisonous *halogen* or silicon compounds that are not quickly *biodegradable* into harmless materials, substances believed to cause cancer and any persistent, floating synthetic substances. Ocean dumping of solid wastes was practised around the USA until 1933, when a court decision made it illegal, and around the UK until the early 1950s. However, many other sorts of ocean dumping continue—of sewage *sludge,* demolition debris, colliery waste, etc. The UK trend now is to ban all dumping except under licence from the

Minister of Agriculture, Fisheries and Food. *See Dumping at Sea Act 1974, marine pollution*, and below.

ocean incineration of toxic wastes *Organo-chlorine compounds* and similar wastes that are difficult to dispose of have to be incinerated at sea, by special permission, in *incinerator* ships. An *EPA* special permit to Shell Chemical Co. allowed it to burn some 16 000 t of organo-chlorine wastes in 2.5 years on an incinerator ship 190 sea miles south of Cameron, Louisiana, in the Gulf of Mexico. EPA filed its *environment impact statement* for this site in July 1976, stating that it proposed to use it for 5 years.

ocean temperatures *See bathythermograph.*

Odonata *Dragon flies* and *damsel flies.*

odour control Water and *sewage* usually do not smell if they remain aerated. In water supply, smell may be removed by *ozonation* or *chlorination* or by adding a little *activated carbon.* *Sewer* smells are ordinarily reduced by venting through a scrubber which may contain a strong *alkali* or active carbon. *Hydrogen peroxide*, pure *oxygen* and air have all been added to sewage to maintain aeration and stop smells, as well as to prevent *sulphide corrosion.* Various perfumed masking chemicals can be sprayed as *aerosols* but often the smell of rotting *sludge* persists over a greater area than the masking chemical.

odour threshold A 'smell limit' below which a substance cannot be detected by the human nose. Odour thresholds for different people can differ a thousandfold. They are therefore decided by a group of people, a trained odour panel, and the threshold is the lowest concentration at which half the members recognise the smell. In parts per million by volume, some common odour thresholds are: acetaldehyde 0.2; acetic acid 1; acetone 100; *ammonia* 47; *bromine* 0.05; butyric acid 0.001; ethyl alcohol 10; ethyl *mercaptan* 0.001; formaldehyde 1; *phenol* 0.05; *sulphur dioxide* 0.05. *See threshold odour number.*

Office of Solid Waste Management Programs *See* OSWMP.

oil coalescer A device for separating tiny oil particles from water. The oil–water mixture passes through a special porous medium which holds back the oil drops. They coalesce as they pass through, become larger and so rise rapidly on the exit side.

oiled birds *See marine pollution.*

oil films Films of crude oil hinder the passage of light into water and probably also hinder gas exchange.

oil interceptor A type of *oil separator*, an *intercepting trap* designed to collect oil, especially from garages, to prevent it fouling the *sewer.* To make sure that the essential maintenance (emptying) is done regularly and correctly, it has been suggested that oil interceptors should be emptied by a representative of the sewer authority.

oil pollution Under the Oil in Navigable Waters Act 1955, a discharge is not 'polluting' in certain UK waters unless it contains more than 100 mg/litre of oil. However, river or water authorities can require higher standards for discharge into *receiving waters* and often insist on 'no visible oil'. *See also dispersants, marine pollution.*

oil reclamation *See sump oil.*

oil refinery effluent Waste waters from oil refineries may be treated by *coagulation, sedimentation* or *flotation* and *biological treatment*. This can usually fulfil the strictest European requirements, namely 3 mg/litre of oil, 25 mg/litre of *BOD$_5$* and 0.2 mg/litre of *phenols* before discharge to a *receiving water*. More than 80% of the water from refineries is *cooling water* contaminated by leaks and spilt oil, which may reach 3% of the oil treated.

oil separator A tank which separates oil from water or *sewage*, as in an *inclined-tube settling tank, flotation tank, grease trap, oil coalescer* or *oil interceptor*.

oil tanker washings, slops Every oil tanker on its return voyage empty has to wash out its tanks somewhere. Under the load-on-top system, a voluntary arrangement made in 1964, ships agreed not to discharge any washings into the sea and the receiving refineries agreed to receive the slops. Not all refineries were willing, but 80% agreed to take on the extra expense. If all the oil tanker washings of 1970 had gone into the sea, this would have amounted to 3 million tons of pollution, reduced by load-on-top to some 600 000 tons.

olfactory Related to the sense of smell.

oligochaete, Oligochaeta Segmented worms having a few bristles (chaetae), often found in polluted rivers, *trickling filters* and occasionally in *activated sludge*—for example, *Lumbriculidae, Nais* and *Tubificidae.*

oligosaprobic river A river with little pollution, in the *saprobic classification*. It has high *dissolved oxygen*, little or no *organic* pollution and a wide variety of plants and animals. There is an appreciable proportion of *freshwater shrimps, caddis flies, mayflies* and *stone flies.*

oligotrophic Description of a water with little life because of the low content of *nutrients*. The water is clear and contains *game fish*. It is aesthetically pleasing and there are no *algal blooms. Compare eutrophic.*

once-through cooling, single-pass c., direct c. A cooling system for industrial plant or for domestic air conditioners in the USA, wasteful because it uses the water only once before discharge, thus spending about 100 times as much water as an *open recirculation system*. For seaside power stations it has not been shown to be harmful.

Onchocerca volvulus A *nematode* worm with a 15-year life, which is a *parasite* on humans, breeding under the skin and eventually causing *onchocerciasis*. It is transmitted to people by another *host, Simulium damnosum*.

onchocerciasis, river blindness A type of *filariasis* that affects people in Africa, South America and Central America with a debilitating and incurable skin disease that leads eventually to blindness, caused by the bites of *Simulium*. The bites inject *Onchocerca volvulus* into the patient. The fly *larva* lives in well-oxygenated river rapids. In the 1970s there were 30 million sufferers in Africa.

one-pass clarifier *See flocculator-clarifier*.

on-site compaction The use of *static compaction* at apartment blocks, office buildings, etc. The compacted volume may be as small as 20% of the original or as large as 60% of it. If on-site compaction is used before *incinerators*, the compacted refuse must be broken up again before it enters the incinerator. This extra expense is not needed with direct *landfill*. On-site compaction of garbage strongly discourages *mosquitoes*, flies, rats, cockroaches, etc., and makes the storage area neater.

on-site incinerator An *incinerator* installed at a block of flats, hospital, etc., usually gas-fired to ensure smokeless operation. On-site incineration is difficult if not impossible to apply to small or low buildings in the UK, partly because the *Clean Air Acts* require chimneys to discharge above possible future nearby buildings. The law also strictly limits smoke emissions, thereby implicitly requiring competent skilled supervision.

opacity Opaqueness. The amount by which vision is obscured, often based on *Ringelmann chart* numbers.

opacity monitor A *smoke-density meter*.

open burning Burning of refuse on a tip, a forbidden practice that pollutes the air.

open-channel flow Flow in a pipe that is not running full is regarded as flow in an open channel because any free surface of water is normally at atmospheric pressure (*see Figure O.1*). *Compare pipe flow*.

open recirculating system An industrial cooling system, mainly for power generation, in which the water is cooled in towers and recirculated with only a small loss of water to atmosphere by evaporation. It consumes much less water than a *once-through system* but more than a *closed recirculation system*.

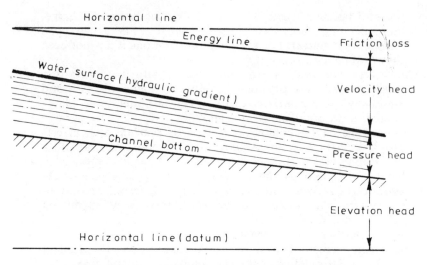

Figure O.1 Open-channel flow (*compare* pressure-pipe flow)

Opercularia Stalked *ciliate protozoa* that live in colonies and are common in *trickling filters* and healthy *activated sludges*. They may also be found in polluted streams.

optical diameter A description of particle size which implies that the particle has been seen and its diameter determined under a *microscope* by comparison with graduations on the reticule. *See Stokes diameter.*

optical microscope, light m. A viewing instrument that can magnify up to about 1000 times but no more. Much greater magnification can be had with the *electron microscope*. The light microscope has been used for the *bacteriological examination* of drinking water since about 1850. *Fungi, algae* and colonies of *bacteria* can be seen but it may be difficult to see single bacteria.

optical separation A glass industry separation method that involves a photoelectric cell detecting each piece of glass that is the wrong colour, followed by the activation of an air blast to push it out of the flow.

order In *taxonomy*, a division below class, not formally accepted.

organic Concerned with compounds of carbon, especially (so far as water is concerned) those from *sewage*. *See* below.

organic compounds, organics Compounds of carbon (excluding carbonates, CO_3^{2-}; bicarbonates, HCO_3^-; *carbon dioxide*; and *carbon monoxide*) which form the basis of living matter. In domestic *sewage* the organics are mainly metabolic wastes from the *faeces* or urine plus grease, oil, *detergents*, etc., from the kitchen. About 50% of the organics in *raw sewage* are solid. Some of these settle in *primary sedimentation*, but the finely divided or colloidal organic solids are adsorbed, absorbed and metabolised in aerobic *biological treatment*. During *anaerobic* treatment (e.g. in *anaerobic sludge digesters* or *septic tanks*) some of the solids will be hydrolysed by *bacteria* to form simpler soluble organics. *See trace organics, hydrolysis.*

organic load, o. loading The rate of application of BOD_5 to a treatment process, defined differently for different processes. Thus, for *trickling filters* or *anaerobic sludge digestion* and sometimes for *activated sludge* it is expressed in kg BOD_5 per day per m^3 of tank (*volumetric loading*). However, the *sludge loading rate* also is often used for activated sludge. For *rotating biological contactors,* organic loading is expressed in kg BOD_5 applied per day per m^2 of disc area.

organic load in a stream Dissolved and suspended *organic compounds* (e.g. from *sewage*) can be biodegraded by *bacteria* which consume thereby the *oxygen* dissolved in the stream. The organics thus create a demand for oxygen from the stream's oxygen supply. If the load is high and the stream is sluggish, its water is likely to be excessively de-oxygenated and may become *anaerobic.* A stream with *turbulent flow* will be able to take up oxygen from the air many times more rapidly than a sluggish stream and will suffer less severely from organic pollution. *See oxygen balance, oxygen demand.*

organic nitrogen *Nitrogen* in *sewage* exists in combined form, the main *organic compounds* being *amino acids, proteins* or *urea*, which are determined by the *Kjeldahl technique. Bacteria* can hydrolyse most organic nitrogen to *ammonia.* Ammonia and its compounds, as well as *nitrates*, are not *organic.*

organics *Organic compounds.*

organics in refuse The organics are paper (including cardboard), plastics, vegetables and other foods, totalling about 60% by weight of UK municipal refuse.

organo-chlorine compounds, organo-chlorides A general name for *organic compounds* containing *chlorine*, and including *chlorinated hydrocarbons.* They are often *pesticides* or antiseptics. Some are only slightly soluble in water but very soluble in the fat of humans, animals or fishes. Many are not easily *biodegradable* and therefore may concentrate in aquatic life (e.g. mussels) at a level more than 10 000 times their concentration in water. They include several groups: *insecticides* such as aldrin, BHC, chlordane, DDE, DDD, *DDT, Dieldrin,* endosulfan, endrin, heptachlor, heptachlor epoxide, Kelthane, methoxychlor, Perthane, Sulphenone; *fungicides* such as

211

hexachlorobenzene and pentachlorophenol; and chlorinated aromatics, including chlorinated naphthalenes and *polychlorinated biphenyls*. Human body fat in Israel was found to contain 30 p.p.m. of *DDT* and in India 19 p.p.m., because in those countries much DDT was used. In West Germany in 1958–59, however, only 2 p.p.m. were found because little DDT was used there. In hot countries organo-chlorines are used for *malaria* control but in Europe their use is restricted. Their major problem is that they persist in the environment, and so accumulate in *food chains* (*biomagnification*). Strains of *insects* that resist them are formed and become more difficult to control. They may also affect harmless *fauna*. Many bird species, especially birds of prey, have been decimated by *pesticides* that have entered their food and hindered their ability to reproduce, often by weakening the eggshell so that the eggs crack during incubation.

ORP   Oxidation–reduction potential, *redox potential.*

ORSANCO, Ohio River Valley Water Sanitation Commission   An inter-state agency that has published much information on methods of analysis of water and waste water, water quality and water quality criteria for various purposes as well as on the treatment of industrial liquid wastes. *See river basin commission.*

Orsat apparatus   A gas-analysis apparatus that measures CO_2, CO and O_2 by passing the gas mixture through various liquids that absorb them. Addition of an oxidative assembly enables it to measure *methane* or to analyse *sludge gas.*

orthophosphate   A salt whose formula ends in PO_4—i.e. simple salts of the acid H_3PO_4, not a *polyphosphate*, or a metaphosphate (e.g. $NaPO_3$) or a pyrophosphate (e.g. $Na_4P_2O_7$).

Oscillatoria   A filamentous *alga* of the phylum *blue-green algae* common in muddy pools or waters with high *organic* pollution. Water containing it tastes muddy.

Oslo convention   *See ocean dumping.*

osmosis   The tendency of water from a dilute solution to diffuse through a *membrane* into a stronger *solution*, with the effect that the concentrations tend to equalise. A *semi-permeable membrane* is used for this purpose. To stop the diffusion of water into the stronger solution, a pressure must be applied to it, known as the osmotic pressure. The force of osmosis is considerable and with capillarity it enables the sap in trees to mount as high as 50 m or more in tropical rain forests. *See reverse osmosis* (and illustration).

osteomalacia   A bone disease, in one type of which the *calcium* of strong bone is replaced by *cadmium* that weakens the bone. The

The cadmium can enter the body through, for example, rice that has been irrigated with water containing less than 1 mg/litre of cadmium, which concentrates in the rice plant to a dangerous level. This disease, often fatal, occurred in the Zintsu river valley, Japan, in the late 1950s, downstream of the beneficiation plant of a lead–zinc–cadmium mine, and slightly resembles the *mercury* poisoning at *Minamata Bay, Japan.*

OSWMP   The Office of Solid Waste Management Programs. One of its first aims was to replace the 5000 dumps in the USA by *controlled tipping* in its 'Mission 5000'. OSWMP originated in 1965 with the passing of the Solid Waste Disposal Act, as part of the Department of Health, Education and Welfare, but passed to the *EPA* after its formation in 1970.

outage   *Down time.*

outfall   The opening at the end of a *sewer* from which *effluent* or *sewage* is discharged into a river, lake, sea, etc. One or more *diffusers* near the outfall may help the sewage to disperse. At sea, sewage is liable to rise to the surface because sea-water is denser and often also cooler. *See sea outfall.*

overband separator, suspended magnet s.   A *magnetic separator* in the form of a short cross conveyor belt with an electromagnet inside it just above the bottom strand of the cross belt (*Figure O.2*). The magnetic field of the electromagnet lifts iron and steel

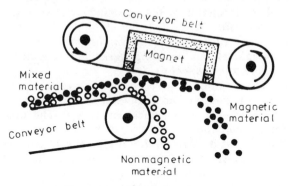

Figure O.2   Overband separator

pieces off the refuse belt below it and the cross belt pulls them off it sideways, dropping them into a chute beside the refuse belt. They automatically disconnect from the short belt of the separator as they approach its pulley.

overburden pressure, lithostatic p., cover load p.   The rock pressure in a mine caused by the weight of the overlying rocks.

overfire air, overgrate a., secondary a.   Air supplied above the

213

grate, either as high-pressure secondary air to improve turbulence and help complete the burning or in an *incinerator* as dilution air to cool the *flue gas*. Overfire air in completing combustion helps to prevent *dark smoke*, and in reducing the flow of *underfire air* helps to reduce *carry-over* of solids in the flue gases. Dilution air may be added periodically when a thermostat switches it on.

overflow rate, upward-flow velocity   The *surface loading rate* of a *sewage* or *water treatment* process having some upward flow, such as a *clarifier* or *sedimentation tank* but not a downflow process such as a *sand filter*. It is the mass flow divided by the water surface area of the tank and has the dimension of speed, m^3 per unit of time per m^2 = metres per unit of time.

overhung latrine   A latrine from which the excreta fall directly on to a tidal flat, river or canal, often causing a *public health* problem.

overland flow, surface runoff   Water from rain, etc., which travels over the ground surface to a stream, channel, conduit, etc.

overland flow land treatment   The use of *grass plots*.

overpumping of wells   Removal of too much water from a well may excessively lower the *water table*. This can alter the structure of the *aquifer* or cause *saline intrusion* if it is near the coast.

oversize   The material that cannot pass through a *screen*. Nevertheless, depending on the screening effectiveness, it always contains some *undersize*.

overturn of lakes, turnover   *See thermal stratification.*

oxidant   In general, an *oxidising agent* but specifically in air pollution a *photochemical oxidant*.

oxidation   The addition of *oxygen* or the loss of electrons. The purification of *sewage* often involves oxidation using *bacteria* or other microbes—i.e. biochemical oxidation. The opposite of oxidation is *reduction*.

oxidation ditch, ring d., Pasveer d.   A type of *extended aeration*, a Dutch development of the Haworth system, due to A. Pasveer, in which *brush aerators* such as the *Kessener brush* aerate and circulate the *mixed liquor* through a channel dug to the shape of an oval in plan (*see Figure O.3*). The flow rate is about 0.3 m/s. Operation may be continuous or intermittent. For continuous operation an *activated sludge* settling tank is provided, with a pumped re-cycle of sludge to the ditch. For intermittent operation the rotors are periodically stopped to allow the sludge to settle in the ditch. A batch of *effluent* is then discharged, and a corresponding volume of *raw sewage* is brought in. Continuous operation is preferable but more costly. Ditches for small

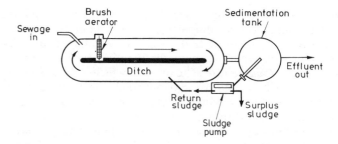

Figure O.3 Oxidation ditch, continuous system. In the intermittent system the circulation is periodically stopped so that the ditch can be emptied, there being no sedimentation tank (Whitehead and Poole, Manchester)

communities are 1.5 to 2 m deep but this may be increased to 4 m for large plants. *MLSS* is maintained at about 3000 to 5000 mg/litre. *Organic loadings, sludge production* and effluent quality have the values stated for *extended aeration* plants. Oxidation ditches have been used to treat the sewage from a *population equivalent* above 100 000. *See also Carrousel system.*

oxidation pond   A *waste stabilisation pond.*

oxidation prevention   An *acid mine drainage* pollution prevention measure, involving the injection of cement mortar into fractures in sulphide-rich zones to prevent them oxidising further. Drainage is allowed around the grouted zones but the grout prevents water and air reaching the richest ore. Mining is not precluded as it may be when underground dams are built in *discharge prevention* methods. Oxidation prevention also helps the mining engineer by reducing the corrosion of his equipment.

oxidation–reduction potential, ORP   The *redox potential.*

oxidising agent   A substance that loses *oxygen* or gains electrons in an *oxidation* reaction, thereby itself suffering *reduction.*

oxygen, O_2 The colourless, odourless gas that forms about one-fifth of normal air and is essential to the life of vertebrates, including humans, and much other life. It is provided by *photosynthesis*, but is also the most abundant element in the earth's crust, forming 45% of it, in combination with other elements. *See* above and below; *see also ozone.*

oxygen-activated sludge   Injection of *oxygen* instead of air into the *aeration tanks* of an *activated sludge* process. Oxygen is only 21% of air; consequently, pure oxygen can maintain higher *dissolved oxygen* levels and cope with higher *oxygen demands.* In other words, with the same *aerators* and other conditions, theoretically four times as much oxygen can be dissolved as with air. The aeration tanks are covered to enable the oxygen

215

atmosphere to be re-cycled into the *mixed liquor* and finally vented from the last compartment with about 50% CO_2. The *MLSS* in the aeration tanks can be maintained at 6000 to 8000 mg/litre, which reduces the aeration tank volume for a given *sludge loading rate*. The oxygen, which is usually produced on site, can be expensive and outweigh the lower cost of the aeration tanks and savings in blower energy. Many claims are made for oxygen aeration in comparison with air: the activated sludge settles better in the *secondary sedimentation tanks*; high sludge loading rates are possible; *bulking* is prevented. Oxygen-activated sludge plants (such as that illustrated in *Figure O.4*) are becoming popular in the USA, where cheaper energy favours oxygen production.

oxygenation In *water treatment*, but particularly *sewage treatment*, this means getting oxygen to dissolve in water, often so as to prevent the smells of *stale sewage*. *See also dissolved oxygen, game fish.*

oxygenation capacity The rate of absorption of *oxygen* per unit volume of de-oxygenated liquid expressed in kg per m³ per h at 10°C and 760 mmHg, resulting from the use of an *aerator*.

oxygenation efficiency (1) The *oxygenation capacity* for a liquid, per kWh. *Surface aerators* (brush or cone aerators) have an efficiency of about 2.5 kg/kWh in clear, de-oxygenated water, but this drops to 1.8 kg/kWh in the practical situation of aerating a *mixed liquor*. For *fine-bubble aeration* in clear, de-oxygenated water, the efficiency is about 4 kg/kWh but for mixed liquor this falls to about the same level. The amount of *oxygen* required in an activated sludge aeration tank varies from 1 kg O_2 per kg BOD_5 removed at a *sludge loading rate* of 0.3 per day to 2 kg O_2 per kg BOD_5 removed at a sludge loading rate of 0.1 per day. For *nitrification*, 4 to 5 kg of O_2 are required per kg of *ammonia* oxidised to *nitrate*. (2) The *oxygen* dissolved compared with that supplied, expressed as a proportion. With clear, de-oxygenated water, about 15 to 20% is dissolved by fine-bubble aeration and 5% or less by *coarse-bubble aeration*.

oxygen balance of natural waters *Oxygen* is available in *solution* in a river, lake or sea from three main sources: *dissolved oxygen*; *re-aeration* when turbulence at the water surface dissolves more oxygen; and oxygen released into solution from *algae* or other water plants by their *photosynthesis*. Oxygen is removed from solution by *bacteria*, fish and many other forms of water life and the overall result of the uptakes and consumptions is called the oxygen balance. It can be affected by many factors. The speed of the chemical and *biochemical* reactions increases with temperature but the *air saturation value* decreases as temperature

216

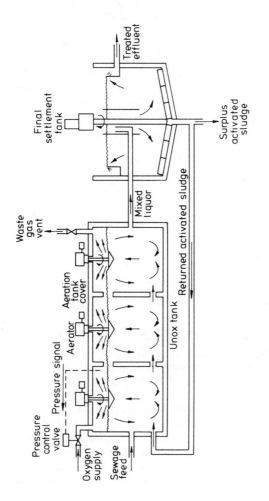

Figure O.4 Oxygen-activated sludge, Unox system (Wimpey-Unox Ltd.)

217

rises. At high river flows, bottom mud may be re-suspended and exert an extra *oxygen demand* but the extra turbulence increases the *aeration*. At low flows, *organic* solids may settle out and reduce the oxygen demand of the river. Oil films may hinder re-aeration at the water surface. Sunlight encourages photosynthesis. Toxic substances may reduce the biochemical activity of micro-organisms, etc.

oxygen deficit The amount by which the *dissolved oxygen* is less than the *air saturation value. See oxygen sag curve.*

oxygen demand *See biochemical oxygen demand, chemical oxygen demand, respirometer, total oxygen demand, ultimate oxygen demand*; *see also* above and below.

oxygen sag curve A graph of *dissolved oxygen* (DO) against time, plotted for a watercourse from the point of addition of pollution

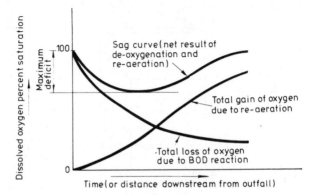

Figure O.5 Oxygen sag curve (the variation of dissolved oxygen in a river saturated with oxygen, that then receives domestic sewage)

(*see Figure O.5*). It shows the drop in DO, the sag in the curve picking up again after an interval, which indicates *self-purification* in the river and the change in the *oxygen balance* in this distance downstream. When excessive pollution is added, the DO may drop to zero for a period. Many mathematical models have been made to simulate the effect of pollution on the *oxygen deficit* of a river.

oxygen saturation value (1) The concentration of *oxygen* in water in equilibrium with an oxygen atmosphere. (2) *Air saturation value.*

ozonation The injection of an *ozone*–air mixture into clean, filtered water as a final *disinfection. Since ozone* (O_3) is unstable and completely decomposes to *oxygen* (O_2) in an hour or less, it is not possible to have excess ozone for long. Ozone is the fastest and most powerful disinfectant known and kills *viruses* with certainty if the water is clean. A 5 to 10 min contact time is needed if there is 1% ozone in the ozone–air mixture injected. The main disadvantage is the cost of making the ozone. For sewage *effluents*, ozonation is preferable to *chlorination* because there is no residual disinfectant. At an *activated sludge* site, ozonation becomes less expensive if pure oxygen is used in *aeration*. Some of the oxygen can be fed to the *ozoniser*, which then works at higher efficiency. No plants are working in the UK but ozonation is popular in France.

ozone, O₃ (1) A poisonous blue gas used for disinfecting water, which decomposes so quickly to *oxygen* (O_2) that it has to be made at the waterworks in an *ozoniser* immediately before use. Its *half-life* is only 20 min. It leaves no residual but is a very strong oxidising agent and can therefore remove colour, smell or taste where other methods fail. The volume of ozonised air to be injected into the water can reach one-third of the volume of the water. A real advantage is that colour control can be achieved without making sludge. (2) From the viewpoint of *air pollutants*, ozone is a poison formed in *photochemical smog* when sunlight causes *nitrogen dioxide* (NO_2) to decompose:

$$NO_2 + O_2 = O_3 + NO$$

The ozone would quickly react with nitric oxide (NO) to re-form NO_2 but hydrocarbons in the industrial atmosphere, present in unburnt oil or petrol, react with the NO and therefore the ozone accumulates during the daytime. Some of the unburnt oil, NO and ozone combine to form *PAN* or other organic *oxidants*. Polluting concentrations of ozone can be formed by the ultra-violet radiation from electric arc welding. *See smog alert*.

ozoniser An electrical discharge unit operating at 5000 to 20 000 V and 50 to 500 Hz, which converts some of the *oxygen* in clean, dried air temporarily to *ozone* (O_3). It produces ionised air with 1% O_3. Since ozone is so unstable, a very small *chlorine residual* is often also injected into the supply. *See ozonation*.

ozonising *Ozonation*.

P

p symbol for pico-, or 10^{-12}.

packaged sewage treatment plant A complete, small treatment plant made at a factory, able to function as soon as it is placed on the site and connected to its plumbing inlet and outlet. Packaged plants are often *activated sludge* plants, with *contact stabilisation*, *extended aeration* or *rotating biological contactors*.

packed tower, packed-bed scrubber An early type of *wet scrubber*, consisting of a hollow tower packed with *media* down which the reagent (scrubbing or wash water) trickles. Packed towers can have upward (*countercurrent*) gas flow, downward (concurrent) flow or cross flow. In countercurrent flow the structure slightly resembles a *trickling filter*. Gas pressure drops are low, only about 1 cm water gauge per 24 cm height of tower, but packed beds cannot clean very dusty gases or the bed blocks up. They can, however, be used as *entrainment separators*. *See moving-bed scrubber*.

packing *See medium* (2).

PAH *Polycyclic aromatic hydrocarbons*.

Paladin bin A refuse bin of 900 litres capacity, with small wheels that enable it to be pushed to the collecting lorry and there emptied by a powered lifting device on the lorry. It is used at apartment blocks and other buildings in the UK where the volume of refuse is large enough.

palaeolimnology The study of past *limnology* from the records provided by sediment cores extracted from a lake bed.

palatability of water *See taste and palatability of water*.

PAN, peroxyacetyl nitrate An *oxidant* that irritates the eyes, and results from reactions between unburnt oil or petrol, *ozone* and nitric oxide in *photochemical smog*. It can reach 0.3 p.p.m. but is typically 1 to 2% of the ozone concentration.

paper and pulp manufacturing wastes Wood is prepared for papermaking either mechanically by grinding with water into a pulp or chemically by heating wood chips with calcium bisulphite (sulphite process), or with a mixture of sodium sulphide, sulphate, hydroxide and carbonate (*kraft* process). Chemical pulping produces a highly polluted waste water with a BOD_5 that may exceed 500 000 mg/litre. However, the pulping wastes are often diluted by the paper finishing processes to give a mixed *effluent* of about 200 to 2000 mg/litre BOD_5.

paper salvage The saving of paper and its re-sale to paper merchants. Some 2 million t of paper are salvaged and re-used in the UK every year. Printer's ink can be poisonous and is therefore a hazard in re-cycled paper, even in paper towels. Most old papers are, however, entirely suitable for making building paper or light wallboard for partitions, or cupboards. The *protein* in paper enables it to be fed to cattle, horses and pigs if it is shredded and not poisoned by printer's ink. Mixed with *compost*, paper has much the same value for conditioning soil as straw. If heat or any other energy is to be recovered from the refuse, the extraction of the paper will reduce its value for this purpose. *See shredding.*

paper stock   Waste paper.

Paramecium   'A free-swimming *ciliate protozoon* which may occur in *activated sludge* or organically polluted waters where it has many *bacteria* to eat.

parasite (adjective **parasitic**)   An organism that obtains food from another, normally larger, one called the *host*, in or on which it lives, for the whole or part of its life. *See obligate parasite.*

parasitic worms   *See helminthic diseases.*

paratyphoid   A disease slightly less serious than *typhoid*, due to a bacterium, *Salmonella paratyphi*. It may be waterborne or come from infected food, but is passed in the *faeces* of the infected person. *See waterborne diseases.*

partially separate system   A *sewerage* system in which each street has two sets of *sewers*, one *foul sewer* and one for rainwater. Unlike the rigidly *separate system*, the rainwater from the back yard and house back roof drains into the foul sewer.

particle size distribution, grain size d.   The proportions of grains of different sizes that make up a particular sand, gravel, refuse, etc., usually expressed in percentage by weight. *See Coulter counter, effective size, uniformity coefficient.*

particulate   A particle, whether liquid or solid, especially in *smoke*, although many US legal definitions by *EPA* and states exclude uncombined water from particulates. It includes *aerosols, dust, fume, grit,* etc., but not gases. Some experts, however, think that *sulphur dioxide* may be adsorbed on to smoke particles. It is certain that, combined with smoke, sulphur dioxide is especially harmful to health. Typical annual urban average levels of particulates are around 30 μg/m^3.

particulate emission   Under the US Air Quality Act 1970, the allowable emission of *particulates* from a new boiler or furnace of 73 MW heat input (about 10 t/h of coal) is 45.4 g per million B.t.u. (10^6 B.t.u. = 293 kWh). Old plants may emit double this amount.

parts per billion, p.p.b.   In Western Europe this meant parts per million million but in the USA it means parts per thousand million, μg/litre, and this is coming to be accepted in Europe.

parts per million, p.p.m.   One part per million is 1 g/t or 1 mg/kg. 1 p.p.m. is often taken to be 1 mg/litre but strictly this is true only when 1 litre weighs 1 kg. For pure water this is true only at 4°C. Concentrations in liquids, therefore, should be expressed in mg/litre in all instances where high accuracy is needed.

221

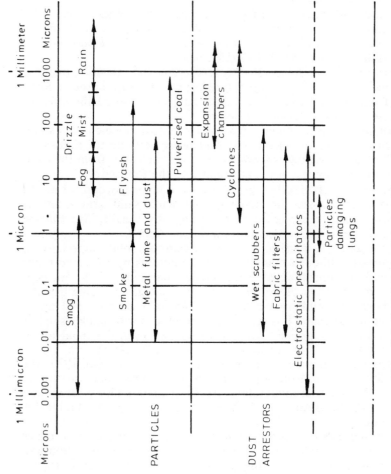

Figure P.1 Particle sizes and dust-catching equipment

Pasteurella tularensis One type of the *bacteria* that cause *tularaemia*.

Pasveer ditch An *oxidation ditch*.

pathogen (adjective **pathogenic**) A microbe that causes disease. It is impracticable to test water for all possible pathogens and most of them are less likely to survive· outside the body than *Escherichia coli*. Consequently its absence is taken to show that pathogens are also absent, but this does not apply to *cysts, endospores* or *viruses*. *See disinfection*.

PCB *Polychlorinated biphenyls*.

PCT *Physico-chemical treatment*.

pebble-bed clarifier, Banks filter A rectangular, upward-flow, *secondary sedimentation tank* with a layer of pea gravel hung near the outlet (*see Figure P.2*). The *effluent* is forced to flow up

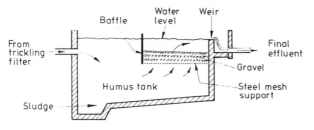

Figure P.2 Pebble-bed clarifier

through the gravel before it passes over the outlet *weir*. The 150 mm deep gravel bed acts as a coarse filter after the *humus sludge* has settled. The top of the gravel is 150 mm below the top water level. Periodically the water level in the tank is lowered so that the gravel layer can be hosed down and cleaned. It is most suitable for *polishing* the effluent from a small *sewage* works. *See sand filter*.

pegging The jamming of a particle in a *screen* opening (like a peg in a hole). It can be reduced by using *wedge-wire screens*.

pelagic Concerned with the open sea, not shallow water.

pelletiser, densifier A machine that is fed with pulverised *waste-derived fuel*, which it extrudes as a solid cylinder of 10 mm or so diameter, broken off in convenient lengths or pellets. *See also* below.

pellet reactor, pelletiser for sludge, Gyractor, Spiractor (*Figure P.3*) A cone with its apex down, resembling a *wet cyclone,* which precipitates *calcium carbonate* during the *lime softening* of water. The water with added lime enters the cone tangentially to give it a swirl, and grains of sand or marble (0.2 to 1 mm diameter) are dropped in at the top. The grains form nuclei

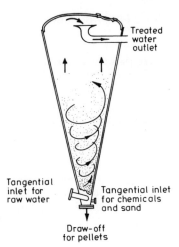

Figure P.3 Pellet reactor (Spiractor) (Permutit-Boby Ltd.)

which attract calcium carbonate precipitate, grow and settle out when they are large enough. Design loadings may be 20 to 50 m³/h per m² surface area of tank. The process is restricted to waters of low *turbidity* and low *magnesium* content.

pellicular water Water held against the force of gravity in the *zone of aeration* between the *soil water zone* and the *capillary fringe*.

penstock (1) A device to control the flow of fluid, containing a vertical gate perpendicular to the flow in an open channel. (2) In a hydro-electric scheme, a pipe, tunnel or channel leading water from a dam to turbines.

peptide link *See proteins*.

perched water table *Groundwater* held temporarily or permanently above the main groundwater, by a relatively small layer of impervious strata in the *zone of aeration*.

percolate *Leachate*.

percolating filter A *trickling filter*.

percolation Movement of water through the soil, usually downwards.

peripheral weir, launder A *weir* around the outside edge of a circular or square *sedimentation tank*, over which the *effluent* flows.

periphyton Plants or animals that cling to rooted water plants, above the bottom mud.

peristaltic pump A pump with three rotating arms that alternately compress and release a flexible tube that is bent into a U-shape, forcing fluid along the tube. It is used to pump small, accurate volumes of water in some *samplers*. Larger peristaltics move highly corrosive fluids because the fluid does not touch metal.

permanent hardness *Non-carbonate hardness.*

permanganate value, PV The oxygen absorbed from one-eightieth normal (N/80), occasionally N/8, potassium permanganate in dilute sulphuric acid at 27°C. Two test periods exist. The 3 min test measures readily oxidisable inorganic matter (*nitrites*, sulphides, sulphites, *ferrous salts*) and *organic* salts. The 4 h test measures more of the organic impurities but oxidation is less complete than in the *BOD* test. The test is quick and easy but because it oxidises only certain organics its use is limited. *See Kubel test, McGowan strength.*

permeability The ability of a substance to allow fluids to pass through it: in particular, for a soil, its ability to let water through; and for an air *diffuser*, its ability to let air through. *See Darcy's law.*

peroxyacetyl nitrate *PAN.*

persistent pesticides Pollutants that biodegrade slowly or not at all. The best-known are the *organo-chlorine compounds.* They may undergo *biomagnification* in a *food chain.*

pesticide A chemical that kills pests. It includes *algicides, fungicides, herbicides, insecticides, larvicides, molluscicides, nematicides, piscicides* and *rodenticides.* There should be at least one centre in each country or region, capable of investigating pesticide residues in drinking water (WHO). *See organo-chlorine compounds.*

petro-chemical industry waste *See oil refinery effluent, plastics industry waste.*

PFA, pulverised fuel ash *See flyash.*

pH, pH value A measure of the *acidity* or *alkalinity* of water, based on its concentration of hydrogen *ions*, H^+. The pH value is the logarithm of the reciprocal of the concentration of hydrogen ions in the water. A pH of 7 is neutral. A very strong acid has a negative pH—e.g. a pH of −1 means a H^+ concentration of 10 mole/litre. A very strong alkali produces a high OH^- (hydroxyl ion) concentration with a corresponding low H^+ concentration and a pH of 14 or more. The range of pH values that allow life to continue is narrow and the *sulphur-oxidising bacteria* that may thrive in acid water at pH below 3 are quite exceptional. Very few if any *bacteria* survive above pH 11. Meters exist that measure pH continuously and some of them record it.

phaeophytes, Phaeophyta The *brown algae.*

phage A *bacteriophage.*

pharmaceutical industry wastes Wastes from pharmaceutical production vary with the product. Many have a high *organic* content and may contain extremely toxic chemicals.

phase separation, PS *Concentrating viruses* by putting them in aqueous solution into contact with two *polymers* such as dextran sulphate and polyethylene glycol. These form two phases. The viruses separate into the phase of the smallest particles and concentrate to about one-hundredth of their former volume. Using two steps of PS, even higher concentrations can be achieved. The method is useful for detecting viruses in *sewage*, drinking water or sea-water.

phenol, carbolic acid, C_6H_5OH The word 'phenol' applies also to a group of aromatic compounds with one or more hydroxyl (OH) groups attached to the benzene ring (cresols, naphthols). Ten micrograms of phenol per litre in drinking water can make it unpalatable, especially after *chlorination*, which forms *chlorophenols*. Phenols may be found in water that has run off tar-sprayed roads or from a tar distillery or from *coke plant wastes*. They may be biologically oxidised in *sewage treatment* at concentrations up to 500 mg/litre if the *biological treatment* has been acclimatised to them. Taste and smell are the real criteria for this large class of compounds, which are some of the many *micro-pollutants*. *See* Table of Allowable Contaminants in Drinking Water, page 359.

phenol adsorption test A test for *activated carbon*, in which a *phenol* solution at about 5000 p.p.m. is circulated through an *adsorption* bed. The percentage phenol adsorbed is measured. If the drop in adsorption from the value with fresh carbon exceeds about 40%, the carbon should be changed. This test has the advantage of using the carbon in its working state. *See also iodine number*.

phenolphthalein alkalinity *Caustic alkalinity*.

pheromone A chemical secreted by an animal, which transmits a communication to another one—e.g. a chemical attractant.

Phormidium A *blue-green alga* which may form olive-green sheets growing on the surface of *trickling filters*.

phosphate *See orthophosphate, polyphosphates*.

phosphorus, P Phosphorus is a major *nutrient* for *bacteria* and animals, and occurs naturally in water, usually as *orthophosphates*. Human activity adds these and other forms of phosphorus to waste waters. Fertilisers containing phosphates are washed off farmland into rivers and lakes. Domestic *sewage* usually contains 5 to 15 mg/litre of phosphorus, of which up to 70% comes from the *polyphosphates* that are *builders* in *synthetic detergents*. The major effect of phosphorus in natural waters is *eutrophication*. Consequently, there is an increasing emphasis on the limitation of phosphorus concentrations in sewage *effluents* or industrial discharges. Phosphates (sodium hexametaphosphate) are occasionally added to water in doses of 1 to 2 mg/litre to inhibit *corrosion. See* below.

phosphorus removal *Phosphorus* normally occurs in *sewage* as *phosphates*. In domestic sewage about 20% is removed by conventional *sedimentation* and a further 20% assimilated by *bacteria* in *secondary treatment*. More complete removal can be obtained by forming relatively insoluble precipitates through the addition of *calcium,* aluminium or ferric compounds such as *lime, alum* or *ferric chloride*. If they are added before *primary sedimentation,* the lack of phosphorus may limit the efficiency of *biological treatments*. If the chemicals are added during biological treatment, the precipitate settles out in the *secondary sedimentation tank*. In *activated sludge* plants this may concentrate insoluble phosphate in the system because of the continuous return of sludge but the effects of such a concentration are not known. Adding the chemicals to the effluent from the secondary tanks will not interfere with the processes but it will involve a further stage of sedimentation or *sand filters*. More than 95% of the phosphorus can be removed by these methods. It can also be removed by *activated alumina, algal harvesting, electrolytic treatment of sewage* or *luxury uptake of phosphorus*.

photoautotroph An organism that uses light for an energy source and *carbon dioxide* as a carbon source.

photochemical oxidant Any strong oxidising agent in *photochemical smog*, mainly *ozone, nitrogen dioxide* or *PAN*. Oxidant levels measured, as usually they must be, in the presence of *sulphur dioxide* (SO_2), may seem smaller than in reality they are, because SO_2 is a reducing agent. The 1971 *EPA* ambient air quality standards require oxidants, corrected for SO_2 and NO_2, not to exceed 0.16 mg/m^3 (0.08 p.p.m.) more than once yearly for a 1 h average. However, all US cities where measurements were made had regular maxima of double this value or even more.

photochemical smog, Los Angeles smog A *smog* caused by the effect of sunlight on motor vehicle exhaust gases. The resulting light haze is unpleasant to breathe and causes smarting of the eyes, coughing and chest soreness. It is believed to originate in bright sunlight with *nitrogen dioxide* (NO_2) dissociating to nitric oxide (NO) and an atom of oxygen (O) which reacts with the oxygen in the air to form *ozone*. Some 80 reactions are believed to occur, forming *oxidants* and irritants such as *PAN* or formaldehyde with hydrocarbons in the air, quite apart from common pollutants such as *sulphur dioxide*. Severe photochemical smogs in Los Angeles are caused by the stable *inversions* protected by nearby hills. Photochemical smog occurs in summer in most large US cities, also in Melbourne, Australia, and London, England, Continental European cities and Tokyo, Japan. *See smog alert*, and above.

photodegradable A description of something that decomposes under the action of light. *Compare biodegradable.*

photometer A light meter. *See smoke density meter.*

photosynthesis An activity of plants and *algae*, one of their great contributions to all animal life. It is the creation of *organic* matter from water and *carbon dioxide* in sunlight with the liberation of *oxygen*. *Compare chemosynthesis.*

photosynthetic Description of an organism or process that uses *photosynthesis*.

phototaxis *See taxis.*

phototroph An organism that uses *photosynthesis* to produce *organic* matter and *oxygen*. *Compare chemotroph.*

phreatic surface US (and French) term for water table.

phreatic zone The *zone of saturation*.

phreatophyte A plant with deep roots that reach to the *water table*, as long as 10 m.

phylum, sub-kingdom The division in *taxonomy* that comes below *kingdom*.

physical pollution Pollution of water because of colour, *suspended solids*, temperature (*thermal pollution*), *radioactivity* or *foaming*.

physico-chemical treatment, PCT (*Figure P.4*) *Water treatment* for homes or industry has always involved physics and chemistry. Present-day PCT is, however, a considerable

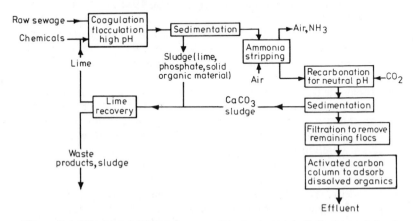

Figure P.4 Physio-chemical treatment of raw sewage including removal of phosphorus and nitrogen, flow diagram

improvement on earlier efforts and is becoming a realistic alternative to conventional *sewage treatment* with its physical and *biochemical* processes. In PCT, chemicals are used to coagulate and flocculate the *raw sewage* just before *primary sedimentation.* The remaining dissolved *organic* matter can be removed by oxidation with *chlorine* or *ozone* or by *adsorption* on *activated carbon.* But it is still thought cheaper and more practical to remove it by *biological treatment.* Many *advanced wastewater treatments,* including the removal of *ammonia, phosphates* or *trace organics,* are based on PCT. Examples of PCT are at *Coleshill AWT plant,* England, and *Lake Tahoe,* USA. *See coagulation.*

phytoflagellates, phytoflagellate protozoa Photosynthetic *flagellate protozoa* (Mastigophora) which may be considered either as *algae* or as *protozoa*—for example, *Euglenophyta.*

phytoplankton *Plankton* which are *photosynthetic,* mainly *algae.* *Compare zooplankton.*

phytotoxic Description of a substance that poisons plants or algae.

picket-fence stirrer Vertical bars hanging from a slowly rotating central vertical shaft, used for *sludge thickening.* The rim speed is 150 mm/s ± 50%.

piezometer tube An open-topped tube in which water rises because of the pressure in the *confined groundwater* or soil where its lower end is submerged. The water level in it indicates the pressure. The tube can be inserted into the earth of an earth-fill dam to determine whether it is safe to fill the *reservoir.*

piezometric surface The surface of the *standing-water level,* shown by the level in *piezometer tubes* whose lower ends pass below the surface of *confined groundwater* and indicate the level to which the water from an *artesian well* would rise. (The *hydraulic gradient* refers to moving water.) *Compare water table.*

pig, go-devil A scraper or brush or foam-plastics swab that is pushed along a pipe by the fluid pressure behind it, used for cleaning a water *main, sewer,* or pipeline. A modern pig, for detecting leaks in oil pipelines, can detect the small noise of escaping oil. It is pumped through during normal working.

pig feed The feeding of garbage to pigs can cause disease which may spread to humans. The garbage should therefore first be boiled before it is given to the animals.

piggery wastes The weight of *BOD* per day from one pig is about twice as much as from one human. Copper salts, sometimes present in significant quantities because of their use as feed additives, may restrict the disposal of the manure on land.

pilot plant   A small-scale treatment plant which simulates the full-size plant before this is built, so as to optimise its design or engineering conditions.

pinworm, *Enterobius vermicularis*   *Nematodes* that cause human illness and are passed in the *faeces* and transmitted without an intermediate *host* via soiled clothes, food, *sewage*-contaminated water, etc.

pipe flow   Flow through a pipe that runs full. Several types of pipe flow exist. In a *siphon* the *hydraulic gradient* is below the pipe.

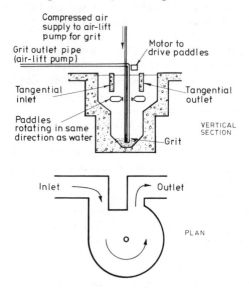

Compressed air supply to air-lift pump for grit

Grit outlet pipe (air-lift pump)

Motor to drive paddles

Tangential inlet

Tangential outlet

Paddles rotating in same direction as water

Grit

VERTICAL SECTION

Inlet

Outlet

PLAN

Figure P.5   Pista grit trap (Jones and Attwood Ltd.)

In pressure flow the hydraulic gradient is above the pipe (*see Figure P.7*). See *Hazen–Williams formula, open-channel flow.*

pipeline transport of solid wastes   *See pneumatic transport of solid waste, slurry transport of solid waste.*

piscicide   A *pesticide* that kills fish.

Pista grit trap   A circular tank that removes *grit* from *sewage*. It has a stepped conical bottom, of two different diameters, diminishing downwards (*see Figure P.5*). Power-driven rotating paddles maintain the correct horizontal velocity in the sewage for the grit to settle and the *organic* solids to be carried on. The grit sinks to the central sump and is removed through a pipe by an *air-lift pump.*

piston flow   *Plug flow.*

Pitot tube   A tube that measures the static pressure as well as the kinetic energy of a flowing fluid. Ordinarily two tubes are used, connected at their far ends to a U-tube. One opening faces upstream and measures the total head. The other faces at right angles to the flow and measures the static head only. The U-tube thus shows the difference in pressure which is due to the velocity—the velocity head, from which the velocity can be calculated. Many variations exist.

pit privy A pit several metres deep. It is a low-cost latrine but needs a lid to exclude flies.

planarians, Turbellaria Free-living flatworms. Some species may be found in water that is clean or mildly polluted with sewage—e.g. *Dendrocoelum lacteum. Flukes* and *tapeworms* are the main ones.

plankton Micro-organisms that float or swim freely in oceans or fresh water. They neither attach themselves to the bed nor live near it, unlike the *benthon*, and include algae, tiny animals, rotifers, protozoa and crustaceans, which form part of holoplankton or meroplankton. Many types of water life exist, including *nekton, neuston, phytoplankton, seston* and *zooplankton.*

plaque An area of a bacterial *culture* that has been destroyed by *virus* infection. The number of viruses present is assumed to be related to the number of plaques that have formed.

Plasmodium A *parasitic, spore-forming protozoon* that causes *malaria,* transmitted by *mosquito* bites.

plastics Man-made *polymers* most commonly used for packaging, e.g. polythene (polyethylene), vinyls (*polyvinyl chloride*) polypropylene, polyesters. The weight of plastics in UK refuse was only 2 to 3% in 1976. If we except PVC, the plastics all burn with a high heating value, more than double that of paper. The USA used twice as much plastics packaging as the UK and Germany 50% more. *See thermoplastics.*

plastics filter medium Media made of *plastics* and used to fill a *high-rate trickling filter.* They may be semi-rigid, corrugated PVC or individual plastics shapes. All such media have large voids that prevent *ponding* at high *organic loads.*

plastics industry wastes The liquid wastes are mainly from boilers or coolers that do not come into contact with the process. Process water, however, may be contaminated with toxic chemicals. Waste gases also from the plastics and rubber industries can be obnoxious unless air pollution control equipment is installed.

plate and frame press A *filter press.*

plate count The number of colonies of *bacteria* or other microbes that have developed on a plate of nutrient *agar-agar* inoculated in the laboratory with the diluted sample, and counted after a certain number of hours *incubation* (e.g. 48 to 72 h) at a certain temperature. Incubated at 20°C the plate count will tend to indicate bacteria that grow in natural waters, and at 37°C those from mammals. As *organic* pollution in a river increases, the ratio between the counts at 20° and 37° normally decreases below ten to one.

plate count technique, pour-plate method of counting bacteria After making dilutions of the water or waste water to be examined, at 0.1, 0.01, 0.001 and any others needed, samples of 1 ml of each dilution are pipetted into sterile Petri dishes about 90 mm diameter. The 1 ml samples are then covered with 15 ml of molten *agar-agar* at 45°C, well mixed with it by circular and sideways motions and allowed to set. A further thin layer of agar-agar is then poured over the top, to prevent a profusion of large colonies growing on the surface and hiding those below. The agar-agar contains various salts or other chemicals so that certain types of *bacteria* develop. Yeast extract agar-agar should be used for coliform bacteria. When the agar-agar has set, the Petri dishes are inverted and placed in the incubator at the appropriate temperature. The colonies that develop in each dish after this incubation period are counted, and those with more than 300 or fewer than 30 are ignored. More than 300 cannot easily be counted; fewer than 30 are subject to large statistical error. Assuming a dilution of one-hundredth, they would be reported, respectively, as 'over 30 000 per ml' and 'under 3000 per ml'.

plate diffuser A *diffuser* (1).

plate precipitator The usual *electrostatic precipitator* which collects solid particles, unlike the *tube precipitator*.

plate press A *filter press*.

plate scrubber An *impingement scrubber*.

plating wastes *See metal-plating wastes*.

Platyhelminthes *Flatworms*.

Plecoptera *Stone flies*.

plug flow, piston f., slug f. Flow of fluid in an unmixed mass. Thus, if a slug of *tracer* is injected at the inlet to a tank, it will issue from the tank also as a slug after the *detention period*. *Compare arbitrary flow, complete mixing system*.

plug valve A bored plug that can be rotated to stop the flow through a pipe. Although more expensive than a *sluice valve*, it is more easily turned by hand when the pipe is under high pressure.

plumbism Lead poisoning. One of the causes of the decline of the Roman Empire may have been poisoning from the lead in the water carried by the new leaden conduits or from that in the wine drunk out of leaden vessels.

plumbosolvency The ability of drinking waters from, e.g., peaty moorland to dissolve *lead* pipe and thus to cause lead poisoning. (Such waters are slightly acid and have low *carbonate hardness*.) This property of the water can be reduced or eliminated by raising the *pH* or the carbonate hardness. Lead is less soluble in slightly *alkaline* waters and the carbonate is deposited as a protective layer inside the lead pipe. Alternatively, lead pipes should be eliminated from drinking water circuits. In the west of Scotland, plumbosolvent water left overnight in lead-lined cold water tanks has resulted in a lead concentration equal to 400 times the WHO standard.

plume, vapour p. The visible, occasionally invisible, *smoke* from a chimney.

plume rise *See efflux velocity.*

pneumatic ejector A pump for raising *sewage* or *sludge*, developed in 1878 by I. Shone. It admits the sludge into an airtight chamber through a non-return valve. The chamber is then filled with compressed air, dispatching the sludge under pressure through another non-return valve. Pneumatic ejectors are economical for lifting small flows of sewage up to about 400 litre/min and are convenient because they do not clog, but they need a supply of compressed air or steam.

pneumatic transport of solid waste A pipe system through which solid waste travels, drawn along by air pressure, in capsules or bags or loose. In vacuum transport the air is pumped out at the destination. In pressurised pneumatic transport the air is blown in at the point of origin of the waste. *See Centralsug*; *see also* below.

pneumo-slurry transport of solid waste A concept of *pneumatic transport of solid waste* for a short distance from the dwelling to a *shredding* and *mechanical sorting plant* where sewage *effluent* or *sludge* or water would be mixed into the shredded refuse for long-distance *slurry transport of solid waste*.

poikilotherm Description of fishes and most water animals whose body temperature follows that of the environment, being slightly warmer than the water. *See homoiotherm.*

poisoning In connection with *catalysts, ion-exchange resins* or *membranes*, poisoning is the destruction of their properties—e.g. arsenic is a chemical poison for a platinum catalyst. *Organic* material, even at 5 mg/litre, is a poison for membranes in *electro-dialysis*.

polarography An electro-chemical technique that can be used to measure small concentrations of *lead, cadmium, chromium* and other *heavy metals*. The voltage across two *electrodes* is increased until the *solution* polarises and passes current. This minimum voltage is specific for each metal and the value of the current can be related to the concentration of the metal in the water.

poliomyelitis, infantile paralysis   A disease caused by a *virus* that attacks the nerves controlling human muscle. The virus is present in the excreta from an infected person and is transmitted through food or water.

polishing   A final purification of water, not always used. For sewage *effluent* it may be called *tertiary treatment*.

pollution control   The reduction of pollution. It includes both the law with its regulations for reducing pollution and the various equipments that make it possible. For *air pollutants* there is a division in responsibility between the central *Alkali Inspectorate*, on the one hand, and the local councils with their *environmental health officers*, on the other. This does not occur with water, *sewage* or *solid waste*. In England and Wales water authorities have sole responsibility. In Scotland, however, *river purification boards* control river pollution, regional authorities control water and *sewage treatment*, and district councils control solid waste. In the USA pollution control is the duty of many federal and state agencies, the bulk of the federal activity passing through the *EPA*. *See river basin commission.*

Pollution Inspectorate   The *Royal Commission on Environmental Pollution*, in its fifth report, 1976, entitled 'Air Pollution, an integrated approach', recommended that a new body, Her Majesty's Pollution Inspectorate, be set up, of highly qualified people, to be concerned with the environment as a whole—all water, air and land. It would supersede the *Alkali Inspectorate*.

polyacrylamide   A *coagulant aid*. If it is used in drinking water, care must be taken that toxic components such as the *monomer* are reduced to a safe level.

polychaete, Polychaeta   A class of *annelid*—small worms that abound in seas and estuaries. They can amount to 70% of the macro invertebrates in tidal mud and are often chosen for *bioassays* of coastal regions—e.g. *Capitella capitata* and *Neanthes*.

polychlorinated biphenyls, PCB   *Organo-chlorine compounds* that degrade slowly. They have been used as fire retardants, in insulating materials, paints and high-temperature lubricants. They precipitate in *sludge* and are adsorbed by clays to such an extent that they are usually barely detectable in drinking water. They are soluble in oil and *lipids* but not in water, sometimes concentrate in fish and may therefore be found in human tissue. As little as 0.5 g harms humans. Dead sea birds in the Irish Sea in 1969 and 1970 contained significant concentrations of PCB in their bodies. The PCB probably originated from *sewage* sludge contaminated with PCB that had been dumped in the sea.

polycyclic aromatic hydrocarbons, PAH Cancer-producing compounds in tar, mineral oil, plants, the soil and air polluted by exhaust fumes or other *smoke*. Their concentration in drinking water should be less than 0.2 μg/litre. One centre, at least, in each country or region should be able to test for and investigate PAH, according to *WHO 1971.*

polyelectrolyte A *polymer* that carries ionisable groups. When placed in water, they acquire many positive or negative charges (cationic or anionic polyelectrolytes). As a *coagulant aid*, a polyelectrolyte is added to water in small quantities to help in the flocculation and *coagulation* of *colloids* and other fine *suspended solids*, reducing their surface charges, so that they sink and can be separated from the water. Natural polyelectrolytes are often based on *proteins* or *polysaccharides*. Modern ones are synthetic. They are used in *sewage treatment* to help de-water *sludge*, and in *water treatment*. Since *sewage* particles are negatively charged, cationic polyelectrolytes are used as the primary coagulants. Most of them are *polyacrylamides* or compounds of polyacrylic acid. A few are polyamides or quaternary ammonium compounds. Non-ionic polyelectrolytes are a misnomer. They are either *ampholytic* or do not ionise at all, and should not be called polyelectrolytes.

polyfloc A polyelectrolyte *flocculant.*

polyglycol A substance used in industry as a lubricant for cutting metals, etc. It can cause *foaming* of a river or of *activated sludge* or of a *digester*.

polymer A substance composed of long-chain molecules, sometimes with more than 1000 atoms, and built up from simple molecules (monomers) linked together. Many are synthetic, but some, such as *cellulose*, are of natural origin. Different polymers may be formed from different unions of one monomer, so that they have the same formula but different linkages. In *water treatment, polyelectrolytes* are the commonest polymers used.

polymer treatment for concentrating viruses See *hydro-extraction, phase separation.*

polynuclear aromatic hydrocarbons See *polycyclic aromatic hydrocarbons.*

polypeptides *Polymers* of amino acids, of simpler structure than *proteins.*

polyphosphates Complex molecules that include the *orthophosphate* (PO_4) with additional atoms of *phosphorus,* hydrogen and *oxygen.* They hydrolyse slowly to the orthophosphate. See *hydrolysis, sodium hexametaphosphate,* *sodium tripolyphosphate.*

polysaccharides *Polymers* of monosaccharides or simple sugars (glucose, xylose, etc.) or sugar-based structures. Starch and *cellulose* are polymers of glucose but with different links between the glucose molecules. *Agar-agar* also is a polysaccharide, and much bacterial *slime, humus* or *activated sludge* is based on polysaccharides.

polysaprobic river In the *saprobic classification,* a river with heavy *organic* pollution. It is *anaerobic,* with only simple forms of life. Red worms, *Tubificidae* and the rat-tailed maggot *Eristalis tenax* are common in such rivers.

polyvinyl chloride, PVC, $(CH_2CHCl)_n$ A common material of which *plastics* packaging film is made. When burnt it gives off hydrochloric acid to the extent of about 40% of its weight. It could therefore damage *incinerator* linings. But even with a high PVC content reaching 1% of municipal refuse, the air pollution caused by HCl from PVC is expected to be only one two-hundredth of that from the *sulphur dioxide* from chimneys.

ponding The blocking of a *trickling filter* with the formation of pools on its surface, common in winter when it is too cold for the *grazing fauna* to feed on the surface *biological film* and thus to break it up. The blocking is caused by excessive growths of *bacteria, fungi* or *algae.* Ponding stops the efficient operation of the filter and may occur with too high an *organic loading.* It is overcome in several ways. One way is a natural one. With the return of warm weather the grazing organisms revert to the surface of the film and break it up, with the result that large quantities of *humus* are sloughed off the filter. *Alternating double filtration* helps, as does *chlorination* of the *influent,* and possibly inoculation of the filter with an *insect* such as the water *springtail* which feeds on fungi that may be clogging the filter. *Recirculation* of the *effluent* or resting the filter for several days or slowing the speed of the dosing arm may help by reducing the microbial population adhering to the *media. See sloughing off.*

pondweed *Potamogeton.*

population dynamics The study of the variations in numbers of various species in an *ecosystem.* It shows up the interrelationships between species, and their changes.

population equivalent Industrial waste waters are often expressed in terms of equivalent population, based in the UK on the assumption that the average person exerts a BOD_5 load of 60 g per day. Thus, if the BOD_5 of the waste is 600 kg/day, the waste is equivalent to a population of $600 \div 0.06 = 10\ 000$. In the USA 80 g/day may be used to calculate population equivalents.

pore space In soils, the cavities that are filled with fluid—i.e. gas, water, air, occasionally oil, or any mixture of them.

Porifera   The sponges.

porosity   The ratio of the volume of the voids in a soil or *medium* to the total volume. *Compare void ratio.*

porous air diffuser   A *diffuser* (1).

porous bed   A *sludge drying bed.*

Porteous process   *See heat treatment of sludge.*

potable water   *Drinking water.*

Potamogeton, **pond weed**   A water plant that is a *biological indicator of pollution,* unable to grow in water with 2.5 mg/litre of *detergent.* It is less sensitive to mild *organic* pollution.

potamology   The study of streams and rivers. *Compare limnology.*

potassium, K   Sewage *effluents* contain about 10 to 20 mg/litre of K and it can be an indication of their presence. Potassium-39 is not radioactive but contains about one part in 10 000 of the radioactive *isotope* ^{40}K (potassium-40). If a water contains 5 mg/litre of the element at this concentration, it will have a minute *radioactivity* of about 4 picocuries per litre.

potassium permanganate, KMnO₄   An oxidising agent used for precipitating iron or *manganese* or for measuring *oxygen demand* by the *permanganate value* test or for oxidising industrial wastes containing *cyanide compounds.* It has also been used as a disinfectant for drinking water. For removing the taste of *algae,* from 0.25 to 0.5 mg/litre has been added with success. As a 1 to 2% (by weight) *solution* in water at pH 7 to 9, it is helpful in *wet scrubbers* for reducing smells in air, removing *hydrogen sulphide* and *sulphur dioxide* easily and carbon disulphide with difficulty. It also removes the smells of some evil-smelling *organic compounds* (*aldehydes*). *See Kubel test.*

poultry house wastes   The liquid waste from washing down a poultry house is high in *nitrogen,* phosphate and *potassium,* compared with other livestock wastes. About eight hens produce the same *BOD* per day as one human. Chicken manure may be spread on land or treated in an *anaerobic lagoon.*

pour plate method   The *plate count technique.*

p.p.m.v.   Parts per million by volume.

p.p.t.   In the USA, parts per trillion: 1 p.p.t. = 10^{-6} p.p.m.

pre-aeration   A treatment of *sewage* before the *sedimentation tanks* by blowing air through it. It may improve *sedimentation,* certainly helps to keep it fresh and to remove greases and fats by *flotation,* and may form part of grit removal in the *aerated grit chamber. Effluent* can be aerated also after sedimentation and before *biological treatment.*

pre-chlorination   A preliminary *chlorination* before *coagulation* and *filtration* of *raw water*, in order to restrict microbial growths in the *sand filter*. When a *turbid* water is chlorinated, there can be no assurance of good disinfection; consequently, a final chlorination also is needed.

precipitation (1) Rain, dew, snow, hail, sleet or any other form of water from the sky, measured as millimetres of rain. It can also mean a *fallout* of polluting dust or *electrostatic precipitation.* (2) Release from *solution* of a substance by its appearance in the *solvent* as a fine powder or crystals. For removal from water, the precipitate must either flocculate for settlement or be filtered. *See also coagulation.*

pre-coated filter For de-watering a difficult *sludge*, the expense of pre-coating *filter* cloth in a *vacuum filter* or *filter press* may be justified by the increase in efficiency of the filter. The precoating, of *diatomite, flyash,* etc., is applied 50 to 100 mm thick by a *slurry* which is allowed to flow through the filter under vacuum. After the whole area has been coated, the *sludge* cycle is started. *Filter cake* is discharged with a thin slice of the pre-coat. *See filter candle.*

pre-comminution *Shredding.*

pre-composting, preparation for composting The work done on municipal refuse before it goes to the *digestion* process—removal of some or all of the glass and metals. It may include single or *double shredding.*

predator A being that attacks and kills another for food.

prefilt US term for a *sludge, slurry* or other feed, before it enters a *filter.*

pre-filtration, primary filtration, rough filtration The use of *rapid gravity sand filters* or *micro-strainers* for *raw water* without chemical additions, ahead of the *slow sand filters* to increase the length of their *filter run.*

preliminary treatment Any treatment of *sewage* that precedes *primary sedimentation.* It is partly aimed at protecting equipment in later treatment. Mainly it is the removal of large solids, rags and grit by *screens, comminutors, grit channels*, etc. Oil, fat, grease, etc., may be removed in *pre-aeration.* For sewage works fed by *combined sewers*, preliminary treatment includes the proportioning of storm sewage—i.e. *stormwater overflows.*

prescribed flow, minimum acceptable flow The absolute minimum river flow which must not be diminished by *abstraction* and is maintained by *compensation water.* It is the minimum quality and quantity of water that will safeguard *public health*, including the dilution of *effluents*, and meeting the needs of other abstractors.

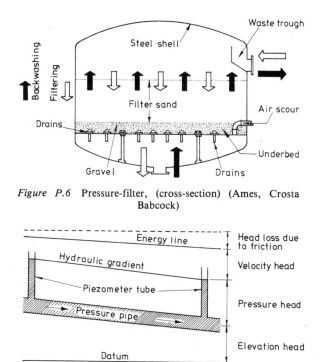

Figure P.6 Pressure-filter, (cross-section) (Ames, Crosta Babcock)

Figure P.7 Pressure-pipe flow

pressure filter (1) A *filter press*. (2) A rapid *sand filter* that is not open to the weather but is contained in a steel shell (*see Figure P.6*), can therefore work at a pressure higher than the gravity head and may be an *in-line* treatment. The sand is 0.45 to 1 m deep, with an *effective size* between 0.5 and 1 mm and a maximum *uniformity coefficient* of 1.7. *Surface loadings* are from 4 to 10 m³/day per m² of filter area, although up to 20 m³/day per m² has been used. *Compare gravity filter.*

pressure flotation *See dissolved-air flotation.*

pressure flow, p. pipe *See Figure P.7. See also pipe flow.*

pressure surge in pipe flow, water hammer A pressure wave caused by an adjustment to a valve, a change in pumping rate, etc. Such waves may be partly reflected at pipe fittings or changes in cross-section. The pressure generated during the surge greatly exceeds that in steady flow, and resonance may result in pressures being multiplied even more. No major pumping scheme or hydro-electric scheme should be designed without an analysis of pressure surges. In a French hydro-electric scheme

239

pipes have exploded because of them. Water hammer is so called because of the noise caused in the pipes. *See surge prevention.*

pressure zones Zones into which water-supply distribution systems are normally divided, each with its separate *service reservoir.* The supply within each zone should be on a *ring main.* The difference in pressures between adjoining zones should not be less than 15 m or more than 75 m head of water. Too small a pressure difference is expensive in service reservoirs and too large a difference produces excessive pressures in low-lying buildings. Tall buildings need their own booster pumps to fill their cold-water *cisterns.*

pre-straining The use of *pre-coated filters.*

presumptive coliform count A *coliform count* made statistically, as in the *most probable number* and *membrane filtration* methods, not by a direct *plate count.*

pre-treatment (1) Or in-house treatment. For industrial waste water, treatment of an *effluent* at the works of origin which reduces the cost of the treatment at the *sewage* works. (2) For *sludge,* any *conditioning* to improve its settling quality or *filterability.*

primary clarifier US term for *primary sedimentation* tank.

primary digestion *See anaerobic sludge digestion.*

primary effluent *Settled sewage.*

primary emission US description of *air pollutants* from an identifiable source (the emitter), often a chimney. *Compare secondary emission.*

primary filtration *Pre-filtration.*

primary material A raw material obtained from natural resources such as plants, trees, animals or minerals, unlike a *secondary material.*

primary production Term used in *ecology* to describe *organic* substances such as material produced in *photosynthesis* by *algae,* plants, etc.

primary sedimentation The *sedimentation* in *sewage treatment,* that immediately follows grit removal, releases an *effluent* (top water) for *biological treatment* and removes *organic* solids as *sludge. Radial-flow sedimentation tanks* or *horizontal-flow sedimentation tanks* may be used. Design may be based on a *detention period* of 1.5 to 2 h at maximum flow, usually 3 times dwf. Alternatively, the *overflow rate* at maximum flow should be 30 to 40 m^3 m^{-2} day^{-1}, which should ensure a 50 to 70% removal of *suspended solids* and 25 to 35% removal of *BOD* from domestic *sewage. Weir loadings* should be less than 250 m^3 m^{-1} day^{-1}, although this is less critical than the overflow rate. *See primary sludge, sludge hopper.*

primary shredding, single s. The first passage of refuse through a *mill*. It can be followed in *two-stage shredding* or *multi-stage shredding*, by *secondary shredding*.

primary sludge The *sludge* from the settlement of *sewage* in *primary sedimentation*. It contains from 6 to 8% solids, unless it is a *mixed sludge*, which will be 3 to 4% solids. *See sludge production*.

primary treatment In *sewage treatment, primary sedimentation*.

priority ranking unit, PRU A classification used in the USA for the *ranking* or *rating of toxic wastes* in order of harmfulness. A number between 0 and 40 is assigned arbitrarily to variables of first priority (e.g. harm to animals), up to 24 for a second-order priority matter or up to 16 for a third-order priority. These are summed, giving, for example, 102 to dilute potassium cyanide, KCN, and 104 to *primary sludge* from *sewage* works. Thus, a total ranking above 80 is considered hazardous, from 61 to 80 is moderately hazardous, from 31 to 60 is slightly hazardous and from 0 to 30 is non-hazardous.

privy vault A bucket latrine connected to a ventilated concrete chamber which is periodically emptied by a tanker truck.

procaryote A lower *protist*.

production In an *ecological* sense, breeding and the bodily growth that accompanies it. *See primary production*.

propeller pump An *axial-flow pump*.

proportioning of storm sewage The separation, at a *stormwater overflow*, of large *sewage* flows into what can and what cannot be treated at the sewage works.

proteins Substances, essential to the living cell and to human food, which are *polymers* of *amino acids*. The link between the amino acid groups is called a peptide link. In *sewage treatment* proteins may be hydrolysed to polypeptides and then to amino acids, which may be further degraded to *ammonia* and to simple *organic compounds*. Proteins are the basis of *enzymes*. *See hydrolysis, single-cell protein*.

proteolytic enzyme An *enzyme* that helps to break down *proteins* into simpler compounds.

Proteus vulgaris, P. mirabilis *Bacteria* that can cause *gastro-enteritis*.

protist, Protista A vast group of highly varied types of micro-organism, which includes lower and higher protists. *Bacteria, actinomycetes* and *blue-green algae* are lower protists (procaryotes). The other *algae, protozoa, fungi* and slime moulds are higher protists (eucaryotes). Micro-organisms are no longer divided into plants and animals, because the distinction is not always clear.

241

protoplasm The body fluid of all *cells* of *bacteria*, plants or animals. An approximation to the protoplasm in *protozoa* is $C_7H_{14}O_{13}N$, but this mixture of water, *proteins*, mineral salts, carbohydrates, etc., is in a constant state of chemical change.

protozoology The study of *protozoa*.

protozoon (plural **protozoa**) Mobile, single-celled *protists* from 2 to several hundred microns long. Most of them are *aerobic* and *heterotrophic*. They move by *cilia, flagella* or *pseudopods*, and are classified into *ciliate protozoa, flagellate protozoa, rhizopods* and *spore-forming protozoa*. Protozoa may indicate (by their type) the condition of *activated sludge*, and they are useful also in *trickling filters*. They do not purify the *sewage* but graze on the *bacteria*, and some of them eat *algae*. Most protozoa are harmless or helpful and only a very few cause illness in humans, *Entamoeba histolytica, Giardia lamblia, Trypanosoma* and *Plasmodium* being some of the exceptions. Because protozoa are easily seen under the *optical microscope*, they are valuable indicators of river conditions. In streams with high sewage pollution *Paramecium* and *Colpidium* are often found in the *zooplankton*, while *Vorticella* and *Opercularia* are found in the bottom mud. Often the population densities of protozoa are more important than the species present.

Pseudomonas A widespread *facultative bacterium* with many species which thrives in both *aerobic* and anaerobic *biological treatments*. In humans some can cause obstinate infections of the ear, urinary tract, etc. Others are important de-nitrifiers, or affect the health of fish, or grow rapidly on *membranes* of de-*salination* plants or *sand filters* or *carbon adsorption beds*. Others even can consume Kuwait crude oil at the oil−water interface, provided that *nitrates* are present.

Pseudomonas aeruginosa This *Pseudomonas*, often present in human excreta, resists *chlorination* more than the coliforms. It has therefore been suggested that its count should be used for evaluating the bacterial quality of water for swimming pools or drinking. Although often found on healthy skin, it can cause disease in wounds or the ear.

pseudopod (plural **pseudopods**), also **pseudopodium** (plural **pseudopodia**) A temporary protrusion from a cell (e.g. of an *amoeba*) which engulfs food or moves its parent *protozoon*.

Psychoda Diptera flies that are common on *trickling filters,* particularly *P. alternata* and *P. severini*. Often the commonest *filter fly,* they are *grazing fauna*.

psychrophile A micro-organism that thrives at low temperatures, even at $0°C$ provided that the water is not frozen, although most of them flourish around $20°C$, unlike *thermophiles* and *mesophiles.*

public health, community medicine The mental and physical health of people, both as individuals and as part of the community. It therefore includes the quality of food, air, water, work, home, leisure, etc., and is partly controlled by *demography.*

public health engineering UK term for *environmental control engineering.*

public health inspector The title used by *environmental health officers* before 1975.

puddling Crushing a soil (e.g. clay) with water and then compacting it with a number of passes of a roller so as to make it homogeneous and reduce its *permeability*. The technique has recently been used to make a watertight base 50 cm thick under an intended site for a *controlled tip* which will prevent *leachate* reaching an *aquifer*. Formerly it was not thought possible to make chalk watertight, but trials in the Kentish chalk in the UK have shown that its permeability can be reduced, by puddling, to one-hundredth of its original value.

pulmonary irritants *Chlorine,* oxides of *nitrogen* or sulphur and other substances that affect the lungs.

pulp (1) A mixture of finely ground mineral and water which is subjected to wet *mineral dressing*. It may have only 1% solids but is usually more concentrated. (2) In paper manufacture, a mixture of *cellulose* and water.

pulp and papermaking waste *See paper and pulp manufacturing wastes.*

Pulsator A type of *sludge blanket clarifier* into which water with coagulants is fed in pulses, not continuously. The pulses occur every 30 s, with 10 s to empty the water from the storage bell into the *clarifier*. The coagulated water enters beneath the *sludge*

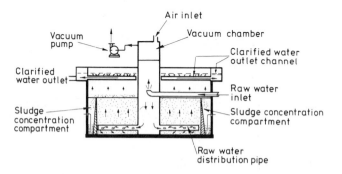

Figure P.8 Pulsator, cross-section ('Water Treatment Handbook' Degremont-Laing)

blanket and the clarified water flows from the top (*see Figure P.8*). Surface loading rates can be several times higher than with other *flocculator-clarifiers*.

pulsed-air cleaning   Removal of the dust that has collected on a *fabric filter* by cutting off the dirty gas supply and reversing the gas flow to the bags, giving them a blast of clean air. The dust drops off into the dust bunker.

pulverised fuel ash, PFA   *See flyash.*

pulveriser   A high-speed, dry *mill*, defined by US makers as having blades with a tip speed above 70 m/s, quite unlike the slow **drum pulveriser.**

pulverising   *Shredding.*

pumped storage   (1) The pumping of water some hundreds of metres above a base reservoir for the storage of energy to drive water turbines that turn electrical generators at the foot of the pipes. Usually the pump is also used as the turbine, since pumping and generation take place at different times. Large pumped storage schemes exist throughout Europe. (2) The pumping of water into a *reservoir* for water supply (*Figure P.9*).

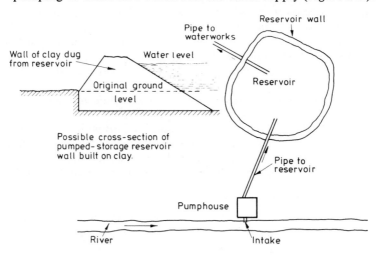

Figure P.9   Pumped-storage reservoir for water supply

The expense of pumping is compensated by the freedom with which the reservoir can be located. Reservoirs for London are filled by pumping from the rivers Thames and Lee. *See raw water storage.*

pump well   A *wet well* or a *dry well.*

pupa   The 'sleeping' stage of an *insect*'s life after the end of the *larval* stage. It develops into the adult (imago) usually in summer or spring.

pure oxygen activated sludge *Oxygen-activated sludge.*

purification In *water treatment* purification is usually the removal of solids, taste, odour and organisms that cause illness. For conventional *sewage treatment* it is often the removal of more than 90% of the solids and *organic compounds*, although the *effluent* may be further purified by tertiary or *advanced waste water treatment*. Allowable limits of impurities have been stated by the *United States Public Health Service*, the US *EPA* and the *World Health Organization*. See *self-purification*, and Table of Allowable Contaminants in Drinking Water, page 359.

putrescible A description of organic material that rots and stinks.

PV *Permanganate value.*

PVC, p.v.c. *Polyvinyl chloride.*

pyramidal sedimentation tank A tank shaped like an inverted pyramid, sometimes used as a *primary sedimentation* or *secondary sedimentation tank* in country *sewage* works. A slope enables the *sludge* to slide down without the need for scraping. The inlet is usually central and the outlet *weirs* peripheral. The maximum practical surface width is 9 m and the *overflow rate* should not exceed 12 m^3 m^{-2} day^{-1}. The inlet may be near the bottom of the tank but above the *sludge blanket*, to produce some *upward-flow sedimentation*.

pyrolysis (verb **pyrolyse**), **destructive distillation** Heating in the absence of air or with a restricted air or oxygen supply. The distinction between pyrolysis and *incineration* is that in a pyrolysis unit the *organic* material or carbon is not fully oxidised and the residue still has a heating value. (Some *total incineration* systems in the USA have been wrongly described as pyrolysis units.) The pyrolysis products differ with the temperature of treatment. Metallurgical coke, for example, because it is made at high temperature, produces much gas; low-temperature coke produces more liquids. For rapid pyrolysis material must be finely ground, but this is expensive. Pyrolysis residue, assuming that all glass, metals, stones, etc., have been removed, can be a char of chemical composition like anthracite. It might be used as a fuel or as a sterile *filter* medium or as a soil conditioner. Pyrolysis units can be heated either externally or internally, the latter by burning a small part of the refuse, as in some US units. The capital cost of pyrolysis may be less than that for incineration but the volume reduction is smaller. In externally heated retorts the amount of heat needed for pyrolysis is some 4000 megajoules per tonne of refuse or about half the heating value of the refuse. Because *rubber tyres* are difficult to burn without black smoke or other pollution and are nearly non-biodegradable, it has been hoped that pyrolysis could be developed to convert scrap tyres into some form of energy.

Q

quaternary ammonium compound A compound in which the four hydrogen atoms next to the nitrogen in NH_4^+ are replaced by four organic radicals. They may be used as dispersants, corrosion inhibitors, antiseptics or cationic *synthetic detergents*.
quenching tower An *evaporative tower*.
quiescent sedimentation *Sedimentation* by a *fill and draw tank*.

R

rack A *bar screen*.
radial-flow pump A pump with the water moving away from the axis of rotation, as in the *centrifugal pump*.
radial-flow sedimentation tank (*Figure R.1*) A circular *sedimentation tank* with the *sewage* or water entering at the centre through an upward-flow pipe into a vertical-walled drum.

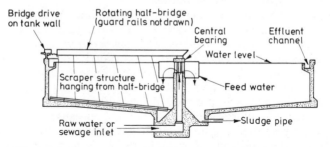

Figure R.1 Radial-flow sedimentation tank with half-bridge scraper
(for raw water or sewage)

The drum distributes the sewage or water through its bottom and sides. *Effluent* flows out over a perimeter *weir*. The flow velocity is greatest at the centre and there is also an upward flow effect. Rotating sludge *scrapers* (either *fixed-bridge scrapers* or *rotating-bridge scrapers*) sweep the *sludge* into a central *sludge hopper*. Scraper speeds should not exceed 25 mm/s, to avoid breaking up the sludge. Radial flow tanks may be used for *primary sedimentation* or *secondary sedimentation* of sewage. Tanks can be large, up to 60 m diameter for primary and 35 m diameter for secondary tanks. Average depths are usually 2.5 to 3 m with a floor slope of 7.5 to 10° to the horizontal. Some tanks may be flat-bottomed, with *vee-blade scrapers*.

radioactive wastes *See nuclear reactor wastes.*

radioactivity, radioactive decay The distintegration of some atoms, called *radio-isotopes,* with the resulting emission of *alpha particles* or *beta particles* or *gamma radiation*, leading to the formation of other *isotopes* (decay products).

radio-isotope, radioactive isotope An *isotope* that disintegrates, with emission of radiation. Some radio-isotopes are used as *tracers*.

radiotracer A radioactive *tracer*.

rain gun A nozzle that sprays water, *sewage,* liquid manure or industrial *effluent* into the air and out on to the land. The jet also makes the nozzle rotate. One rain gun can throw its water jet up to 60 m away, consuming a maximum 2 m^3/min. The gun may be stationary or move automatically along the field.

rainout Removal of particles or gases from the atmosphere during the formation of raindrops on nuclei. Rainout takes place in the cloud or near it; *washout* takes place further down, nearer the earth.

rainwater One of the purest of waters, apart from its gas content, which can be considerable. It is soft and contains *carbon dioxide*; therefore, it can dissolve lead and be poisonous if it passes through lead pipes or over a lead roof. If rainwater is caught from any roof for drinking, the first flush should be rejected down the drain and the gutters and roof should be kept clean.

ram pump One type of *reciprocating pump*.

ranking or **rating of landfill sites** Sites that are in view for *controlled tipping* in the USA have been 'marked' in a similar way to *toxic wastes* by *priority ranking units*. Ten variables are marked, four of them concerned with the *permeability*, etc., of the soil under the tip, four with *groundwater* and two with wind. The markings for the ten variables are summed. The lower the total ranking mark, the better the site for the disposal of hazardous wastes.

ranking or **rating of toxic wastes** *Toxic wastes* intended for disposal on a tip (i.e. not *radioactive wastes*, or those that boil and flash at a temperature below normal air temperature) are listed by US specialists for their hazard to the community according to their human toxicity, *groundwater* toxicity, disease transmission potential, lack of biodegradability and mobility. Harm to animals or humans has a first-order priority, persistence in the *ecosystem* second-order, other matters third-order. *See priority ranking unit.*

rapid gravity sand filter, rapid s.f., rapid downflow s.f. An open

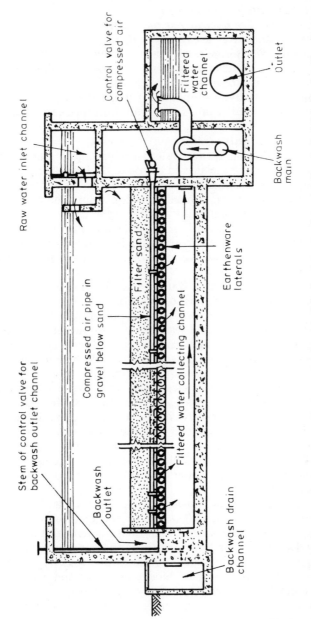

Figure R.2 Rapid gravity filter (Paterson Candy International Ltd) Isolated arrows show the direction of flow during filtration. During backwashing the flow is reversed)

248

tank resembling the *slow sand filter* except that 1 to 1.3 m of sand is covered with 2 m or more of water. Rapid gravity sand filters are widely used in *water treatment* and sometimes for the *tertiary treatment* of sewage. They use sand with a D_{10} *effective size* between 0.35 and 0.55 mm and a *uniformity coefficient* between 1.5 and 2.0. The *underdrains* or pipe manifold beneath the sand, set in pea gravel or shingle, are used both for removing the *filtrate* and for *backwashing* the *filter* with air or wash water. The filtrate is used as *wash water*. Because of the rapid flow rate the sand clogs quickly, but automatic backwashing is now possible. There is little biological activity in the filter, although some *dissolved oxygen* may be used up. The filtrate can be re-aerated by flow over cascades or along a channel into which air is blown. Skilled maintenance and operation are needed. The *surface loading* of a rapid sand filter should be 4 to 6 m^3 m^{-2} day^{-1}, although plants have worked well at 10 m^3 m^{-2} day^{-1}. Higher rates can be achieved if the filter is converted to a *multi-media filter* or *dual-media filter*. (The term 'rapid gravity sand filter' may also mean a *pressure filter* (2).)

rapping *Electrostatic precipitators* need to have the dust rapped off them periodically.

rasp mill, Dutch rasp A slow, wet *mill* used especially in the Netherlands to prepare refuse for *composting*. It resembles a large, vertical-shaft *hammer mill* but rotates much more slowly, at about 5 to 6 rev/min, and the diameter is larger, some 6 to 8 m. Rasping pins on the base plate of the mill break up the refuse as it is pushed round. The crushed refuse falls through 50 mm diameter holes in the base plate. Uncrushable objects are pushed to the rim of the mill to exit holes designed for the purpose.

rating In *hydrology*, the relationship between the water level and the discharge (flow rate) of a stream, well or *aquifer*, or the work of taking the observations and making the calculations that establish the relationship. *See* also *ranking*.

rating curve A *stage/discharge curve*.

rational method The *Lloyd—Davies method* for the estimation of *runoff*.

rat-tailed maggot The *larva* of *Eristalis tenax*.

rave The beam along the back edge of a rear-loading refuse collecting vehicle, on which a dust bin can be rested as it is emptied.

raw sewage *Sewage* before it has received even *preliminary treatment*.

raw sludge *Primary sludge* or *secondary sludge* from sewage, which has undergone neither *aerobic digestion* nor *anaerobic* *sludge digestion*, nor *heat treatment of sludge*. After *mechanical de-watering of sludge* it is called raw *sludge cake*, to distinguish it from de-watered *digested sludge*. Raw sludge contains many enteric bacteria. **Septic** raw sludge often stinks. *See sludge disposal*.

raw water Water intended for drinking or industry which has not been treated and so is stored in an *impounding reservoir* but not in a *service reservoir*. The quality of raw water varies with its source. *See* below.

raw water storage *Surface water* used for a supply needs to be stored before *water treatment*. For *direct-supply reservoirs* this should be enough to maintain the supply to consumers during periods of low rainfall. *River-regulating reservoirs* balance out peak flows in the river and maintain a given minimum flow. Water withdrawn from a river needs *bankside storage*, usually for 7 days. This enables the river intake to be closed if the water quality becomes unsuitable for a short time. Storage of *raw water* influences its quality in several ways. Some harmful *bacteria* die, partly because of *disinfection* by *ultra-violet* light. In warm weather *algae* will grow, particularly if the water is rich in *nutrients*, as in lowland industrial rivers. The algae will increase the *turbidity* of the water, and may influence its taste and smell and affect the working of the treatment processes. *Suspended matter*, other than algae, should, however, settle during storage and reduce turbidity. If the bottom deposits become de-oxygenated, they may produce tastes and smells and reduce ferric and manganic ions to ferrous and manganous *ions*, which are more soluble in water. The organic content of the water may decrease because of bacterial oxidation or *sedimentation*, but algal growth or *anaerobic* decay of the bottom deposits will increase the organic content. In general, if algal growth can be restricted and the organic deposits are not great, raw water improves with storage. *See upland catchment water*.

RDF Refuse-derived fuel. *See waste-derived fuel*.

reactor A container in which chemical or biological changes take place.

re-aeration (1) In *activated sludge* treatment, the addition of air to *return activated sludge* in a second *aeration tank* between the *secondary sedimentation tank* and the main aeration tank. (2) In rivers, lakes, etc., the acquisition of *dissolved oxygen* by turbulence or *photosynthesis*. It is important for maintaining the oxygen balance.

recalcitrant Description of a chemical that biodegrades slowly or not at all.

recarbonation Lowering the *pH* of a water by bubbling *carbon dioxide,* CO_2, through it. In *water treatment* it is needed after *lime-soda softening* for *stabilisation*. In *sewage treatment* recarbonation is needed after ammonia stripping. The CO_2 is supplied either in bulk from brewers or distillers or in the *flue gas* from heating *calcium carbonate* in a lime kiln, etc.

receiving water The stream, river, lake or sea that receives *sewage* or the final *effluent* from a sewage works and by dilution and *self-purification* may render it wholly non-polluting.

recessed-plate press A type of *filter press*.

recharge Refilling of *aquifers* either by *artificial recharge* or naturally by rain percolating through the soil and becoming *groundwater*.

recharge prevention An *acid mine drainage* pollution control method that reduces the infiltration of water by diverting streams away from the mine and from piles of waste rock.

reciprocating-arm grit washer Equipment for separating organic matter from grit in a *Detritor*. The reciprocating arm moves the grit up a ramp at the side of the unit, while *wash water* flows down the ramp back into it.

reciprocating pump, ram p. A pump that displaces the fluid positively, drawing it in and then forcing it out along the pipe. Very high pressures can be produced and *sludges* with as much as 10% solids can be pumped.

recirculation Circulation of fluid. In *sewage treatment* usually either the return of activated sludge from *secondary sedimentation* to the *aeration tank* or the return of sewage *effluent* for further *biological treatment*. In a *trickling filter*, whether high-rate or standard-rate, a minimum flow is needed to keep the *biological film* moist. It is, therefore, sometimes necessary to pump *humus tank* effluent back to the filter as *influent*. A waste that is too strong can thus be diluted, and recirculation may reduce the final *nitrification*. In *two-stage filtration* the final effluent can be recirculated to the first filter—e.g. at night when the feed is low.

reclaim Raw material obtained by *reclamation*.

reclamation of solid waste A first stage of *re-cycling*—the separation, processing and upgrading of materials that otherwise would go to a tip. About 45% of the world's steel, 40% of its copper, 50% of its lead, 30% of its zinc, 25% of its aluminium and 20% of its paper are made from reclaimed, discarded or obsolete material. Energy needs for smelting from ore are always much larger than when making metal from *reclaim*. In megawatt-hours per tonne the figures are roughly: aluminium 80 (from reclaim 5); copper 13.5 (1.7); iron or steel 4 (3);

251

magnesium 91 (19); titanium 125 (52). *See mechanical sorting plant.*

reclamation of water   The treatment of sewage *effluent* or other waste so that it can be re-used directly, not necessarily for drinking. Two of the best-known examples are at *Lake Tahoe*, California, and *Windhoek*, Namibia. In another semi-desert country, Israel, sewage *effluent* is regularly used for *irrigation* and other purposes. In Europe *artificial recharge* of *groundwater* with treated sewage effluent is a form of reclamation, since the groundwater is later pumped out as *raw water*. Often the most expensive pollutant to be removed from sewage is *chloride*, because *de-salination* is needed. Reclamation may involve removal of *nitrogen, phosphorus,* chlorides and *trace organics*, followed by conventional treatment for raw water. *See re-use of water.*

reconditioning tank   A *re-aeration* tank for *returned activated sludge.*

recovery zone of a river   Rivers are capable of *self-purification.* Consequently, a length downstream of an *effluent* outfall is the recovery zone for that *outfall. See oxygen sag curve, saprobic classification.*

rectangular sedimentation tank   A *horizontal-flow sedimentation tank.*

rectification   Separation by distillation of mixed liquids that have different boiling points, the last process in the manufacture of ethanol, for example.

recuperative or **regenerative heat exchange**   Cooling of one fluid by the heating of another fluid (occasionally a powder), usually in a tubular heat exchanger (but *see heat-recovery wheel*). One type, gas-to-gas heat exchange, is sometimes used for heating *flue gas* after *wet scrubbing* and before discharge to the chimney in *stack gas re-heating.*

recuperative incineration   *Incineration with energy recovery.*

re-cycling plant   A *mechanical sorting plant.*

re-cycling of solid waste   Re-use. While 're-use' applies more to the refilling of a bottle or other container, 're-cycling' is usually applied to the material of which the container is made. Direct re-cycling is the re-use of the broken bottle to make similar bottles. Indirect re-cycling is its re-use for a different purpose, usually making something of a lower quality, such as posts and pipes from old *plastics. Pyrolysis* and energy recovery are also sometimes regarded as examples of re-cycling. Some materials, especially plastics and paper, degrade with each re-cycle. Paper fibres shorten at each pulping. Re-cycling begins with *reclamation.*

re-cycling of water or sewage   *See recirculation, re-use of water.*

redox potential, oxidation−reduction potential, ORP, rH, Eh   The electromotive force in millivolts set up in a solution between standard *electrodes,* due to the concentrations of oxidising or reducing substances present. *Aerobic* (oxidising) systems need positive redox potentials of + 200 to + 600 mV for the best results. *Anaerobic* processes have negative potentials of − 100 to − 200 mV. Several types of redox meter exist. Continuous redox meters are often used in chemical treatment to control automatically the amount of reagent added. Redox potentials also are important for controlling *corrosion.*

reduction   In chemistry, loss of oxygen or gaining of electrons by a compound, the opposite of *oxidation.*

reduction of refuse   *See refuse reduction.*

red water   *See ferruginous discharges.*

re-feeding of wastes   Poultry excreta have been fed to other poultry, cattle and sheep without ill effect, after drying by heat. The milk of dairy cows did not suffer, nor did the cows themselves. Sheep also have been fed with excreta from cattle, pigs and poultry, to the extent of 70% of their diet.

refinery waste water   *See oil refinery effluent.*

reflectometer   An instrument that measures the proportion of the light that is reflected from a surface—an indication of its dirtiness or otherwise.

reflux valve   A *non-return valve.*

refractory   Obstinate; not damaged by heat. A refractory is a stony, often artificial material that withstands heat and *flue gases*, typically of firebrick or of concrete made with aluminous cement, and used for lining the inside of *incinerators* and other furnaces, where there is no *water wall.* Refractory organics are *organic compounds* that resist *biological treatment.* They biodegrade slowly or not at all.

refuse   *See* below; *see also solid waste.*

refuse chute   A vertical or nearly vertical pipe, 30 to 90 cm diameter, through which refuse can be dropped. Chutes should be isolated from the dwelling structure so as to prevent disturbance of neighbours by the noise of refuse falling down a chute. Chutes can be a fire hazard if they are not built with traps to prevent them acting as chimneys and if they are not fire-resistant enough. *See flue-fed incinerator.*

refuse compression   *See high-density baling.*

refuse-derived fuel, RDF   *See waste-derived fuel.*

refuse disposal   *See solid waste treatment.*

refuse reduction   (1) Size reduction in a *mill.* (2) Reduction of the mass of refuse to be tipped. This can be achieved in many ways, by *incineration* or *composting* or by salvaging material such as iron, steel, paper and glass in a *mechanical sorting plant*. (3) Reduction of the volume of refuse by *compactors*.

regeneration Generally refers to the recovery of chemical reagents. Spent *activated carbon* grains are regenerated by passing steam through them. Caustic soda that has been converted from NaOH to Na_2SO_4 is regenerated by the action of *lime*, $Ca(OH)_2$. *See zeolite*.

regional water authority A *water authority* (England and Wales).

registered works, scheduled w. Any works that is controlled for pollution purposes by the *Alkali Inspectorate*. They number only about 3000, compared with the 300 000 or so that are not registered and so are supervised by *environmental health officers*. If the means of controlling the pollution from a registered works has become so thoroughly understood that it is standard engineering practice, it is usual to remove the works from the list—to de-schedule it.

regulating reservoir A *river-regulating reservoir*.

re-heating of flue gas *See stack gas re-heating*.

Reinluft process Absorption of *sulphur dioxide* from *flue gases* by *activated carbon*.

relative humidity, RH The ratio between the amount of water vapour in the air and the maximum which it could hold at the temperature concerned. The RH is a fraction, never more than 1. At 100% RH rain or dew is already falling or can soon be expected.

rem, roentgen equivalent man (or **mammal**) A measure of ionising radiation, that which has the biological effect of a roentgen of X-rays. The UK Parliament laid down in 1960 that the population as a whole should not be exposed to more than 1 rem in 30 years, preferably much less. *See also millirem*.

re-oxygenation Addition of oxygen to water from any source, including *photosynthesis, re-aeration* or the entry of *dilution* water.

reservoir A large tank or lake to contain either *raw water* or treated water. *See direct-supply reservoir, impounding reservoir, pumped storage, river-regulating reservoir, service reservoir*.

reservoir by-wash channel A *by-wash channel*.

reservoir storage *See raw water storage*.

reservoir yield *See yield of a water source*.

residence time *Detention period*.

residual chlorine *See chlorine residual*.

residuals, residue Ash, clinker, iron and steel, etc., that have passed through an *incinerator*.

resolving power The *limit of resolution*.

resource recovery Obtaining what is useful. Often it means getting out of refuse what can be got. It can therefore imply a *mechanical sorting plant* to extract metals, glass, *cellulose, compost* or merely the heat from burning. Some of these aims exclude one another. All the heat cannot be recovered if the bulk of the paper is salvaged as fibre or converted to compost.

resource recovery facility US term for a *mechanical sorting plant*.

respirable dust Dust so small (usually between about 10 and 0.25 μm) that it is carried into the lungs, may be caught there and harm them. The bulk of the dust caught is smaller than 5 μm. Larger dust is caught in the nose and throat and never reaches the lungs, so does no harm. Some 40% of the 2 μm size is caught in the lungs. Some particles caught in the lungs may be as small as 0.02 μm, but most of the very small dust floats out again with the outgoing air.

respiration The consumption of *oxygen* and release of *carbon dioxide* that happen when *organic* material is broken down (digested). In *microbiology* it is more precisely defined as a *catabolic* process by which substances containing energy (carbohydrates, etc.) are oxidised by *enzymes* into simpler substances with less energy. Respiration is usually *aerobic*, involving molecular oxygen, although under *anoxic* conditions inorganic compounds can act as the source of oxygen (e.g. *denitrification*). This is still considered to be respiration, as opposed to fermentation.

respiration rate For *activated sludge* the rate of absorption of *oxygen*, expressed in mg O_2 per gram of dry solid in a unit of time.

respiratory quotient The ratio of the volume of *carbon dioxide* produced by *respiration* to the volume of *oxygen* consumed.

respirometer The Warburg respirometer (*Figure R.3*) is an airtight, constant-temperature flask used for measuring the *oxygen demand* of a water sample in it. A manometer connected to the flask shows the drop in pressure at constant volume caused by *oxygen* consumption. Any *carbon dioxide* produced is absorbed by potassium hydroxide held on a tray in a separate central well above the water surface. Respirometers may also be used at constant pressure, by adding the required volume of air or oxygen to the flask. The apparatus is used both for research and for *BOD* measurements. Many other types exist, including some that measure oxygen uptake directly by electrolysis—electrolytic respirometers.

retention period A *detention period*.

retort pyrolysis *Pyrolysis* in a retort heated from outside; nothing burns inside the retort.

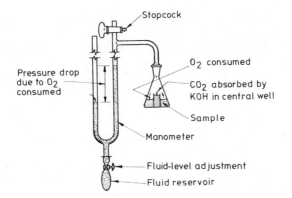

Figure R.3 Respirometer, Warburg type

retrofitting As air pollution legislation in the USA has become stricter, *dust arrestors* have had to be fitted to older plant, and this is called retrofitting. The newer dust arrestors are *electrostatic precipitators, fabric filters,* high-energy scrubbers. Retrofitting also describes the application of *fluidised bed combustion* to boilers that originally were not designed for them.

returnable bottle, r. container The 10% of UK refuse that is glass could be appreciably reduced, with benefit to the cost of living, if the public were to return all returnable bottles and throwaway bottles were eliminated. Large, expensive steel containers are regularly cleaned and re-filled by manufacturers, whereas small containers, tin cans, etc., are usually non-returnable and are a waste of resources.

return activated sludge, returned s. *Activated sludge* that is returned to the *aeration tank* to mix with the influent *sewage,* unlike the *surplus sludge.* The *suspended solids* in the return sludge is about twice that in the *MLSS.* The flow rate of return sludge is usually 50% of the *raw sewage* flow, or it may be kept constant and equal to the *dry-weather flow.*

re-use of water In industrialised or arid regions water supplies are so small in relation to the population that the treated *sewage* of an upstream town, discharged into a river, has to be used as part of their water supply by towns downstream. Such indirect re-use is common in the UK and the rest of Europe, where five or ten re-uses are known. It is limited mainly by increase of *salinity.* Direct re-use of *effluents* is usually known as re-circulation or *reclamation of water.* Industries may re-use their process or cooling water.

reverse current filter A *bi-flow filter* with a layer of gravel below the lowest sand. The *raw water* is partly filtered by upward

256

passage through the gravel. Some of the flow is then diverted on to the top of the sand and the rest flows up through the sand to the outlet in the centre of the sand bed.

reverse-flow cyclone A conventional, tangential-inlet *cyclone*, usually for gas cleaning.

reverse osmosis, RO A *de-salination* method for sea-water or *effluents*. In osmosis pure water passes from a weak solution through a *semi-permeable membrane* into a stronger solution. In RO the direction of flow is reversed by the application of a pressure, to the surface of the stronger solution, which is larger than the osmotic pressure. Relatively pure water is thus 'squeezed' out of the stronger solution. Even with an applied pressure that is double the osmotic pressure, the flow rate is very low. Pressures can be very high and may rise to 100 atm, which is

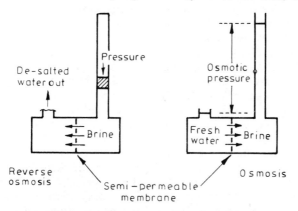

Figure R.4 Reverse osmosis and osmosis

costly in energy. The semi-permeable membrane may be of *cellulose acetate* or other *polymer*, with a mechanical support behind it. Like *electro-dialysis,* RO is more suitable for *brackish* water than for sea-water. In 1976 about 30 plants existed in Japan, but Europeans consider it an expensive way of de-salting sea-water.

reversible fan Vehicles that travel over a *controlled tip* should be fitted with a reversible fan. This enables paper and *plastics* to be dislodged from the radiator at will.

revolving screen A *trommel.*

Reynolds number, *Re* A dimensionless number that describes pipe flow. It is equal to

$$\frac{v \text{ times } d}{\text{kinematic viscosity}}$$

257

in which v = mean velocity and d = diameter of pipe or depth of flow in channel. *See viscosity.*

rH The *redox potential*, measured in millivolts.

rhizopod, Rhizopoda A class of *protozoa* including *amoebas*, which move and take in food by their *pseudopods*.

rhodamine B A pink dye which has been used as a chemical *tracer* although it causes cancer. It tends to *adsorb* on to solid particles.

rickettsiae Very small *bacteria* that resemble *viruses* because they are *obligate parasites*. They cause typhus and other fevers, for which the *vectors* are often lice, but they are not transmitted by drinking water.

ridge and furrow aeration An old method of forming an *aeration tank*, with *diffusers* in the bottom of the furrows to eliminate points where *sludge* might settle and putrefy.

ridge and furrow irrigation *Furrow irrigation.*

rigidly separate system *See separate system.*

ring ditch An *oxidation ditch.*

Ringelmann chart A chart showing five grades of darkness, through the palest gray to pure black, by which *smoke* density can be assigned a Ringelmann value or number, from 0 to 5. Under the UK *Clean Air Acts* in smokeless zones, dark smoke (No. 2) is forbidden except for occasional short periods, which may include *soot blowing*. Ringelmann 0 is white, 1 is 20% opacity, 2 is 40%, 3 is 50%, 4 is 80% and 5 is 100%, dead black. The British Standard chart has the disadvantage that two people are needed to make an observation, one of them holding it at 15 m distance from the other, who observes it and the smoke. This disadvantage is overcome by the *micro-Ringelmann chart, telesmoke, carboscope* and other devices. In some US states Ringelmann has been superseded by *particulate emission* standards.

ring main A *main* connected to the water supply at two points—i.e. with a ring layout so that any consumer is served by at least two routes. This minimises inconvenience to consumers in the event of a burst pipe. A ring main also has the advantage that it eliminates dead ends in the pipe and the resulting deterioration in water quality that can occur with *stagnation*. The pipes in a ring main tend to be smaller and the pressures lower than with a *branch main*.

riparian rights The water rights of those who own land on the banks of rivers, lakes or streams.

ripening *See maturing.*

rise rate of a clarifier *Overflow rate.*

rising-current separator A device used in *mineral dressing* or a *mechanical sorting plant*, in which a rising current of water

258

removes the lighter material from a mixture of light and dense materials. *See elutriation.*

rising main, force m. A *main* up which water or *sewage* is pumped. Lack of oxygen can occur in rising *sewers*, causing *sulphide corrosion* at the outlet from the rising main.

rising sludge *Sludge* that has risen from the bottom of a *sedimentation tank* to float on the surface, lifted either by *methane* or *nitrogen*. Nitrogen is formed if a nitrified *activated sludge* effluent de-nitrifies in the activated sludge settling tank, releasing nitrogen gas. This is unlikely if the sludge spends less than 2 h in the settling tank. The major remedy is to increase the rates of sludge collection and return from the settling tank. Sludge collection can be increased by having three or four scraper arms in the tank (*see rotating-bridge scraper*). In *primary sedimentation* tanks, sludge rises because it decomposes anaerobically, releasing methane. In warm weather de-sludging of the tanks may be needed more often because the oxygen is used up more quickly. *See de-nitrification.*

river authority, r. board English and Welsh river authorities and boards were taken over by the new *water authorities* in 1974 under the *Water Act 1973*. Scotland, however, still has *river purification boards. See Scottish water industry.*

river basin commission In the USA a body formed by the federal government with one or more states, to co-ordinate the development of a river and related resources. The commission aim at broader planning of water resources than is possible for the individual state agencies, but only a few exist. *See ORSANCO.*

river blindness *Onchocerciasis.*

river classification Various classifications of river water quality exist, such as the *saprobic classification* and the *Trent biotic index*. Another classification, used by the Department of the Environment and the Scottish Development Department, divides rivers into four classes:

(1) Clear water with BOD_5 below 3 mg/litre, well oxygenated and without evidence of pollution.

(2) Slightly de-oxygenated and known to receive some pollution.

(3) Less than 50% saturated with *dissolved oxygen*, or with substances believed to be at toxic concentrations.

(4) Grossly polluted water, offensive in smell or appearance, or with a BOD_5 of 12 mg/litre or more, or completely de-oxygenated in any reach, or with toxic substances, incapable of supporting fish life.

In addition, the Department of the Environment (D. of E.) recommends four pollution classes based on the *fauna* present, broadly the same as the saprobic classification. Classes A, B, C, D of the D. of E. are equivalent to the *oligosaprobic, β-mesosaprobic, α-mesosaprobic* and *polysaprobic* subdivisions.

river purification board, RPB Scottish bodies responsible for the quality of rivers and coastal waters. *See Scottish water industry.*

river-regulating reservoir, regulating reservoir (*Figure R.5*) An upstream *impounding reservoir* for *raw water storage* that can

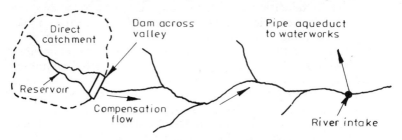

Figure R.5 River-regulating reservoir

either reduce floods or release water when the river level downstream is too low for water to be withdrawn from it. It eliminates the cost of a long *aqueduct* and allows for multi-purpose use, including amenity, recreation, sailing and economy of water, quite unlike the old single-purpose *direct-supply reservoir*.

river water Rivers furnish two-thirds of drinking water supplies in England and Wales, largely from lower reaches. In Scotland most of the water is from upland catchments. *See groundwater, oxygen balance, raw water.*

RO *Reverse osmosis.*

road drainage *Runoff* from streets enters the *sewers* through *gulleys*. For an 8 m wide street, a gulley every 35 m should be sufficient. The runoff may contain much *grit* and salt in winter.

rodding Cleaning out blockages in drains or *sewers* with *drain rods*.

rodenticide A *pesticide* to kill rodents—e.g. rats, mice, squirrels.

rotary-drum pulveriser A *drum pulveriser*.

rotary-drum screen A *trommel*.

rotary kiln, rotating k. A continuous furnace that can be used as an *incinerator*. It consists of a nearly horizontal steel cylinder lined inside with firebrick. The burning gases or other fuel are injected at the lower end. The material to be roasted or incinerated is poured in at the upper end and slowly descends as

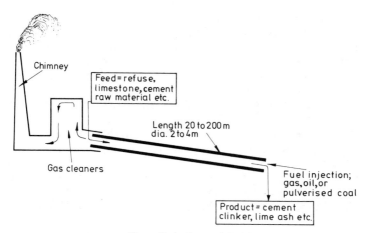

Figure R.6   Rotary kiln

the kiln rotates at about 12 revolutions per hour (*see Figure R.6*). A kiln 18 m long and 2.4 m diameter has been used for burning sewage *sludge* with 96% water. For sludge burning, to avoid smell, the gas temperature should not be below about 750°C. Rotary kilns can burn any burnable liquid as fuel, provided that it is atomised by steam or air through a burner nozzle. Solids or tars, even substances that liquefy in the kiln, can be burnt without difficulty, and the long *detention period* ensures complete combustion for the slowest-burning objects unless they are spherical or cylindrical or airborne.

rotary regenerative air heater   A *heat-recovery wheel* used by the UK Central Electricity Generating Board to recover heat from *flue gases* and transfer it to the incoming air. The flue gases traverse one half of the wheel, which cools them, passing the heat to the incoming air as it rotates through it. For a 500 MW boiler, two wheels about 10 m diameter and 2 m deep are needed. One difficulty has been the fouling of the gas passages with *flyash* because the *electrostatic precipitators* that remove it come later. This difficulty has been largely overcome by re-design of the gas passages.

rotary screen   A *trommel*.

rotary screen magnet   A short *trommel* with an electromagnet outside the riding edge. This edge therefore lifts iron and steel pieces and drops them at the point where the electromagnet ceases to be effective—at the top of the trommel. The metal falls from there into a chute inside the trommel that takes it outside. The trommel also functions as a *screen* in the ordinary way, thus yielding three products—an oversize, an undersize and magnetic

261

material. *See magnetic separator.*

rotating-arm distributor A *distributor* rotating over a circular *trickling filter*. The arms (pipes) are pushed round either by the reaction of the water flowing out of them or, preferably, by an electric motor. Two or four arms are carried on the central column. Holes in different arms are staggered so as to distribute the *sewage* uniformly. The arms and holes must be cleaned regularly.

rotating biological contactor, RBC, rotating biological disc, rotating disc An *aerobic* sewage treatment in which a horizontal shaft just above the surface of the *sewage* carries closely spaced discs up to 3 m diameter or random plastics media in circular wire cages, that revolve with the shaft (*see Figure R.7*). Some 25 to 45% of the discs or other media are submerged.

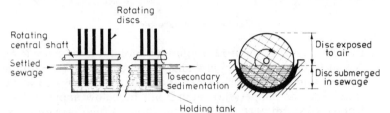

Figure R.7 Rotating biological contactor

The slow rotation develops a *biological film* which oxidises the sewage. When the slime layer becomes too thick, it sloughs off and is settled in a separate tank or compartment. In some plants, especially those for small communities, the discs are contained in the same tank as the zones for *primary* and *secondary sedimentation* and *sludge* storage (Biodisc). Some *anaerobic sludge digestion* will occur and the plant is de-sludged two to four times yearly. Little maintenance is needed. Alternatively, primary and *humus* settlement and sludge treatment may all be in separate tanks. Such plants may treat the sewage from populations of 500 to 10 000 or so. RBC plants have the advantage of low power consumption and have only one-tenth of the land demand of a *trickling filter*. Design is based on kg *BOD* per m^2 of disc area per day. To obtain 90% reduction of BOD in plants with settlement and storage of sludge in one tank, 7 kg BOD_5 should be allowed per m^2 per day for populations smaller than 50 and up to 10 kg m^{-2} day^{-1} for populations of 300 or more. *Sludge production* is about 0.6 kg of dry sludge per kg of BOD_5 applied in the raw sewage. Activated sludge plants have been upgraded by adding rotating discs to the aeration tank. Higher organic loads per unit volume of tank are possible because bacteria can grow in the mixed liquor as well as on the surfaces of the discs.

rotating-bridge scraper A mechanical *scraper* for *sludge*, used in *radial-flow sedimentation tanks* in the UK, consisting of a bridge pivoted at the centre of the tank and carried along round the rim wall on a trolley. In the USA *fixed-bridge scrapers* are commoner. For *primary sedimentation* a bridge covering half the diameter of the tank is usual (half-bridge scraper). For *secondary sedimentation* one or two full-diameter bridges may be used or three half-bridges. The greater the number of bridges in a tank, the faster the sludge will be removed. This may be important to stop sludge rising because of *de-nitrification*. A chain scraper may be used instead of the usual scraper blades.

rotating-drum air separator A rotating horizontal drum through which air is blown to separate heavy from light refuse—a principle that has been further developed with the use of a conical drum.

rotating-drum furnace A *rotary kiln*.

rotifer, Rotifera, wheel animal Tiny *aerobic* creatures 50 to 250 μm long, the simplest of the multi-cell invertebrate animals. They have *cilia* around the mouth and can engulf *bacteria* or other *organic* matter. Their presence in an *effluent* indicates highly efficient aerobic *biological treatment*.

Rotodisintegrator A type of *comminutor*.

rotodynamic Description of a rotating mechanism that imparts movement—e.g. in pumps to water.

Roto-plug sludge concentrator De-watering equipment for *sludge* which is fed into a series of drums lined with nylon filter cloths, turning at about 2.8 rev/min. The sludge de-waters slowly and then passes through a *filter press*, concentrating finally to 20 to 30% solids.

roughing filter (1) A coarse true *filter*, formerly used for straining *effluent* before discharge into a *receiving water* or on to land. (2) A *high-rate trickling filter*. (3) A unit for *pre-filtration*.

routing *See flood routing.*

Royal Commission effluent The Royal Commision on Sewage Disposal was set up in 1898 and was not disbanded until 1915. A Royal Commission effluent has a *BOD_5* below 20 mg/litre and *suspended solids* below 30 mg/litre, provided that the *receiving water* can dilute it at least 8 times. This standard of sewage *effluent* was intended to ensure that a clean river with BOD_5 of 2 mg/litre would not exceed 4 mg/litre when mixed with the effluent. However, a Royal Commission standard has tended to become synonymous with a 20:30 effluent, regardless of the available *dilution*.

Royal Commission on Environmental Pollution A body that in its fifth report (1976) proposed that the *Alkali Inspectorate* be forthwith removed from the Health and Safety Executive and placed in the Department of the Environment, with the other organisations concerned with the environment. After the appropriate legislation it would be re-named H.M. *Pollution Inspectorate.* The report also proposed appointing some *environmental health officers* as agents of the Alkali Inspectorate.

rubber crumb One of the products of the *cryofragging* of *rubber tyres*, it can be used in the rubber industry. The other products are fragments of textile and of wire reinforcement from the tyres.

rubber tyres Some 27 million tyres from cars and lorries were scrapped or reconditioned in the UK in 1974. Of these, 5.3 million provided reclaimed rubber, 4.7 million were re-treated and 1.4 million were exported. There were few if any uses for the remaining 15.6 million. If tyres are burnt, this has to be done with special care because of the sulphur and other substances they contain. Their calorific value is high, and if a safe way of burning them existed, this way of disposal could quickly pay for itself. *See cryogenic processing.*

runoff Rainwater, molten snow or other *meteoric water* that collects to follow a stream channel. It has several components, including *overland flow* and channel precipitation (falling directly on the stream). In *sewers* runoff is overland flow plus *infiltration* (2).

runoff coefficient The *impermeability coefficient.*

S

sacrificial anode In *cathodic protection*, an *anode* that is eaten away before the metal that it protects. It is electrically connected to this metal, which is lower in the *galvanic series*. The protected metal thus becomes cathodic, the sacrificed metal anodic.

sag curve The *oxygen sag curve.*

saline Salty. *Nitrates, chlorides* and phosphates are the commonest salts in *sewage* that may have to have their concentrations reduced by *advanced waste water treatment* or *de-nitrification. See salinity.*

saline intrusion, s. encroachment, sea-water intrusion Movement of salt water underground from the sea coast inland. Since salt water is denser than fresh, it passes below the fresh water, forming a 'saline wedge'. Excessive pumping of deep wells may extend the wedge inland to produce saline well water.

salinity The amount of *salts* (not only common salt but all salts) dissolved in a water. The *total dissolved solids* added for each domestic *re-use of water* (because it becomes *sewage* each time) can reach 250 mg/litre, and they can be removed only by *desalination*. Salt water can contain less *dissolved oxygen* than fresh. The salinity of the Atlantic Ocean varies from 3.3 to 3.7% and of the Dead Sea from 19 to 26%.

Salmonella Enteric bacteria closely related to *Escherichia coli* and *Shigella*. *Salmonella* may cause enteric fever, *gastro-enteritis* or salmonella septicaemia. The diseases are spread by food or drink that has been contaminated by the *faeces* of an infected person. Flies (which travel from faeces to food) may transmit the *bacteria*. Many animals, particularly pigs or poultry, may be naturally infected with *Salmonella* and their meat should therefore be well cooked before it is eaten. *See typhoid fever*.

salmonellosis Acute *gastro-enteritis* with severe diarrhoea, occasionally a *waterborne disease* but usually foodborne, caused by *Salmonella*.

salt Salts are formed when an *acid* (*see acidity*) reacts with an *alkali*. They form *ions* in water. Organic salts are called esters. Table salt is sodium chloride, NaCl. *See saline*.

saltation Jumping. In river *hydraulics*, intermittent movement downstream of particles along the bed.

sample A small part of a large quantity, one that truly represents the whole.

sampler A device that collects a sample for analysis. For accurate, consistent analyses, much depends on satisfactory sampling. Important variables include the position from which the sample is taken and the frequency of sampling. Samples can be single, random, single at specified intervals, flow-proportional (for fluid), *composite* or continuous. Considerable errors can be made in sampling. For example, obtaining a sample of water, to determine the concentration of oil in it, is difficult because the oil lies on the surface of the water. Any intermittent sampling is totally unsuitable for a water polluted by batch industrial discharges. *See automatic sampler*.

sampling train, s. assembly A sequence of equipment for the sampling of *flue gases* which includes an intake nozzle on a probe, a filter, collectors for polluting gases, a condenser, a gas flow meter, and a pump or aspirator to suck the gas through the train. The flow rate and other gas conditions are designed to be the same as in the flue, and if this is achieved, the two flows are described as isokinetic.

sand filter A layer of sand supported on a bed of gravel that usually contains *underdrains* to draw off the water quickly.

Many types exist, the *slow sand filter* being the earliest. They may be contained in open concrete tanks or even within earth walls, as in *land filtration*. Downflow or *rapid gravity sand filters* are common in *water treatment*. *Upward-flow sand filters* include the *Immedium sand filter* and the *pebble-bed clarifier*. *See also filter, moving-bed sand filter*.

sand re-circulation filter A *moving-bed sand filter*.

sand river A dry river bed beneath which there is water, common in the wadis of African and Arabian deserts, which enjoy running water only occasionally, after heavy rain. Underground dams have occasionally been built in the bed of a sand river to store water.

sand-wash filter A *moving-bed sand filter*.

sanitary engineering The US term for water and *wastewater engineering*.

Sanitary Inspectors' Association The name of the *Environmental Health Officers' Association* before 1957.

sanitary landfill US description of *controlled tipping*, probably first used in Fresno, California, in the 1930s, although the practice may have been older.

sanitary sewer US term for the UK *foul sewer*.

sanitation truck US term for refuse collecting lorry.

saprobe An organism that feeds on dead or decaying *organic* matter.

saprobic classification The *Kolkwitz–Marsson saprobic system*, a pollution classification for rivers, dividing the *recovery zone* into *oligosaprobic, mesosaprobic* and *polysaprobic*, with subdivisions. *See also river classification*.

saprobicity The effects of organic matter on life in rivers.

saprophyte A *saprobe* that feeds on soluble *organic* matter. Many *bacteria* and *fungi* are saprophytes.

saprozoic Descriptive of *saprobes*, which feed on solid *organic* matter, like most animals.

SAR *Sodium absorption ratio*.

Sarcodina *Rhizopods*.

satellite treatment The arrangement of small *sewage* plants around a city to treat mainly domestic sewage, possibly with the return of the *sludge* through a *sewer* to a central treatment plant.

saturated solution A *solution* in which the *solvent* has dissolved as much as is possible of the *solute* at the temperature in question. When a saturated solution is cooled, some solute may come out of solution and be deposited, either immediately or later. If the deposition is delayed, the solution is 'supersaturated'.

saturation flotation *See dissolved-air flotation*.

saturation index *See Langelier saturation index.*

saturator A pressure chamber in which air can be injected into water at (say) 4 atm pressure—e.g. as a part of *dissolved-air flotation.* A saturator can be either unpacked (empty) or packed with stones or plastics medium. Packing helps to break up the bubbles and flow pattern. The *detention period* is normally from 30 to 60 s.

scale, fur The whitish crust formed inside kettles or hot water systems when the *carbonate hardness* of water is deposited by boiling. It is mainly *calcium carbonate*, $CaCO_3$.

scales US term for a *weighbridge.*

scalping (1) Removal of large lumps at an early stage of *mineral dressing.* Sometimes in a *mechanical sorting plant* it may mean the early removal by *magnetic separators* of iron and steel pieces larger than about 20 cm. (2) Removing the larger particles of *dust* in an *expansion chamber* or other inexpensive device before passing the dust-laden gas to the units that remove fine dust—*electrostatic precipitators, fabric filters,* high-energy scrubbers, etc.

scavenging The use of voluminous absorbents such as sawdust, hay, straw or tannery waste (leather shavings) to soak up floating oil spills. The solids then have to be collected for burning on the beach. In the usual sense of the word, scavenging over *controlled tips* should be prevented where possible, because it is dangerous to the people who do it and, like *totting,* is illegal in the UK.

scheduled works A *registered works.*

schistosome, *Schistosoma* The main species of *fluke* which cause *schistosomiasis*—namely *Schistosoma mansoni, S. haematobium* and *S. japonica.* The eggs of the schistosome are excreted by an infected person but die if they do not reach water. They hatch to a *ciliate* larva, which must quickly enter certain freshwater snails that live in slow-moving warm water. The *larvae* develop in the snail and eventually swim about, penetrate unbroken skin and infect a human in the water. The flukes live in the blood vessels, particularly around the bladder or intestine. *Irrigation* schemes in hot climates can substantially increase their population. The life of a fluke averages 3 to 8 years but may reach 25 years. They are normally reduced by a *molluscicide* that kills the *host* snails.

schistosome dermatitis, swimmer's itch A skin trouble acquired by humans from the larvae of *schistosomes* that affect rats, birds and mice.

schistosomiasis, bilharzia A debilitating infection in which the male and female *schistosomes* inhabit the veins of humans, resulting in blood in the urine or faeces. In severe cases the *flukes* damage the liver, kidneys or urinary tract. The disease is believed to affect some 200 million people, including more than half the population in some areas, and is as much a socio-economic as a medical problem.

Schmutzdecke (German for 'dirt cover') Over a *slow sand filter*, the reddish-brown sticky coating formed on top of the sand, consisting of micro-organisms, partly decomposed *organic* matter and salts of iron, manganese, aluminium and silica. Below the Schmutzdecke for a few millimetres is an autotrophic zone where *nitrogen* and *carbon dioxide* are consumed and the oxygen emitted is taken up usefully by the water. For a depth of about 0.3 m, however, there is a further most useful *heterotrophic* zone where the organic matter is broken down by *bacteria*. It is a type of *biological film*.

Scottish Development Department, SDD The SDD is responsible in Scotland for many of the functions of the *Department of the Environment* in England and Wales.

Scottish water industry In Scotland the water industry was re-organised in a different year, 1975, and in a different way from the English and Welsh reorganisations of 1973. Water and *sewage* services are operated by the Scots regional authorities. River pollution is the concern of the seven *river purification boards*, and the sewage works must meet their *effluent* standards.

scouring organisms *Grazing fauna.*

scouring sluice, scour pipe A *washout valve* through the lower part of a dam, by which accumulated *sediment* can occasionally be expelled to prevent the *reservoir* being filled by it.

scouring velocity of sewers The *self-cleansing velocity*.

SCP *Single-cell protein.*

scrap Metals or other waste materials, usually after *reclamation*. The terminology of scrap is too vast to be covered here, but the annual handbook of the British scrap trade journal, *Materials Reclamation*, includes glossaries of classified scrap terms.

scrapers Power equipment provided at *sewage* works to scrape the *sludge* along the bottom of *horizontal-flow sedimentation tanks* or *radial-flow sedimentation tanks* to the sludge outlet. Hopper-bottom tanks of pyramidal or conical shape should not need scrapers, since the sludge gravitates to the central outlet. *See chain scraper, fixed-bridge scraper, flight scraper, rotating-bridge scraper, scum weir, travelling-bridge scraper, vee-blade scraper.*

scrap processing *See high-density baling, fragmentiser, de-tinning.*

scrap rubber *See rubber crumb, rubber tyres.*

screen (1) A sieve made of a flat sheet of wire mesh or punched steel plate for separating granular material into sizes (*classification*). It can also be made of parallel, wedge-shaped wires or bars or round wire either flat or formed into a cylinder, as in the *trommel*. In a *mechanical sorting plant* trommels are an inexpensive first step and may also be used intermediately. (2) In the treatment of *sewage* or water, a device that strains some of the solids out of the mixture of solids and water. At its most refined it is a *micro-strainer*. In a sewage works the use of a *comminutor* eliminates the need for handling the solids. At *reservoir* intakes screens are usually racks of massive bars. They should be installed parallel to flow lines, not perpendicular to them. This is expensive in screen area, but it is less likely to result in blocking, overtopping or collapse of screens. At sewage works

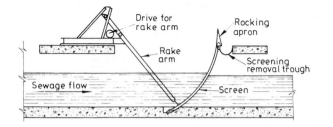

Figure S.1   Sewage screen showing mechanical cleaning arm

screens are part of *preliminary treatment*. *Coarse screens* may be used to remove large solids from *raw sewage*, to protect from damage the pumps at the works inlet. They are followed by *fine screens* to separate smaller solids before the grit removal tanks. Screens are often cleaned by a rake mechanism. They are usually designed for a velocity through the bars or mesh of 1.2 to 1.4 m/s at maximum flow. *See also band screen, cup screen, disc screen, drum screen, vibrating screen.* (3) An embankment or a wire mesh fence, up to 3 m high, built around a *controlled tip*, especially on its lee side to catch windborne refuse.

screening   The use of a *screen*.

screenings   Coarse solids removed by *screens*. In *sewage* works the screenings quickly putrefy, especially in hot weather. They are macerated (*see* below) or disposed of by tipping or *incineration*. The volume of screenings is very variable but should not exceed 5 cm³ per person per day for screens with 5 cm spacing and 25 cm³ per person per day for 1.5 cm spacing. Screenings are only 20 to 30% solids.

screenings disintegrator, disintegrator pump, macerator   A pump that chops up sewage *screenings* after removal from the flow of *sewage* so that they can be returned to it without causing obstruction. A *comminutor*, on the other hand, is in the sewage flow.

screw classifier A *mineral dressing* device resembling a *screw pump*, with a sloping trough in which an *Archimedean screw* rotates, separating the light (or fine) material in the overflow from the dense (or coarse) material in the *underflow*.

screw pump An *Archimedean screw* in a semi-circular or circular trough set at an angle of 30 to 40° to the horizontal (*see Figure S.2*). Screw pumps are efficient over a wide range of flows, and

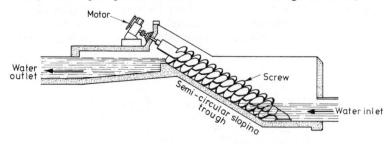

Figure S.2 Screw pump (longitudinal section)

are ideal for pumping sludge and similar substances. The greatest lift for one screw pump is about 10 m.

scrubber A *wet scrubber*.

scum, skimmings Any material that floats to the top of still water, especially in *sedimentation tanks*. In the UK *sewage* scum contains plastics, polythene and terylene film, fats, oils, greases, soaps as well as *rising sludge*. Scum should be removed regularly because it smells and may attract birds which make an even worse mess. It can block pipes and the *distributors* of *trickling filters*. Scum is usually disposed of with *sludge*. Scrapers may be designed to drag it over a *scum weir*. The insoluble precipitates formed by soap in hard water are also scum—calcium and magnesium salts of the soap substance.

scum board, s. baffle, inverted weir A board that prevents scum floating out with the stream of *effluent*. Those provided at the outlet from *sedimentation tanks* in *sewage treatment* are about 0.45 m high, projecting 75 to 150 mm above the surface and dipping about 0.3 m below it across the full width of the channel.

scum trough A trough in or beside a *sedimentation tank*, down which the *scum* flows away. For *horizontal-flow sedimentation tanks* the scum trough should be at the inlet end.

scum weir, skimming w. A *weir* with an adjustable crest that can be lowered to remove *scum*. Often a *scraper* pushes the scum over the scum weir to the *scum trough.*

sea disposal of sludge   *See ocean dumping.*

sea outfall   The Jeger report recommended that crude *sewage* should be discharged to sea only after *comminution* and *screening* and through many *diffusers* at the end of a long *outfall* in deep water, sited after a careful local investigation. In general, the initial *dilution* of the sewage at the outfall should be at least 50:1, and 100:1 if offshore currents are weak. About 12% of the sewage and a high percentage of the *sludge* from England and Wales was discharged direct to the sea with little or no treatment in 1975. Sea outfalls may be pipes on the bed or tunnels in rock. Tunnels have the advantage that ships dragging their anchors cannot break them, which often happens to pipes, but they may be too expensive or impracticable.

sea-water intrusion   *See saline intrusion.*

secator conveyor separator   An *inertial separator* consisting of a rapidly rotating wheel that throws dense objects a long way while objects such as paper fall short.

secondary digestion   *See anaerobic sludge digestion.*

secondary effluent   Effluent from *secondary sedimentation tanks.*

secondary emissions, secondary air pollution   Pollutants, such as *photochemical smog,* which originate from reactions in air polluted by *primary emissions.*

secondary main   A pipe that distributes water from a *trunk main* to street *service mains* but may also provide water direct to an industrial consumer.

secondary material, reclaim   A raw material that has been obtained by *reclamation* from waste.

secondary sedimentation tank   A *humus tank* or any *sedimentation tank* that follows *activated sludge* or other *secondary treatment. Radial-flow tanks* are invariably used except in small works, where *pyramidal sedimentation tanks* or *pebble-bed clarifiers* may be preferred. Their maximum diameter is usually 30 m. Design for radial-flow tanks is usually based on *overflow rate,* which should be about 30 to 40 m^3 m^{-2} day^{-1} unless the *sludge* is expected to be difficult to settle, in which case the overflow rate should be reduced. The *detention period* should not be less than 2 h at maximum flow. *See secondary sludge, surface solids loading rate.*

secondary shredding   A second *shredding,* after the iron and steel have been removed, probably most of the other metals and glass also. The *mill* suffers less wear than in *primary shredding.* Usually secondary shredding is to a maximum size of about 25 mm, although it may be much smaller.

secondary sludge   *Sludge* from *secondary sedimentation tanks* (*return activated sludge* or *humus sludge*). In temperate climates secondary sludge is often returned to the *primary sedimentation* tanks and is there thickened to give a *mixed sludge*. In hot climates this may be unsuitable because the secondary sludge becomes *anaerobic* very quickly and *rising sludge* is seen in the primary tanks. *Flotation* may be better than *sedimentation* for secondary sludges in warm climates. *See sludge production.*

secondary treatment For sewage this usually means *biological treatment* followed by *sedimentation*, but when biological oxidation is inappropriate, it can imply a *physico-chemical treatment*. This is preceded by *primary sedimentation* and may be followed by *tertiary treatment* or *advanced wastewater treatment*.

second-stage biochemical oxygen demand *Nitrogenous BOD.*

sediment, silt Particles eroded from the river bed or nearby land, which are carried suspended by the stream and dropped when the flow is low or in a reservoir or estuary. *See sediment load.*

sedimentation Settlement, sinking. In *sewage* or *water treatment* it means the removal of *settleable solids* in a *clarifier, grit chamber, sedimentation tank,* gravity *sludge thickener,* etc. Chemicals may be used to help sedimentation. *See coagulation, settling regimes.*

sedimentation tank, settling t. Some of the most important units in *sewage treatment*, also used in the mineral industry. They are tanks for settling the solids out of sewage or water without the help of chemicals. They can be used for *primary sedimentation* immediately after the *grit chambers* or for *secondary sedimentation* or for *stormwater. Horizontal-flow, upward-flow* and *radial-flow* types exist, as well as the now rare *quiescent sedimentation*, the modern *inclined-tube settling tank* and, in small works, *pyramidal sedimentation tanks.*

In *water treatment* sedimentation tanks are uncommon compared with *clarifiers*, but the storage of *raw water* before treatment does allow the *settleable solids* to sink to the bottom of the *reservoir*. In the USA sedimentation tanks are often called clarifiers.

sediment load All rivers or streams carry material downstream as suspended *sediment* or *bed-load,* or by both methods. The total of the two is called sediment load. *Reservoirs* are provided with *scouring sluices* to enable their sediment to be washed out periodically.

seeding, inoculation Bringing in suitable *bacteria* and other organisms to help biochemical reactions. The return of *activated sludge* is one example. To start a new activated sludge plant, the *aeration tank* may be seeded with sludge from another plant.

Digested sludge can be added to undigested sludge to seed it before digestion. Samples in the *BOD* test also need to be seeded with appropriate bacteria.

seepage   Leakage of water into or out of or through a soil or other granular material.

seepage pit   An *absorption pit.*

seiche (French for 'tidal wave')   In a lake, a tide caused by wind raising the water level at the lee end and correspondingly lowering it at the windward end. When the wind drops, the water flows slowly back to the windward end and returns with diminishing amplitude. It helps to mix the water.

selective medium   A *medium* which favours the growth of a particular group or species of micro-organisms to the exclusion of others—e.g. *MacConkey's broth* for growing and identifying bacteria of the *coliform group.*

selenium, Se   A *micro-nutrient* that may counteract the toxicity of *arsenic, cadmium* and *mercury.* Above 2 mg/litre in water, however, it can cause the illness selenosis. Selenium is used in almost all types of paper and is therefore present in municipal refuse at 1 to 5 mg/kg. *See* Table of Allowable Contaminants in Drinking Water, page 359.

self-cleansing   Description of the power of the environment to remove pollution. For example, in the atmosphere large particles drop to earth, entering the topsoil, and smaller ones are caught in leaves or buildings or by rain or are removed by wind. *See also self-purification.*

self-cleansing gradient   For each pipe diameter a range of slopes at which *sewage* carries all solids with it. This should occur daily during the peak flow. *See* below.

self-cleansing velocity, scouring v.   The speed of water flowing down a *sewer* which sweeps all solids away. At lower speeds some solids are stranded and the sewer may become *septic. Organic* matter is kept suspended at speeds of 0.6 m/s or more, but sand and grit need a minimum speed of 0.75 m/s. *See* above.

self-digestion   *Autolysis.*

self-purification, natural purification   The ability of natural water to purify itself of *sewage* or other wastes. It is possible because *bacteria* and other micro-organisms in the water feed on sewage and reduce its concentrations while using up the *dissolved oxygen*. If the self-purification capacity is exceeded (i.e. the water is overloaded with sewage), the dissolved oxygen decreases substantially. Fish disappear early in this state of pollution and eventually all the *oxygen* disappears from the water. As the river recovers, fish are among the last creatures to re-appear. *See oxygen balance.*

semi-permeable membrane A skin or sheet (membrane) that allows *solvent* to pass through it but not *solute* (dissolved salts, etc.). It often occurs in nature, and is used in *osmosis, reverse osmosis,* dialysis, etc. *Electro-dialysis* makes use of *ion-exchange membranes*.

separate system, rigidly s.s. Two separate *sewer* systems, one taking rainwater, the other taking domestic *sewage* and trade wastes. Most modern developments are separately sewered so as to reduce river pollution. The rainwater sewer discharges into the nearest stream without pollution from sewage, unlike the *combined system*. In general, separate systems are more expensive than combined systems because of the extra pipes. *See also partially separate systems.*

separation/incineration, s. and i. Salvage of iron and steel and some other materials before *incineration,* superseded by the modern method or *direct incineration*. The method was common when batch incinerators were in use.

separation weir A *stormwater overflow.*

Sepedonium A *fungus* that grows on *trickling filters* and has a lower optimum temperature than most other active micro-organisms in the *biological film*. It may reach a thickness of 2 cm in winter and thus contribute to *ponding*.

septicisation *De-oxygenation* of *sewage* because of *oxygen* uptake by *aerobic, heterotrophic* micro-organisms that transform the *organic* substances and remove all the oxygen while purifying the sewage.

septic sewage *Fresh sewage,* that has become *stale sewage* because it has had no treatment or added *oxygen,* eventually stinks because of the *aerobic* organisms in it that have reduced the *sulphates* and *organic compounds* of sulphur to *hydrogen sulphide* and *mercaptans.* Warm weather and long travel in *sewers* increase the likelihood of sewage becoming septic. *See sulphide corrosion.*

septic tank A covered, horizontal-flow tank (*Figure S.3*) that involves a series of treatments and is now used only for isolated dwellings which are uneconomic to connect to a *sewage treatment* system. The minimum volume of a septic tank in litres is $180N + 2000$, N being the number of people served. Where *garbage grinders* are used, the factor 180 should become 250, and for schools or other intermittent uses it should be 90. A septic tank is normally at least 1 m wide, with a length:width ratio of 3:1 to 2:1 and a minimum depth of 1.5 m. The tank is often divided into two sections, the first being twice as long as the second. The solids settle to the floor and undergo anaerobic digestion. Some of the gas evolved lifts the *sludge* and forms a

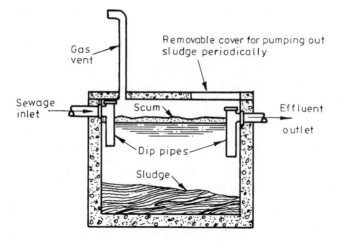

Figure S.3 Septic tank, single compartment type

scum which traps other floating solids. The inlet and outlet *weirs* should be below the scum level. The *effluent* from a septic tank is very variable, with a BOD_5 of 100 to 300 mg/litre and *suspended solids* from 70 to 150 mg/litre. In general, one hopes that the effluent has about 50% less *BOD* and 70% less suspended solids than the *raw sewage* influent. The poor BOD reductions are partly caused by *hydrolysis* of the sludge particles. Further treatment by *subsurface filtration* or *subsurface irrigation* or *trickling filter* may be needed. Septic tanks should be de-sludged once a year and about one-sixth of the sludge should be left to *seed* the new sludge. The sludge usually contains 10% or more solids (BSCP 302).

sequestering *See complexing.*

service main The water (or gas) *main*, buried in the street, from which domestic consumers draw their water or gas.

service pipe The pipe connecting the *service main* to the consumer.

service reservoir, distribution r., clear-water r. A *reservoir* that is supplied, often by a gravity *trunk main,* from the *water treatment* plant. It supplies water to the consumers in one *pressure zone.* Unlike an *impounding reservoir*, it stores treated water and is therefore roofed to prevent pollution by birds or airborne material. A water tower is a service reservoir or tank raised above ground level. Service reservoirs usually contain 1 to 3 days' storage for the zone they serve. This allows for interruptions to the supply from the water treatment plant in the event of a burst trunk main, etc.

seston The smallest *plankton*, living or dead. It adds *turbidity* to water.

settleability Ease of settling. An indication of the ability of a *secondary sludge* to settle in a *sedimentation tank* can be obtained by a *settling column* analysis or the *sludge volume index* or the *stirred specific volume* or the *sludge density index.*

settleable solids The solids in *sewage* which are large or dense enough to settle in the time that is allowed for the sewage to pass through the *sedimentation tank*. The *colloids* or lighter particles remain in the *effluent* and do not settle. Settleable solids are measured in an *Imhoff cone. Compare suspended solids.*

settled sewage The *effluent* from *primary sedimentation tanks.*

settlement *See compaction of landfill, sedimentation,* and below.

settlement plate A square of asbestos cement or a glass microscope slide that is fixed on a jetty pile, buoy or other fixture, with others over a wide area below or partly below water level. They are regularly harvested and replaced. Growths on the plates are recorded and indicate the quality of the water and its life.

settling column A unit of laboratory equipment for measuring settlement rates in a *suspension*, usually a vertical tube with tapping points at intervals. To avoid interference between the particles and the tube wall, the tube diameter should be at least 100 times the diameter of the largest particle.

settling flux *Solids flux.*

settling module *See inclined tube settling tank.*

settling regimes Particles in a liquid settle in four main ways, important in both *sewage treatment* and *water treatment*:

(1) The discrete settling of granular particles that do not flocculate and are not hindered by interaction with other particles. They have a constant settling velocity, calculated from *Stokes' law,* but only if they are spherical.

(2) The settling of *flocculent* particles, in low concentrations. The *flocs* grow and therefore their settling velocity may increase but the forces on the settling floc may break it up. Hence, the floc is continually re-forming.

(3) Zone settling. Individual flocs are close, interfere with one another, and the mass of particles settles as a block.

(4) Compression settling. At high concentrations of solids the particles are in contact with one another and settlement is by compaction of the particles because of the weight of those above them. Any downward movement of floc corresponds to an upward movement of water, which slows down settlement.

The *sludge blanket* in *sedimentation tanks* or *clarifiers* consists of zone settling and compression settling areas. 'Hindered settling' is a loose term applied to all types except (1) above.

settling tank   A *sedimentation tank, clarifier, thickener,* etc.

sewage   The water supply of a town after it has been used. Most municipal sewage is about 99.9% water and 0.1% impurities. In the UK a domestic sewage after *screening* and grit removal contains about 60 g of BOD_5 (in the USA 80 g BOD_5) and 80 to 90 g *suspended solids* per person per day. In calculations the flow of foul sewage is usually considered to be the *dry-weather flow.* The *bacterial count* varies from 100 000 per ml in fresh sewage to 10 or 100 million in *stale sewage. Fungi* and *protozoa* are also present, sometimes in the dormant stages of *spores* or *cysts.* The sewage flows away in *sewers* to the treatment plant or *outfall. See* below; *see also raw sewage, septic sewage.*

sewage disposal   In the nineteenth century *sewage* was often discharged untreated into the sea or the nearest river, which resulted in serious pollution. Towns now normally use *sewage treatment* before disposal, but the ultimate disposal of the *effluent* is always by *dilution* in a river or the sea.

sewage effluent standards   *Effluent* standards at conventional *sewage* works normally quote the *BOD_5* and *suspended solids,* sometimes also the *ammoniacal nitrogen. See advanced waste water treatment, Royal Commission effluent, trade effluent.*

sewage farm   Land on which *sewage* or sewage *effluent* is poured, and where crops are grown, using some *land treatment.*

sewage field   A sewage *slick.*

sewage fungus   A greyish, slimy mass of *filamentous organisms* which can be seen growing on rocks or the banks of water containing *sewage.* The name is misleading, because sewage fungus often includes *bacteria* such as *Sphaerotilus natans* and *protozoa* such as *Carchesium,* as well as true *fungi* such as *Leptomitus* or *Geotrichum*; but it may consist of only one or two species of bacteria or fungi.

sewage sick   A description of land that has had too much *sewage* or *sludge* over it. Its fertility drops and it may also be smelly.

sewage sludge   *See sludge.*

sewage smells   Sewage smells arise partly from the *anaerobic* degradation of *sulphates* in water to *hydrogen sulphide* by the *sulphate-reducing bacteria. Proteins* in sewage also stink when they break down in water to volatile *aldehydes*—e.g. butyric acid, *mercaptans* and skatole. *See odour.*

sewage treatment   (*Figure S.4*)   The separation of the 0.1% of impurities from the 99.9% water in *sewage* so far as it is economically possible. Yet however good the sewage treatment is, none is ever complete. The *effluent* is further treated in the

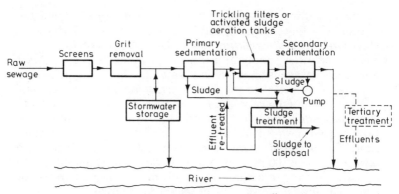

Figure S.4 Sewage treatment, flow diagram

receiving water by *self-purification* or at the waterworks if it is withdrawn downstream. It includes the *preliminary, primary* and *secondary treatments* given at most plants, with *tertiary* or *advanced wastewater treatments* at those with a strict effluent standard. All treatment starts with *solid−liquid separation*. BSCP 302 describes small treatment works.

sewer A *culvert* or buried pipe that leads away domestic and trade waste waters for treatment and disposal. Sewers may be in *combined systems* or *separate systems* or *partially separate systems*. In the USA sewerage is described as waste water collection and a *foul sewer* there is called a sanitary sewer. Sewers in cities were originally only culverts to allow traffic to pass over streams, whether in ancient Rome, Paris, London or elsewhere. In London *sewage* was excluded from them until 1815. Early sewers were of brick or stone, since pipes for sewerage did not exist, and became *combined sewers* as soon as house drains were allowed to be connected to them. Apart from ancient Rome at the time of Christ, waterborne sewage systems existed in India around 4500 B.C. and near Baghdad about 2500 B.C. Modern sewers are usually circular, although very large rectangular sewers have been built. In the UK pipes are normally of concrete, asbestos cement, cast-iron, PVC or vitrified clay. They are laid at a slope which will achieve a *self-cleansing velocity* at least once in 24 h. To avoid excessive wear of the pipe material, the velocity of flow should not exceed 2.5 m/s. Foul sewers should accommodate 4 to 6 times the *dwf* when flowing full, without *surcharging*. Combined sewers should take a once-in-3 years storm without surcharging. *See branch sewer, egg-shaped sewer, intercepting sewer, manhole, trunk sewer.*

sewerage A network of *sewers* or the science of sewers. BSCP 2005 describes sewers, *storm overflows, siphons, sewage*

278

pumping and tidal *outfalls*.

sewer chimney   US term for a *back drop*.

sewer corrosion   *See sulphide corrosion.*

sewer flushing   If a *sewer* is not regularly *self-cleansing*, a flushing device should be built into it, often with a tank and *dosing siphon*.

sewer gas   Any gas in *sewers*, including petrol vapour, town gas, *carbon monoxide, carbon dioxide, hydrogen sulphide, methane* or combinations of them.

sewer ventilation   Ventilation of *sewers* is essential to stop *sewer gas* building up and to prevent air locks in them. In temperate climates the *sewage* is usually warmer than the outside air. In hot climates the sewage is cooler than its surroundings and forced ventilation of sewers is often needed. *Odour control* may be needed at the ventilation shafts.

shaft furnace, s. kiln, vertical s.f.   A furnace resembling a *cupola*, tried in USA for *total incineration* but not used in Britain for *incineration*. Like cupolas, shaft furnaces need re-bricking every year at least at the melt zone and complete re-bricking every 10 years or so.

shale heap   *See spoil heap.*

shallow-depth sedimentation   The use of *inclined-tube settling tanks, lamella separators,* etc., instead of conventional deep *sedimentation tanks*.

shears   *See alligator shears.*

shellfish   Oysters, mussels or other shellfish from *sewage*-polluted water can habour *pathogenic* organisms that may cause, if they are eaten, *typhoid fever,* diarrhoea, *gastro-enteritis* and other *waterborne diseases*. After harvesting from polluted waters, shellfish may be completely purified of bacterial hazards by leaving them first in chlorinated water and then in tanks of clean water. However, it is still possible for them to contain *heavy metals* in proportions much higher than the surrounding water. Some shellfish act as an intermediate *host* to *parasites* that harm humans—for example, *flukes*.

shellfish waters   In *shellfish*-growing waters the *US PHS* requires a median *most probable number* below 70 coliform bacteria per 100 ml of water. Not more than 10% of the samples may ordinarily exceed 230 coliforms per ml.

shigellosis, bacillary dysentery   A common cause of diarrhoea in humans; at least 30 types of the bacterium *Shigella* exist. This enteric bacterium, closely related to *Escherichia coli* and *Salmonella*, is transmitted by *faecal* contamination of food or water.

Shone ejector   A *pneumatic ejector*.

shredder A *mill* that cuts and tears solids put into it—e.g. confidential documents. The term has been loosely applied also to mills for *shredding*, and in the USA specifically is a *comminutor.*

shredding, pulverising Size reduction of the large pieces in municipal refuse by crushing, beating, grinding, milling, rasping, shearing, etc., either wet or dry in some sort of *mill.* Shredded refuse can be acceptable where a tip for raw refuse would certainly be rejected. It can be converted to *compost* in areas where topsoil is scarce. Shredding is essential to fast *mechanical composting* and may often precede *controlled tipping.* Decomposition is faster with shredded than with raw refuse. The same compacting force after shredding reduces the refuse about 30% more than before shredding. *Magnetic separators* for iron and steel work more easily on shredded refuse, so they are often used in this way. Shredding is expensive; consequently, only 5% of UK refuse is shredded. Some of the arguments in favour of shredding are countered by the very high compaction claimed for steel-wheeled *compactors.* The size reduction may be to a maximum size of 10 or 15 cm in one stage or to 3 cm or less in two stages or more. Shredding machines are of three main types: fast *hammer mills*, slow *drum pulverisers* and the supplementary *alligator shears* or other powerful machines for breaking up bulky wastes. Some preliminary sorting of refuse is helpful if blockages are to be avoided. Energy consumption increases in single-stage shredding as the final maximum size of the refuse diminishes—e.g. roughly 10 kWh/tonne for 15 cm final maximum size or 15 kWh/tonne for 8 cm final maximum size. *See primary shredding, secondary shredding, two-stage shredding.*

shredding machine A *mill.*

shrimps *See Gammarus pulex.*

sick activated sludge An *activated sludge,* that does not react as it should, appears dark brown or black and may be difficult to settle. It may become sick because of toxic chemicals in the *sewage* or too little air going into the *aeration tanks.* A sick activated sludge may suffer from *bulking.*

side-weir overflow A common *stormwater overflow* with a *weir* parallel to the pipe or *sewer.* When the weir crest is below the horizontal diameter, it is a 'low' side-weir overflow; above, it is a 'high' side-weir overflow.

siemens, mho Two names for the unit of *electrical conductance.*

sieves *See mesh, screen.*

sigmoid growth curve A graph showing the numerical rate of growth of a species of *bacteria* for a given food input (*see Figure*

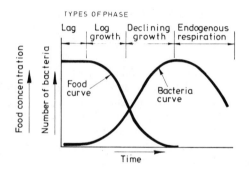

TYPES OF PHASE

| Lag | Log growth | Declining growth | Endogenous respiration |

Food concentration · Number of bacteria

Food curve

Bacteria curve

Time

Figure S.5   Sigmoid growth curve

S.5). Sigmoid means ∫-shaped. Although this curve can describe the growth of a single species of bacteria for a given food input, it is inaccurate to use it to describe population growth in complex microbial systems such as *activated sludge*. *See growth phases of bacteria.*

signature methods   Methods of separation of minerals or refuse which involve the use of a mini-computer to recognise the signature of the material in question. *Impact deceleration* and *infra-red spectroscopy* are two methods being developed, but they were not beyond the research stage in 1978.

silica   Silica is found in most hard waters up to about 40 mg/litre, at about 1 mg/litre in moorland water, and in river waters intermediately. In boiler *feed water* it is undesirable because of its very hard *scale.*

silicon carbide   A *refractory* that provides good protection to a combustion chamber wall at *grate* level in an *incinerator.* It can also be used to protect a *water wall* that is liable to be excessively corroded.

silt, silting, siltation   The fine material deposited from the *sediment load* of a stream.

silver   In the USA silver has been deposited on grains of *activated carbon.* The *bacteria* are killed by the silver as the water passes through the pores. The method may perhaps be used where *chlorination* is doubtful. The limits to silver stated in the Table of Allowable Contaminants in Drinking Water (page 359) are probably set to prevent argyrosis, an infirmity which disfigures the complexion. *See filter candle.*

Simater sand filter   A *horizontal-flow sand filter.*

Simcar aerator   A *cone aerator* which injects air into *activated sludge,* somewhat resembling the *Simplex system.*

281

Simplex system A vertical-shaft *surface aerator* for *activated sludge,* developed in 1920 by J. Bolton at Bury, Lancs. The *mixed liquor* is drawn up to the surface by a Simplex *cone aerator.*

Simuliidae Blackfly such as *Simulium.*

Simulium A genus of blackfly or buffalo gnat that both transmits *Onchocerca volvulus* to humans and receives it from them when it bites and drinks human blood. The *Simulium* larva lives on rocks in torrential water in warm climates. When a river is dammed, many of the rapids are drowned, thus reducing the breeding grounds of *Simulium*, but increasing the likelihood of *schistosomiasis.*

single-cell protein, SCP Foods synthesised by *yeasts, bacteria, algae* and in chemical engineering. Bacteria are single-celled, as also are many algae and *fungi*. Some of them can become food—for example, yeast and *Spirulina maxima*, the edible alga from Lake Chad.

single filtration, single-stage f. One pass of waste water through a *trickling filter*, without *re-circulation* or *two-stage filtration.*

sink and float treatment *Float and sink treatment.*

sinking of oil slicks *See hydrophobic chalk.*

sinuous flow *Turbulent flow.*

siphon A pipe shaped like a 'U' upside down (∩). *Inverted siphons,* U-shapes, are used in *dosing siphons* and the *siphon overflow* and in a *sewer* to ensure *sewer flushing.*

siphon overflow A type of *storm overflow* through an inverted *siphon.*

site licensing *See licensing of sites.*

site ranking *See ranking of landfill sites.*

size distribution of particles *See particle size distribution.*

skimming Removing floating material, not only *scum,* from the surface of water, such as solids (wood) or liquids that do not mix with water or semi-liquid greases, which are removed in a *flotation tank* or skimming tank.

slabbing Slicing flattened cars into slices, producing a contaminated steel, of less value than a shredded scrap car.

slag Ash that has completely melted, unlike *clinker.*

slagging incineration *Total incineration.*

slaked lime *See lime.*

slaughterhouse wastes *See meat-producing wastes.*

sleek *See slick.*

sleeping sickness, trypanosomiasis A disease caused by trypanosomes, an order of flagellate *protozoa* that includes *Trypanosoma zambiense* (or *gambiense*) and *T. rhodesiense*. It is widespread in Africa and is transmitted by the bite of the tse-tse fly, which lives in waterside vegetation. The protozoa live in the bloodstream of the *host* and produce *toxins* that affect the central nervous system. The disease can be controlled by clearing the bush round waterholes so as to destroy the *habitat* of the fly.

slick, sleek A patch of oil, *scum, sewage* or other flotsam on a water surface. To avoid sewage slicks, a *diffuser* is fitted to a modern *outfall.*

slime *See biological film.*

slope In *microbiology,* a small quantity of *medium* that has solidified in a stoppered incubating bottle tilted so that it is nearly horizontal, which results in a large surface for *cultures.*

sloping-trench landfill *See trench method.*

sloughing off, unloading Dropping of *humus* from *trickling filters* or of *biological film* from *rotating biological discs,* common in spring because of the revival of *insects* that have been dormant.

slow sand filter (*Figure S.6*) One of the earliest *filters* for water treatment, developed in 1829 by James Simpson. It has a sand layer about 0.7 m deep over a layer of gravel about 0.1 m deep, increasing in grain size towards the bottom. Perforated concrete or clay underdrains laid in the gravel remove the filtered water. The sand is drowned under about 1.2 m of water. The filter is cleaned every few weeks or months by draining the water and scraping off the top 15 to 25 cm of sand. This is repeated until filtration is no longer efficient, and the sand bed is then renewed. In *sewage treatment* slow sand filters can remove from a *Royal Commission effluent* about 40% of the BOD and 50% of the *Escherichia coli*, provided that the flow rate does not greatly exceed 1 m/day. Double to treble this rate can be achieved with a filtered *raw water* for drinking. The D_{10} *effective size* of the sand is 0.2 to 0.4 mm, with a *uniformity coefficient* of 1.6 to 2.5. *Sand filters* are valued because they can remove taste and smell as well as *bacteria. See Schmutzdecke.*

sludge Solids settled out from water, but containing from 55 to 99% water. Apart from the many industrial and waterworks sludges, there are several types of *sewage* sludge, most of which

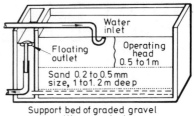

Figure S.6 Slow sand filter (From 'Water Treatment and Sanitation; simple methods for rural areas' Intermediate Technology Publications Ltd., London)

283

have over 95% water before thickening. *See activated sludge, digested sludge, humus sludge, mixed sludge, primary sludge, raw sludge, secondary sludge, sludge diposal* and below.

sludge age   The *mean cell residence time* of *activated sludge.*

sludge blanket   The mass of *sludge* in a *sedimentation tank* or *clarifier.* The bottom of the blanket may rest on the bottom of the tank or it may be a *suspension,* as in the *sludge blanket clarifier.* The *settling regime* of the sludge blanket is either zone settling or compressive settling.

sludge-blanket clarifier, upward-flow clarifier, upward-flow floc-blanket clarifier   A type of *solids contact clarifier* in which the inlet flow passes up through the suspended *sludge blanket,* as in a *Pulsator* or *hopper-type clarifier.* The particles from the coagulated water are 'filtered' out by and amalgamate with the blanket. When the sludge level rises too high, some sludge is wasted from the tank. The top of the sludge blanket is at least 1.0 m beneath the top water level. The clarified water flows over *weirs* at the surface of the clarifier.

sludge cake   *Sludge* that has been thickened to 80% water ceases to flow, can be picked up with a shovel and is called cake. *Filter presses* usually can reduce *sewage* sludge to 65% water, *centrifuges* to 75%, but *vacuum filters* to only about 80%.

sludge circulation clarifier   A *solids recirculation clarifier.*

sludge conditioners, s. conditioning   *See chemical conditioning, conditioning.*

sludge density   *See specific gravity.*

sludge density index, Donaldson s.d.i.   A measure of the *settleability* of *activated sludge.* The index is equal to 100 times the reciprocal of the *sludge volume index*—i.e. 100/ sludge volume index.

sludge de-watering   *See de-watering of sludge.*

sludge digestion   Usually *anaerobic sludge digestion* (*see Figures A.6 and S.7*), a treatment that stabilises *raw sludge.* Fully *digested sludge* has little readily *biodegradable* organic matter. It

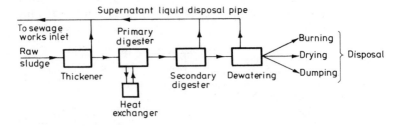
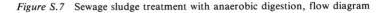

Figure S.7   Sewage sludge treatment with anaerobic digestion, flow diagram

is not smelly and about 50% of the solids are inorganic. Sludge can also be digested aerobically. *See aerobic digester.*

sludge disposal The five common ways of disposing of *sewage* sludge are:

(1) On farm land to improve the soil.
(2) As fill on low-lying land or lagooning on land.
(3) Dumping into the sea.
(4) Use in industry.
(5) *Incineration.*

Sewage *sludges* usually improve land. The dry solids in liquid *digested sludge* or *raw sludge* contain about 7% *nitrogen* and 5% *phosphorus.* Thickened digested sludge contains less *nutrients* than thickened raw sludge because more of the nitrogen and phosphorus are lost in the liquor from thickening. The raw sludge also contains fibres which may help to condition the soil. *Heavy metals* or harmful organisms in sewage sludge may hinder its use on land. Heavy metals precipitate in the primary tanks and accumulate in the sludge. Although many heavy metals are essential *micro-nutrients,* they are poisonous in excess and may render the soil infertile. Therefore, admixtures of industrial wastes may spoil a domestic sewage for farmland. The risk of transmitting microbes in sludge on farmland is not fully understood. In general, the land should lie fallow for at least 6 months after it has been treated with raw sludge. Fully digested sludge, however, can be spread on arable land after harvesting crops. There should be no smell from digested sludge but raw sludge can be foul-smelling. If the sludge can be tipped on land, it is cheapest to tip thickened raw sludge for populations around 20 000, although for cities of 500 000 or more it is probably more economic to buy de-watering equipment and thus to reduce the volume and mass of wet sludge to one-third, and, hence, transport costs. *Ocean dumping* is practised by many large British and US cities. The sludge may be pumped several miles from the sewage works to the sludge boat docks. The dumping ground must be far enough from the shore for the sludge not to be washed on to the beaches. The main pollution problem is the accumulation of heavy metals or persistent *organic compounds* in the plants or animals in the dumping zone. *See polychlorinated biphenyls.*

sludge drying *See* below, and *de-watering of sludge, heat drying of sludge.*

sludge-drying bed An open area for drying *digested sludge,* consisting of rectangular, concrete-floored *lagoons* surrounded by walls 0.9 m high and provided with *land drains* over the

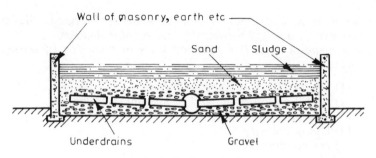

Figure S.8 Sludge-drying bed

concrete (*Figure S.8*), sometimes also with a **decanting valve.** The bottom layer of the bed, covering the drains, is 0.3 m of **clinker** or stone of 25 to 38 mm. A *filter* is formed by covering it with 5 to 7.5 cm of finer clinker or stone of 6 to 13 mm. This may be overlain by 10 cm of sand. A depth of about 25 cm of **sludge** is allowed to flow in. When the drying sludge is removed by spade, at up to 55% solids, the top layer of fine sand or clinker is removed with it and this has to be replaced. In the UK each bed is used only about six times a year because of the time taken for the sludge to dry, particularly in winter. Drying beds occupy a large area of land, which may be expensive. The area needed can be from 0.15 to 0.25 m² per person. Consequently, *mechanical de-watering of sludge* may be preferred. However, recently in the USA sludge drying beds have returned to favour for small **sewage** works. The dried sludge is removed mechanically, with a rubber-tyred front loader with a fork to minimise the loss of sand or clinker during loading. To enable the weight of the power shovel to be carried without crushing the **underdrains,** either concrete strips must be laid into the bed at a spacing equal to the track of the loader or the machine must be fitted with balloon tyres. *Raw sludge* is too smelly for open drying beds.

sludge flotation   *See dissolved-air flotation.*

sludge freezing   *See freezing* (2).

sludge gas, digester g.   Gas produced from *anaerobic sludge digesters* contains **methane,** CH_4, *carbon dioxide,* CO_2, and sometimes small quantities of *nitrogen,* hydrogen or **oxygen.** The gas generated is usually collected and used on the plant in heat exchangers to heat the digesters or the offices. In summer more gas may be produced than is needed, so the excess gas is often flared. Sludge gas is stored in a gasometer or in the space above the anerobic digestion tank under the floating roof. *See methane fermentation.*

sludge growth index   *See sludge production.*

sludge hopper   The lowest part of a *sedimentation tank*, where the settled *sludge* collects, either by flowing or being scraped in. This deep part is at the centre of an *upward-flow tank* or *radial-flow tank* and usually at the inlet end of a *horizontal-flow tank*. Its capacity, in *primary sedimentation* in the UK, should be 1.4 litres per person, which should provide enough space for 24 h *sludge production*. With sides sloping at 60° to the horizontal, it should be deep enough to ensure some consolidation of the sludge by compression settling. *See settling regime, sludge withdrawal.*

sludge lagoon   A pond where (waterworks or) *sewage* sludge is allowed to digest and thicken, without its gas being collected. If land is plentiful and cheap, the method is economical but lagoons can be extremely smelly if sewage *sludge* is undigested. Some water evaporates but *underdrains* are not usual.

sludge loading rate, SLR, food-to-micro-organism ratio, F:M ratio, organic loading   The number of kg of BOD_5 per day applied to an *activated sludge* plant per kg of (dry) activated sludge under *aeration*. It is:

$$\text{SLR} = \frac{\text{BOD}_5 \text{ of sewage (mg/litre)} \times \text{ flow of sewage (m}^3/\text{day})}{\textit{MLSS} \text{ (mg/litre)} \times \text{ tank volume (m}^3)}$$

Thus, the dimension of SLR is kg kg^{-1} day^{-1} or simply per day. SLR is a common design criterion for activated sludge plants. At SLR less than 0.4 per day, more than 90% of the BOD_5 should be removed. To obtain *nitrification*, a lower SLR is needed. Conventional plants are designed at 0.3 to 0.35 per day and *extended aeration* plants at 0.05 to 0.15 per day. In the USA the SLR is often calculated from the *MLVSS* and not from the MLSS. *Compare mean cell residence time; see sludge production.*

sludge pressing   The use of *filter presses* or *filter-belt presses* for de-watering sludge.

sludge production   (1) The mass of *primary sludge* accumulated per day in *sewage treatment* is estimated by multiplying the *raw sewage* flow by its concentration of *suspended solids* and by the percentage removal of solids in the *primary sedimentation* tank (typically 60%). In the UK, primary sludge production is about 50 g per person per day. (2) *Secondary sludge* production is often stated as the number of kg of sludge produced per kg BOD_5 applied to the *secondary treatment*, a figure sometimes known as the sludge growth index. For standard-rate *trickling filters* it is about 0.7. For *activated sludge* it is about 0.75 at a *sludge loading rate* of 0.4 per day and drops to 0.4 at a SLR of 0.1 per day. Standard-rate trickling filters produce about 30 g of humus sludge per person per day. Activated sludge production varies from 30 g per person per day for conventional plants to 15 g per person per day for *extended aeration*. (All sludge figures are now stated as dry solids.)

sludge pumping *Sludges* with 4% solids, when pumped, involve friction losses in the pipes of 50% more than when clean water is pumped. At 10% solids, pumping becomes impracticable because of high friction losses. *Reciprocating pumps* are used for moving thick sludges.

sludge re-circulation clarifier A *solids recirculation clarifier*.

sludge re-cycle line The pipe connection between the secondary settling tank and the *aeration tank*, along which the *return activated sludge* passes.

sludge respiration The absorption by *sludge* of *oxygen* from solution in the water around it. *See respiration rate*.

sludge return ratio, s. re-circulation ratio In an *activated sludge* plant, the rate of flow of the *return activated sludge* expressed as a percentage of the average flow rate of *sewage* through the plant.

sludge stabilisation *Sludge digestion*.

sludge thickening, gravity t. The difference between two *sludges* that are, respectively, 98 and 94% water is that the first, at 2% solids, occupies three times the volume of the second for the same weight of solids. Therefore, thickening gives generous reductions in volume for apparently small increases in solids content. Sludge may be thickened by sinking (gravity) or by *flotation*. Gravity thickeners usually have a *picket-fence stirrer*

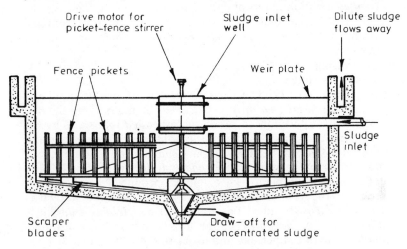

Figure S.9 Sludge-thickening tank with picket-fence stirrer

288

(*Figure S.9*) which slowly rotates, helping the sludge to settle and thicken. Thickener design is usually based on solids loading (mass of dry solids per day per m² of surface area). Typical gravity thickening figures for domestic sewage are as given in the table.

	Percentage solids		
	inlet sludge	thickened sludge	Solids loading (kg m^{-2} day^{-1})
Primary sludge	3 to 6	7 to 10	100
Humus sludge	2 to 4	6 to 8	40
Waste activated sludge	0.5 to 1	2.5 to 3.5	20
Mixed 50:50 primary and activated sludges	3 to 4	6 to 8	50

Gravity thickeners are tanks 3 to 5 m deep, with a *detention period* from 1.5 to 2.5 days. Therefore, they can act also as storage tanks for subsequent treatments. In warm climates sewage sludges may produce so much gas that thickening is prevented. Waterworks sludges of metal hydroxides can be thickened up to 2.5 to 3% solids at a solids loading of 10 kg m^{-2} day^{-1}. *See deep-cone thickener, dissolved-air flotation.*

sludge treatment cost About half the total cost of *sewage treatment* is incurred in treatment and disposal of *sludge—sludge thickening,* digestion and *de-watering of sludge.*

sludge volume index, Mohlman s.v.i., SVI A measurement of the *settleability* of *activated sludge* made in a one-litre graduated cylinder:

$$SVI = \frac{\text{settled volume of sludge in 30 min (\%)} \times 10\ 000}{\text{MLSS(mg/litre)}}$$

For conventional activated sludge plant, the SVI should be less than 150 and not more than 300, otherwise the sludge will not easily settle in *secondary sedimentation.* However, for plants running at a high *MLSS* these values are reduced. For example, if the MLSS is about 10 000 mg/litre and the sludge does not settle at all in 30 min (i.e. settled volume is 100%) the SVI is still 100. Therefore, comparisons of SVI values can be very misleading. *See stirred specific volume.*

sludge withdrawal Sludge is withdrawn from *sedimentation tanks* under its hydrostatic head through a pipe, often using a *telescopic valve.*

slug *See plug flow.*

sluice valve, gate v. A plate or wedge perpendicular to the line of the pipe, which restricts or stops the flow through it. It is opened or closed by screwing in or out.

slurry A fluid mixture of water with a powder such as cement, or a *tailings* slurry from *mineral dressing*.

slurry transport of solid waste A US proposal to pump solid wastes in a mix containing about 6% dry solid, carried in water or *sewage*. Air, injected at intervals along the pipes, would convert the pipe into a biological reactor. Such waterborne transport could be economic for quantities and distances like those needed from a large city to a remote landfill.

smell *See odour control, odour threshold.*

smog Smoky fog. *See London smog, photochemical smog.*

smog alert In Los Angeles county, USA, there are three levels of smog alert, to warn against *photochemical smog*. The first is a forecast of smog. A stage 2 alert is announced when the *ozone* concentration reaches 0.35 p.p.m. At this stage generating stations are asked to burn natural gas instead of oil, and people are asked to reduce electricity consumption and to share cars going to and from work. Important sources of pollution have to close down. At 0.5 p.p.m. there is a signal for emergency action but this level had not been reached by 1978. A maximum background concentration for ozone is 0.05 p.p.m., although at the altitudes usual for supersonic flight the air contains 2.5 p.p.m. O_3.

smoke *Flue gas* after it has left the chimney. It includes suspended *dust, grit, flyash,* tar, vapour, soot, *carbon dioxide, sulphur dioxide, oxygen, nitrogen,* etc., and rarely a coloured gas such as *nitrogen dioxide. See gas-cleaning plant.*

smoke-density meter, opacity m., smoke photometer An instrument that can be installed at a *flue* to record the darkness of the *smoke* that passes through it. The law may require one to be installed. A light ray passes through the smoke on to a photo-electric cell.

snow fence A light fence of vertical pales nailed to alternate sides of the fence, thus leaving gaps for wind to pass through but holding back snow, windblown paper, plastics, etc. It is a convenient *screen* for putting on the lee side of a *controlled tip* to catch windborne refuse.

SO_x Any mixture of *sulphur dioxide*, SO_2, and sulphur trioxide, SO_3. In power station *flue gases* usually only one part of SO_3 is formed per 40 to 80 parts of SO_2. SO_2 has a short lifetime in air, for many reasons. It can be washed out directly by rain, caught on vegetation or, in the presence of metal dust, converted to SO_3.

In damp air it reacts with *ammonia* to form ammonium sulphate, $(NH_4)_2SO_4$. All these substances are very soluble and easily taken up by water.

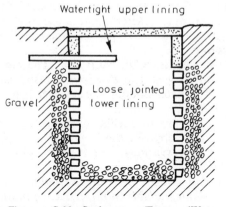

Figure S.10 Soakaway (From 'Water Treatment and Sanitation; simple methods for rural areas' Intermediate Technology Publications Ltd., London)

soakaway   *See absorption pit, subsurface irrigation.*
soda ash   Commercial sodium carbonate, used in *water softening* during the *lime-soda softening of water.*
sodium absorption ratio, SAR   A pointer to the effect of the sodium in a water on the soil and crops:

$$SAR = \frac{Na}{0.5(Ca + Mg)}$$

where Na, Ca and Mg are the concentrations in milli-equivalents per litre of sodium, calcium and magnesium in the *irrigation* water. For crop irrigation the SAR should be less than 6, while an SAR above 15 implies that the water is unsuitable.
sodium bicarbonate, $NaHCO_3$ A common chemical, which has been added to *sewage* to improve its *alkalinity* and *buffering* capacity, especially in *anaerobic sludge digestion.*
sodium chloride, NaCl   Common salt. *See chlorides, brine.*
sodium hexametaphosphate, $(NaPO_3)_6$, Calgon   A cyclic *polyphosphate* used as a *corrosion* inhibitor and added to boiler *feed water* in doses of 1 to 2 mg/litre. It is also claimed to prevent pitting of iron and steel and the precipitation of *calcium carbonate* and iron compounds.
sodium-restricted diet   The American Heart Association

291

recommends a maximum 20 mg/litre of sodium in water for heart sufferers, but this level is often exceeded when *water softening* is used.

sodium thiosulphate, $Na_2S_2O_3$, hypo A substance which, at a concentration of least 100 mg/litre, has a stabilising effect on a population of coliform bacteria and is used for this purpose in bacterial sample bottles. Hypo also *de-chlorinates* any *free residual chlorine.*

sodium tripolyphosphate, STP, $Na_5P_3O_{10}$ The most widely used *builder* in modern *synthetic detergents,* responsible for up to 70% of the phosphates in domestic *sewage.* It has been blamed for the *eutrophication* of lakes.

soffit The under surface of a ceiling or of the top of an arch inside a *sewer*, pipe, etc., opposite the *invert*.

soft detergent A *synthetic detergent* which is readily *biodegradable.*

softening, softening sludge *See water softening.*

soft pesticides *Biodegradable* (non-persistent) *pesticides.*

soft water Water with little or no *carbonate hardness* or *non-carbonate hardness.* In Britain it usually originates in the hard rock areas of the west and north, especially in Wales and Scotland, or from peaty moorland, but not from chalk, nor from limestone formations. Soft water may be corrosive, as in *plumbo-solvency. See corrosiveness of water, sodium-restricted diet.*

soil defence Natural methods whereby harmful *faecal* bacteria are killed by dispersion through the soil because of a combination of lower temperature and hostile natural *bacteria* in the soil. Dry, sandy or gravelly soils with free air circulation are best. Harmful bacteria are generally greatly reduced within the first 7 m of underground travel, unless the rate of travel is fast, above 7 m per day. Harmful bacteria may survive for years in tropical soils but only a few months in a temperate climate. *See virus detection.*

soil water zone The zone, extending from the surface, which is penetrated by plant roots. The water in this zone is soil water, not *groundwater. See field capacity, zone of aeration.*

sol A suspension of *colloid*-sized solid particles dispersed in a liquid, sometimes illogically called a colloidal solution.

solid−liquid separation Solids are usually removed from *sewage* by *screens* or *sedimentation.* In *sludge* treatment, separation processes include gravity *sludge thickening, flotation* or *mechanical de-watering of sludge.* In *raw water* treatment the *tertiary treatment* of sewage, there are also *micro-straining, clarification* and *deep-bed filtration.*

solids contact clarifier, accelerated c. A *clarifier* used in *water treatment*, in which the separated solids or *sludge* are in direct contact with the incoming water. This accelerates *flocculation* and reduces the amount of *coagulants* used. The *surface loading* can be at least twice that of a simple *flocculator-clarifier* for a given *raw water*. *See sludge blanket clarifier, solids recirculation clarifier.*

solids flux, settling f. In *sedimentation tanks* for *sewage*, the product of the concentration of dry solids in mg/litre and the settling velocity. *Compare surface solids loading rate.*

solids loading *See sludge thickening, surface solids loading rate.*

solids recirculation clarifier *(Figure S.11)* *A type of solids contact clarifier* in which the separated solids *(sludge)* are re-circulated and mixed with the *raw water* either within the tank (e.g. Accentrifloc) or in a mixer outside. This re-circulation of

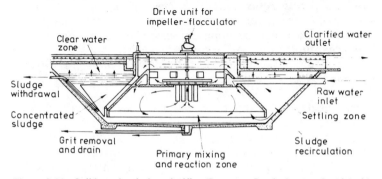

Figure S.11 Solids recirculation clarifier (Paterson Candy International Ltd.)

chemical sludge accelerates *flocculation* and reduces the amount of *coagulants* used. They are usually circular in plan, up to 30 m diameter. The *surface loading* is 60 to 120 m³ per day per m² of surface area, depending on the quality of the raw water. The flow pattern is a mixture of radial and upward flow, and the incoming water does not flow up through a *sludge blanket*, which would be true in a *sludge blanket clarifier*.

solids retention time The *mean cell residence time* in a *biological treatment* process.

solid toxic waste *See toxic waste.*

solid waste House refuse, waste rock, builder's or demolition rubble, steelworks slag, and so on. Mines and quarries produce a greater tonnage of waste than any other industry. Other than mines and steelworks, UK industry produces about 23 million tonnes a year of wastes. Throughout the industrial world, the cheapest and by far the commonest legitimate disposal method is by *controlled tipping*. *See hydrogasification, pyrolysis.*

solid waste analysis Municipal refuse in the UK is generally analysed physically for its components in seven classes, given below with the 1973 national average percentages in parentheses; paper and cardboard (32.0); vegetable and *putrescible* (18.0); undersize below 20 mm (18.8); metals (8.9); glass (10.6); textiles or rags (3.2); *plastics* (2.1); unclassified (6.4). Plastics and paper are steadily increasing their percentages, while the less than 20 mm (ash) class is decreasing.

solid waste arisings The waste generated.

solid waste management Control of the generation, storage, collection, transport, separation, recovery, treatment and disposal of *solid waste.*

solid waste treatment Treatments of solid waste may include *air classification, controlled tipping, incineration, mechanical sorting plants, refuse reduction, ocean dumping* or *shredding.*

solubility The extent to which a substance can dissolve in a *solvent,* measured in grams or *moles* per litre. *See dissolved oxygen, molar solution, saturated solution.*

solute A dissolved substance, forming a *solution* in a *solvent.*

solution A mixture of *solute* and *solvent,* this being usually a liquid. In a solution of solid in liquid the solid disappears completely and cannot be removed by ordinary *filtration,* although it may be removed by *de-salination.* Other types of solution exist: liquid in liquid, gas in solid, solid in solid.

solvent A liquid or other substance that dissolves another to form a *solution.* Thus, in a solution of salt in water, the salt is the *solute* and water is the solvent. Water is the most powerful solvent known.

soot blowing Removal of soot deposits from smoky surfaces in a boiler by blowing compressed air or steam over them. It produces dark smoke which is tolerated under the UK *Clean Air Acts* for limited periods. *See Ringelmann chart.*

sorption All the processes of *absorption, adsorption,* etc. One *advanced waste treatment* for phosphate removal has a sorption column containing *activated alumina.*

sough An *adit.*

sow bug *Asellus aquaticus.*

sparger, sparge pipe A perforated or open-ended metal pipe *diffuser* that produces *coarse-bubble aeration.*

Spartina townsendii A grass that is often found in British salt marshes between high water neap and high water spring tides. It has been killed by *oil refinery effluent* containing 25 to 30 mg/litre of oil, resulting eventually in the scouring away of the marsh soil.

294

species The division, one of the smallest in *taxonomy*, that is below genus.

specific capacity of a well The rate of withdrawal of water from a well in m³ per minute per metre of *drawdown*, as shown in a yield/drawdown curve of depth against pumping rate. *See storage coefficient.*

specific conductance *See electrical conductivity.*

specific gravity, SG, s.g., relative density The ratio of the mass of a substance to that of the same volume of water at 4°C. The s.g. of dry *sewage* solids is about 1.2 and of sewage *sludge* at 5% dry solids is about 1.01.

specific ion electrode An *ion-selective electrode.*

specific resistance, r A measure of the resistance to *filtration* of *sludge* through a filter cloth:

$$ r = \frac{2bPA^2}{C\mu} $$

where P = filtration pressure, normally 49 kPa (49 kilonewton/m²); A = area of filtration m²; C = weight of dry solids per unit volume of unfiltered sludge, kg/m³; μ = dynamic *viscosity* of the *filtrate* in kg m⁻¹ s⁻¹ (= N.s/m²); b = slope of the graph θ/V versus V (units = s/m⁶). V = volume of filtrate collected in time θ. Then the dimension of r is

$$ \frac{s}{m^6} \times \frac{N}{m^2} \times m^4 \times \frac{m^3}{kg} \times \frac{m^2}{N.s} = \frac{m}{kg} $$

The test resembles the **Buchner funnel test**, except that the volume of filtrate is measured at intervals of 1 min or less. The test can be used to optimise the **chemical conditioning** of the sludge and to assess the difficulty of de-watering it mechanically by a pressure or **vacuum filter**. A value greater than 10^{13} m/kg indicates a difficult sludge to de-water, while 10^{11} m/kg indicates one that is readily de-watered mechanically. *Compare capillary suction time.*

specific speed, N_s A concept indicating the design of the runner, used in the choice of pumps and turbines for a known duty. For a pump with a known performance it is

$$ N_s = \frac{N\sqrt{Q}}{H^{0.75}} \quad (\text{rev/min}) $$

in which N = speed (rev/min); Q = discharge (m³/min); and H = optimum pumping head (m).

specific surface A measure of the fineness of a powder, stated as

295

the number of cm² of surface area per gram.

specific yield, effective porosity Of an *aquifer,* the amount of water it yields when it drains by gravity. The value can occasionally reach 45% of the volume of the aquifer but this is exceptional. Ordinarily, a good gravel yields 25% and a clay not more than 3%.

spent liquor The effluent from a *coke plant waste* after the free ammonia has been distilled from the *ammoniacal liquor.* The spent liquor contains *phenols,* thiocyanates, ferrocyanides, thiosulphates and some fixed *ammonia.* Such liquor can be treated by *activated sludge* if the *sludge* is acclimatised—i.e. if the sludge contains the necessary micro-organisms to break down phenols, thiocyanates, etc.

Sphaerotilus natans One of the *filamentous organisms,* a *bacterium* found in *sewage fungus* and in *bulking* sludge. *See iron bacteria.*

spillway, wasteway, waste weir An open channel or pipe that receives the overflow from a *reservoir.* Many different types exist, but all need to dissipate the energy of the overflowing water in a *stilling pond* downstream of the spillway.

spinner Stationary, helically curved vanes, usually at the inlet to a *cyclone* or other *dust arrestor,* which cause the incoming dusty air to swirl (spin).

Spiractor A *pellet reactor.*

spiral classifier A *screw classifier.*

spiral flow aeration tank An *aeration tank* with *diffusers* or *spargers* placed on one side so as to swirl the liquid, giving roughly spiral flow to it down the tank, as in *aerated grit chambers* or the *Inka process* for *activated sludge* treatment.

spirochaete, Spirochaeta Heterotrophic *bacteria* shaped like flexible spirals. They can be 500 μm long but only 0.3 μm across. Spirochaetes are found in muds and water or inside a *host.* Some of them cause diseases (e.g. syphilis and *leptospirosis*).

spirochaetosis *Leptospirosis.*

Spirulina maxima A *single-cell protein,* an *alga* harvested from Lake Chad, Africa, and used as food after drying.

spoil heap A pile of waste, usually rock, from a mine or other industry. Colliery spoil heaps are no longer built as cones with a pointed top. They are compacted during construction by earth-moving vehicles running over them and are therefore flat-topped as convenient for landscaping. This gentle slope reduces polluting *runoff.*

spore A seed or reproductive but dormant *cell.* All *fungi,* some *algae* and a few *protozoa* reproduce by spore formation. Spores resist *disinfection,* heat and desiccation. Bacterial spores are

296

called *endospores.*

spore-forming protozoa, Sporozoa   Parasites, some of which, the four species of *Plasmodium,* cause *malaria,* transmitted by *anopheline mosquitoes.*

spray aerator   An *aerator* in which the water is shot into the air as a fountain with many jets.

spray drying   Conversion of a *solute* to powder by spraying the *solution* into hot air, causing quick evaporation which frees the solute.

spray irrigation   Sprinkling water over land intended for growing crops. This method of *land treatment* originated in the USA and has been successfully used for treating the *effluent* from canneries or chicken-packing plants, with 90% reduction of *BOD.* Spraying growing crops with piped water or even well or river water is allowed in England and Wales only under licence from the local *water authority. Sewage* effluents can be sprayed over crops or grass as a final treatment but not in frost. *See rain gun.*

spray tower   (1) An inexpensive *packed tower.* (2) An *evaporative tower.*

spreading area, s. pit, recharge pit   A pit, usually surfaced with 1 or 2 m depth of sand, used for *artificial recharge* of *groundwater.*

springback   In *high-density baling,* increase in the volume of a bale after it has been released from the baling press. Typical figures 1 to 2 min after release are 40 to 70% volume increase. Holding of the top pressure for only 10 s reduces springback. Bales made at pressures of 150 atm or more are much more stable than those made at lower pressures, which tend to collapse.

springtail, Collembola   Insects without wings, some of which graze on *trickling filters*—e.g. *Hypogastrura viatica.* They can feed on *bacteria, fungi* or decaying *organic* matter.

sprinkling filter   A *trickling filter.*

spun-iron pipe   Cast-iron pipe made by pouring the molten metal into a horizontal, water-cooled mould rotating at high speed. For the same size and strength, the wall thickness of a spun-iron pipe is less than that of a sand-cast iron pipe (BS 4622, 4772).

SS   *Suspended solids* or, occasionally, *settleable solids.*

stability   (1) The 'keeping' quality of a sewage, its ability to stay oxygenated and not to putrefy when it is kept out of contact with air for, e.g., 5 days at 20°C. *Nitrified effluents* usually are stable as measured by the *methylene blue stability test.* An indication of instability is the smell of *hydrogen sulphide.* (2) In meteorology, any rate of change of the air temperature upwards that is equal to or less than the *adiabatic lapse rate* is stable, causing the atmosphere to hold pollutants and not to disperse them.

stabilization (1) In drinking water, prevention of the deposition of solids from the water. Ordinarily, this means that *calcium carbonate*, $CaCO_3$, should not precipitate from it (*see precipitation*). *Carbon dioxide* gas may be lost by heating or *aeration* of a water, resulting in the formation of a *scale* of $CaCO_3$. After water has been softened by adding *lime,* small amounts of $CaCO_3$ may continue to precipitate for some time, forming scale in the water pipes. Addition of CO_2 may prevent this, reversing the reaction (stabilisation by *re-carbonation*). *Polyphosphates* form soluble complexes with calcium or *magnesium* salts and doses of 1 or 2 mg/litre may be used to stablise a water. Strong acids, such as hydrochloric acid, HCl, which form soluble calcium salts may also be used to stabilise the water. Stabilisation with strong acid is called vaccination. (2) As applied to *compost* or *sewage*, putting it into a state in which it will not stink. For sewage this means settlement and *biological treatment* to reduce its *BOD*. (3) As applied to heaps of refuse or *tailings*, preventing them from subsiding or blowing away. This is done by earthing over, watering, compacting with steel-wheeled tractors or bulldozers, seeding with grass, etc. When a tip has ceased to subside, it is said to be stable. This may take as much as 20 years, depending on how *biodegradable* the tip happens to be.

stabilization pond   A *waste stabilisation pond.*

stable effluent   One that does not and will not stink.

stack (1) In *sewerage*, a vertical pipe that collects *sewage* from a building and is connected through a house drain to the *sewer.* (2) A chimney.

stack emission, s. effluent, s. gas   Chimney *emission, smoke* or gas.

stack gas cleaning   The use of *dust arrestors* or *flue gas desulphurisation* or both.

stack gas re-heating   Heating of *flue gas*, usually after *wet scrubbing* so as to ensure *buoyancy of the stack gas* and dryness in the chimney. It also helps to achieve an invisible *vapour plume* and reduces pollution at ground level, although this may be unimportant if the scrubbing has removed all the pollutants. Even so, the descent of a *fog* outside a chimney can be unpleasant.

stack sampling   Sampling of *flue gases*, usually with a *sampling train.*

stage (in USA **gage height**)   The water depth in a river, reservoir, lake etc., measured from a reference level especially chosen to indicate the flow rate.

stage/discharge curve, rating c. A graph of water level (stage) against flow rate (discharge) for a particular point in a stream. It indicates the flow rate there for any water level. Every measuring station should have such a curve drawn for it.

stagnation (1) Stillness, a condition of the air under an *inversion*. (2) Lack of movement of the water in a *main*, usually impossible in a *ring main*. It is undesirable and may lead to an unpleasant taste in the water.

stain In *microbiology*, a dye used to colour microbes selectively—for example, *Gram stain*.

stainless steel Steel with substantial additions of nickel (6 to 12%) and chromium (12 to 18%). The amounts of nickel and chromium can greatly alter the mechanical properties. In *aerobic* conditions a stable, hard oxide film forms on the surface, protecting the metal from *corrosion*. In the absence of air or *oxygen*, once the protective oxide layer is removed by erosion from grit, etc., it is not replaced and the material then corrodes as quickly as mild steel. *Chloride* ions attack the common nickel-chromium steels, which therefore cannot be used in sea-water. Corrosion by chlorides can be controlled by adding up to 1% molybdenum to the steel.

stale sewage *Sewage* that has lost its *dissolved oxygen*. It should quickly be oxygenated or treated, before it becomes *septic*.

stalked ciliates *See ciliate protozoa.*

Standard Methods for the Examination of Water and Waste Water A reference book on water testing, published jointly by the American Public Health Association, the American Water Works Association and the Water Pollution Control Federation. It lists not only standard analyses, but also *bio-assays*. It was first published in 1910 and the 14th edition was issued in 1975.

standard-rate trickling filter *See trickling filter.*

standards for sewage effluents *See sewage effluent standards.*

standards for water *See drinking water standards, irrigation, swimming water.*

stand-by capacity Mechanical or engineering equipment that is not needed for ordinary work but has to be kept in case of breakdown or other emergencies. To reduce expense there must be a balance between the amount of excess plant and the risk of failure.

standing-water level The level at which the water in a borehole or pit stands when it is left undisturbed for some days without pumping. A large number of such levels forms the *water table* of an unconfined aquifer or the *piezometric surface* of *confined groundwater*.

299

standing wave A *hydraulic jump.*

standing-wave flume (*Figure S.12*) A measuring flume with a *hydraulic jump* downstream of its throat:

$$\text{flow rate} = 1.705 \ BH^{1\cdot5} \ (\text{m}^3/\text{s})$$

in which, B = width of flume (m) and H = total upstream head plus kinetic energy head (m). In practice the kinetic energy can be

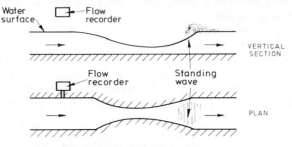

Figure S.12 Standing-wave flume

ignored; thus, H = depth. Flumes should be designed to BS 3680. Such flumes are often used to control the flow in a *constant-velocity grit channel.*

stand pipe A vertical pipe connected to the water *main* and used for water supply in the street. Some 20% or more of city dwellers in poor countries have to use stand pipes to obtain drinking water.

Staphylococcus A spherical *bacterium* (coccus). *Staphylococcus aureus* infects the skin or wounds. Most strains now found in hospitals are immune to penicillin, although they were at first easily treated by it.

starved-air incinerator A *two-stage incinerator.*

static compaction, stationary c. (1) Compression of refuse by a small stationary machine into disposal sacks. (2) Compression of refuse into a steel skip, sometimes with a mechanism that is electrically driven and built into the skip. Other types have an external ram.

statutory control of dams Dams in the UK are controlled by the Department of the Environment. *See ICOLD, impounding.*

statutory nuisance A *nuisance* that is so called because of a law (statute) passed by Parliament, not because of common law.

statutory water company A private water supply company. Many such companies remain in existence after the UK reorganisation of the water industry resulting from the *Water Act 1973.*

steam generation from incineration *See incineration with energy recovery.*

300

stearated limestone dust *See hydrophobic chalk.*

steel Mild steel is a purified cast-iron containing only about 0.1% carbon. Other elements may be added to improve the strength and other properties—e.g. to make *stainless steel.* Steel water pipes should have external and internal protection such as bitumen to prevent *corrosion.*

steel-wheeled compactor *See Figure S.13* and *compactor.*

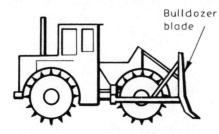

Figure S.13 Steel-wheeled compactor

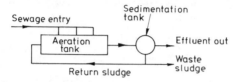

Figure S.14 Step aeration of activated sludge,
flow diagram

step aeration, s. feeding, incremental loading Incremental loading best describes this method of feeding the *influent* sewage into an *aeration tank* in *activated sludge* treatment (*see Figure S.14*). It is added at several inlets spaced out along the tank between the main inlet and the outlet, but all the *return sludge* is added at the main inlet. The *organic* load is more evenly distributed along the tank and there is no need for *tapered aeration.*

sterilisation Destruction of all living micro-organisms. *Compare disinfection.*

sterilising filter A *filter* that removes *bacteria,* unusual in the treatment of drinking water but used medically to sterilise solutions that cannot be heated. In *membrane* filters the pores are too small for the bacteria to pass through, but in packed bed filters the bacteria are adsorbed, usually by asbestos fibre. *Viruses,* being smaller than bacteria, may pass through such filters. *See adsorption.*

Stigeoclonium One of the *filamentous organisms,* a *green alga*

301

found in water polluted by *sewage* which survives in the presence of some toxic metals.

stilling basin, s. pool, s. pond   A mass of water that absorbs the energy of flow downstream of a dam *spillway*, to ensure that the river bed is not scoured.

stilling pond overflow   An overflow downstream of a *stilling basin*. It is designed to have *laminar flow* at the outlet.

stilling well, gauge w.   A chamber that communicates with the main body of water by only a small inlet, thus protecting the instrument within it from waves or surges. Often it contains a gauge that automatically records the water level.

stimulation of a well   Artificially increasing or maintaining the output from a well by methods such as *acidising* an *aquifer* containing limestone or dolomite, or breaking it with high-pressure water or explosives, or any combination of these methods.

stirred specific volume, SSV   Like the *sludge volume index*, a measure of *settleability* for *activated sludge*, except that a 4 litre cylinder is slowly stirred with three vertical 5 mm diameter rods rotating at 1 rev/min:

$$SSV = \frac{\%\text{ settled volume of sludge in 30 min}}{MLSS,\text{ mg/litre}} \times 10\ 000\ (\text{ml/g})$$

stochastic   In statistics, random, having an element of chance.

Stokes diameter   The diameter of a sphere of the same relative density and free-falling speed under *Stokes' law* as the particle under consideration. *See volume diameter.*

Stokes' law, S. formula   An expression for the terminal (free-fall or falling) velocity of a spherical particle dropping freely in a fluid of known specific gravity. If d is the particle diameter in cm and s is the specific gravity, the falling velocity is:

$$\frac{gd^2(s_{\text{solid}} - s_{\text{fluid}})}{18 \times \text{dynamic viscosity}}\ (\text{cm/s})$$

in which g is the acceleration of gravity, 981 cm/s². For particles falling in air at 21°C this simplifies to

$$303\ 000\ d^2(s_{\text{solid}} - s_{\text{air}})\ (\text{cm/s})$$

Stokes' law in water, as in air, applies to separate, not *flocculent*, particles, at *Reynolds numbers* of less than 0.2. *See settling regime, viscosity.*

stone fly, Plecoptera   The *larvae* live in clean, well *aerated* water in upland streams. *See Leuctra.*

stop log A temporary dam, often in the form of one or more concrete or steel or wooden beams, inserted at a water intake to a *reservoir* so as to prevent *abstraction* while *screens* are being repaired or to lower the water level for any other reason.

storage coefficient Of an *aquifer*, the volume of water that it releases when the water level falls by a unit of depth. *See specific capacity of a well.*

storage of water *See reservoir, raw water storage.*

storm drain US term for a *stormwater sewer.*

storm overflow A *stormwater overflow.*

storm relief sewer A *stormwater sewer.*

storm sewage *Storm water.*

storm tanks *Stormwater tanks.*

storm water, stormwater Water from any form of *precipitation.* In waste water engineering it means the *overland flow* that enters the *sewers*, which can be estimated from the *Lloyd−Davies method* or from *unit hydrographs.*

stormwater overflow, storm o., separation weir In a *combined sewer* or *partially separate* sewer, a *side-weir overflow* or similar structure that separates excess water, discharging it into a *stormwater sewer.* A similar overflow, at or near the *sewage treatment* works, separates that part of the *sewage* that, because of the limited capacity of the works, cannot receive full or immediate treatment. Separate sewers do not need them, because they take no rainwater. In the UK usually three times the *dry-weather flow* is taken for full treatment. Anything in addition to this receives *preliminary treatment* only, and is stored in or passes through stormwater tanks. *See also leaping weir overflow, siphon overflow.*

stormwater sewer, storm relief s. (in USA **storm drain**) In a *combined system*, a *sewer* that is dry except after rain, and discharges directly into a watercourse. It is designed to receive overflows from the combined sewers when these are carrying flows equivalent to three, six or even twelve times the *dry-weather flow.* However high the *dilution, stormwater* from a combined sewer cannot avoid being polluted with *sewage.*

stormwater tank, storm t. A *sedimentation* tank provided at a *sewage treatment* works served by *sewers* in a *combined system* or *partially separate system*, which stores *stormwater* so that it can be returned for *biological treatment* when the rain has stopped. If the stormwater exceeds the tank capacity, the *effluent* is discharged without further treatment. Some works use them for equalising the daytime and night-time flow, which restricts their use as stormwater tanks during the day. Their total capacity is often equal to the capacity of the *primary* *sedimentation* tanks, but other methods of calculation exist, including use of the figure of 0.68 m³ per person served. In new works it is often desirable to build the primary and stormwater tanks with the same dimensions, scrapers, etc., and with suitable pipework to ensure that the tanks are interchangeable, allowing flexible operation. *Tank sewers*, used for storing the stormwater in the sewer itself, may be located away from the sewage treatment works.

STP *Sodium tripolyphosphate.*

straight-through cyclone, vortex air cleaner A *cyclone* in which there is no reversal of the air flow. The dust is removed from the wall of the cyclone through a narrow annular slot which bleeds off some 10% of the air towards a dust bunker. The cleaning efficiency can be further improved by wetting the walls (irrigation) to prevent bounce-back and re-entrainment of the dust.

straining Removing particles that are larger than the opening of a *screen* or *filter*. *Compare filtration.*

stratification Separation into layers. *See density currents, de-stratification, thermal stratification.*

stream *See waterway.*

stream flow Flow in a natural open channel. *Compare pipe flow.*

streamflow routine *Flood routing.*

streamline flow *Laminar flow.*

street inlet A *gulley.*

strength of sewage Ordinarily, strength means the content of *organic* or other oxidisable matter in unit volume of *sewage*. This depends on many factors, including the population's water consumption, the trade wastes in the sewage, the *infiltration* water entering old *sewers*, and whether they form a *combined system* or a *separate system*. It is often measured by *BOD₅* or *COD* or *McGowan strength.*

Streptococcus A genus of spherical *bacterium* (coccus) linked to its neighbours in chain form. Some species cause diseases such as pneumonia and scarlet fever.

Streptococcus faecalis A *Streptococcus* that is common in human and animal *faeces,* but scarcer than *Escherichia coli*. It is of less bacteriological interest than *E. coli* except in doubtful cases. It ferments *MacConkey's broth* without producing gas.

stripping tower A tower used for *ammonia stripping* or for removing other unwanted gases from water, such as *carbon dioxide* or *radioactive* gases.

strontium-90, ⁹⁰Sr One of the main *radio-isotopes* in *fallout* and in *nuclear reactor wastes*. With its long *half-life* of 28 years, it could be a hazard to health, particularly since it can replace *calcium* in the bone. Belonging to the same chemical family as calcium, it can also be removed from water by any *water softening* process that removes calcium. In the UK it has rarely if ever been more concentrated than 2 picocuries per litre in drinking water.

subadiabatic *See adiabatic lapse rate.*

subcritical flow Flow in an open channel at a speed below the *critical velocity.*

sub-kingdom A *phylum.*

sub-lethal toxicity Poisonousness that does not kill but nevertheless may be exceedingly harmful in the long term.

submerged-bed aeration *Aeration* in a tank that is loosely packed with plastics *medium,* as in a *high-rate trickling filter,* but differing from it in being fully submerged. The *sewage* is aerated from the bottom, but although the air rises, the sewage may move down or up. The *contact aerator* is one type. The biological action resembles that in the *trickling filter* and the *effluent* goes for *sedimentation* afterwards. The bed may need *backwashing.* It may be used as a *nitrifying filter.*

submerged filter A *high-rate trickling filter* or a *trickling filter* that is fully submerged, as in the *anaerobic filter* or *submerged-bed aeration.*

submerged weir, drowned w. A *weir* with the tailwater above its crest.

submergence The state of being covered by a fluid. In *sewage* sludge de-watering, the submergence of a *vacuum filter* is usually 25 to 40%. *Rotating biological contactors* also are usually 25 to 40% submerged. Thus to keep the outer layers of the slime *aerobic,* 60 to 75% is above the water level.

sub-micron particle A particle smaller than 1 μm (0.001 mm).

sub-natant liquid The bottom water or *sludge,* which sinks below the *supernatant liquid.*

subsidence chamber An *expansion chamber.*

subsidence inversion An *inversion* in which air is warmed by its own downward movement.

substrate A physical background—e.g. one on which a paint or a plaster is laid or a micro-circuit is fastened. In *microbiology* it means nutrients as a whole, the substances on which *enzymes* act.

subsurface filtration A *sewage treatment* used where *subsurface irrigation* is not practicable, using specially prepared sand beds, as in *land filtration.* The *effluent* is introduced to the *sand filter* by *land drains* that are covered with topsoil to keep flies away, and is removed by *underdrains* buried at the bottom of the *filter* (*see Figure S.15*).

Using commercial loose jointed pipe   Using broken rock and
                                                  plastic strip

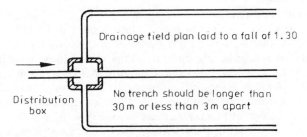

Drainage field plan laid to a fall of 1.30

Distribution box

No trench should be longer than 30m or less than 3m apart

Figure S.15 Subsurface filtration for disposal of rainwater, sullage or different from a septic tank (Reproduced from 'Water Treatment and Sanitation: simple methods for rural areas' published by Intermediate Technology Publications Ltd., London.)

subsurface flow *Interflow* and other *groundwater* flow.

subsurface irrigation Watering of crops by the use of the natural flow of water below ground level or by injection into *land drains*. It is mainly used where sandy soil lies just above an impermeable layer. The *irrigation* water then stays near the surface and can be used by crops. In *sewage treatment* the method provides a type of *absorption pit* for treating *effluent*, usually from *septic tanks*.

suction hood *See dust-extraction hood.*

sugar beet wastes *See beet sugar wastes.*

sullage, grey water Waste water from sinks, baths, washbasins, etc., but not from WCs. Its volume may form two-thirds of domestic *sewage,* though with one-third of the polluting matter and only half its *strength. Compare black water.*

sullage disposal If it is separated from WC wastes, *sullage* can be disposed of in an *absorption trench* or by *subsurface irrigation* without other treatment. In warmer countries *transpiration* areas covered with plants or shrubs having high transpiration rates may be used—e.g. cereals or willow trees.

sulphate-reducing bacteria *Bacteria,* including *Desulphovibrio desulphuricans*, which oxidise *organic compounds,* using *oxygen* from *sulphates* in water, usually producing sulphides or possibly thiosulphates (ending in-S_2O_3). In acid water this may come out as *hydrogen sulphide* gas, H_2S. Sulphides occur in black (*anaerobic*) muds, *septic sewage,* sewage *sludge,* etc.

sulphate-resisting cement A cement to BS 4027 that contains less than 5% tricalcium aluminate, the part of cement that is most easily attacked by sulphates. However, these cements cannot resist the sulphuric acid that occurs in *sulphide corrosion.*

sulphates, -SO_4 Substances whose formula ends in -SO_4, the commonest in water being calcium sulphate, $CaSO_4$. *See* Table of Allowable Contaminants in Drinking Water, page 359; *see also magnesium,* and below.

sulphide corrosion (*Figure S.16*) *Sewers* are eaten away by sulphuric acid produced in them by a series of reactions. If the *sewage* becomes *septic, sulphate-reducing bacteria* reduce *sulphates* to sulphides. In neutral or acid water some of the sulphide will be present as *hydrogen sulphide* gas, H_2S, which will bubble out at turbulent points in the sewer, and some of this

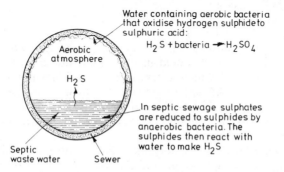

Water containing aerobic bacteria that oxidise hydrogen sulphide to sulphuric acid:

$$H_2S + bacteria \rightarrow H_2SO_4$$

Aerobic atmosphere

H_2S

In septic sewage sulphates are reduced to sulphides by anaerobic bacteria. The sulphides then react with water to make H_2S

Septic waste water   Sewer

Figure S.16 Sulphide corrosion in a sewer

H_2S will re-dissolve in the layer of water inside the pipe above water level. This is an ideal growing area for *Thiobacillus* bacteria, particularly the species *T. concretovorus.* These bacteria oxidise the H_2S to sulphuric acid, H_2SO_4. The *acidity* in the *slime* above water level may drop to pH 1.0. This strong acid attacks the cement in the concrete pipe and its crown eventually falls in. The process depends on the sulphate content and temperature of the sewage. For domestic sewage the production of H_2S takes considerable time below 12°C and so is not a great problem in temperate climates. However, at sewage temperatures above 15°C the sewage becomes septic fast and H_2S is produced even in short sewers. The life of a conventional concrete sewer may be as short as 5 years in a hot climate. *Sulphate-resisting cements* have little advantage over ordinary cement in this strong acid. Possible remedies include the use of *plastics* pipes or epoxy linings inside the concrete. Regularly spaced *back drops* with forced ventilation at each one will bring the H_2S out of solution because of the turbulence, and it is then vented to atmosphere. Alternatively, H_2S production can be stopped by injecting the

307

sewage with air or oxygen. Another possibility is for the sewers to be designed to run full under pressure. *Thiobacillus* bacteria then cannot grow because they are strict **aerobes**. Despite these possible solutions, sulphide corrosion of sewers remains one of the major problems of sewerage in hot climates.

sulphonator Equipment to inject and meter *sulphur dioxide*, SO_2, for *de-chlorination* of water.

sulphur-asphalt paving A road surfacing in which sulphur replaces from one-third to two-thirds of the usual asphalt binder. The sulphur-asphalt material is laid with standard equipment and has some advantages over conventional asphalt, being cheaper and melting at a slightly lower temperature.

sulphur bacteria A general name covering both *sulphur-oxidising* and *sulphate-reducing bacteria*. It often implies only the oxidising bacteria.

sulphur content Municipal refuse generally has a very low sulphur content, well below 1%. Coals and cokes burnt in the UK average 1.6%, most of which is emitted as *sulphur dioxide*. Only 10% of the sulphur is held in the ash. Crude oil also contains sulphur, most of which passes into the residual fuel oil, which can have up to 4.5%S. In general, the lighter the distillate, the less S it contains, but even distillate fuel oil of 35 s viscosity as used in homes can contain 1.5%S, although limited amounts with 0.3%S are available.

sulphur dioxide, SO_2 A poisonous gas produced mainly by the burning of the sulphur in coal and oil and the smelting of the sulphide ores of metals. Some 100 million tonnes of the gas annually are produced throughout the world by human agencies, 90% of it in the northern hemisphere. No law limiting SO_2 emissions exists in the UK, where it has been removed from the *flue gases* of only two existing power stations. Reliance is placed on *dilution* by the *tall stack policy*. After leaving the chimney, the SO_2 dissolves in water droplets and becomes either H_2SO_3 (sulphurous acid) or, if it has been further oxidised beforehand, H_2SO_4 (sulphuric acid). Most of this dilute acid returns to seas, rivers, lakes or underground water. In Norway the effects are particularly severe, because the acid accumulates in the snow all the winter and melts into the rivers in the early summer and spring, when the young fish are the most vulnerable. The culprits are the high chimneys of power stations and factories mainly outside Norway. So far as allowable concentrations of atmospheric SO_2 are concerned, one figure that has been stated as the maximum allowable for continuous exposure is 0.15 mg/m^3, but this is well below its average winter concentration in London air (0.14 in summer, 0.26 in winter).

308

SO_2 is used in raw water treatment for de-chlorinating water after *super-chlorination*. It is exceedingly soluble and the injection can be automated. *See Claus process, de-chlorination, SO_x,* and below.

sulphur dioxide absorber A *flue gas de-sulphurisation* unit that is usually a moving-bed *scrubber*. It can be some 11 m in diameter and 22 m high for a treatment capacity of 150 MW. *Alkaline* reagent is sprayed in from the top and flue gas flows up through the spray. Normally the flue gas first passes through an *electrostatic precipitator* or *fabric filter* that removes 99% of the *flyash.*

sulphur-oxidising bacteria A diffuse group of *bacteria* that oxidise sulphides to sulphur or *sulphates*, or any other form of sulphur to a more oxidised state. The group includes the families of *photosynthetic* bacteria known as *Chlorobiaceae* and *Thiothrix. Thiobacillus* is important to *sulphide corrosion.*

summer stagnation in lakes *See thermal stratification.*

sump oil Used lubricating oil from internal combustion engines is a dangerous pollutant in *sewers* and hinders *sewage treatment*. In the UK 500 000 tonnes of waste oil is recovered every year. Only about 5% is good enough to be re-used as lubricant after passing through the refinery. The remainder is burned to heat soaking pits, brickworks, forges, etc. Usually the smallest quantity bought by refiners is 200 litres, a 45 gallon drum full.

superadiabatic *See adiabatic lapse rate.*

superchlorination The use of comparatively large doses of chlorine in *chlorination* (up to 20 mg/litre) to eliminate objectionable taste or smell in drinking water. It should be followed by *de-chlorination.*

supercritical flow Flow in an open channel at a velocity greater than the *critical velocity.*

superphosphate An important *phosphorus* fertiliser which is a mixture of calcium phosphate, *calcium sulphate*, iron oxide, alumina, silica and water, and contains about 20% phosphorus as P_2O_5.

super-rate filter A *high-rate trickling filter* (1).

supersaturated solution *See saturated solution.*

supplementary fuel *See waste-derived fuel.*

surcharged sewer A *sewer* under internal pressure—that is, at a flow rate higher than when it began to run full. This results in *sewage* rising in the *manholes*. If the manholes are shallow and the surcharge is high, the manhole covers can be forced off and sewage will flow into the street.

surface-active agent A *surfactant.*

surface aeration Mechanical *aeration* usually of *activated sludge*

309

by violent agitation of the surface of an *aeration tank* or *lagoon* with *surface aerators.*

surface aerator Equipment to drive air into the surface of a liquid such as a *mixed liquor* in an *aeration tank,* by a rotating machine with either a horizontal or a vertical shaft. Horizontal shaft aerators—*brush aerators*—include the *Kessener brush,* the *TNO rotor* and the *Mammoth rotor.* Vertical-shaft machines are usually *cone aerators* such as the Simcar, *Simplex system* and Lightnin. Some vertical-shaft machines are carried on floats for use on *aerobic lagoons.* Surface aerators are compared by their *oxygenation efficiencies.*

surface area scrubber A *wet scrubber* consisting of a *packed tower* in which the packing medium can be slats of wood, stones or shapes of ceramic or *plastics.* At a pressure drop in the *flue gases* of some 25 cm *water gauge,* it is claimed that a two-stage scrubber will collect 99% of particles larger than 1 μm.

surface charge Many particles, including *colloids, bacteria* and *algae,* have a surface electrical potential, usually negative. Particles with similar charge repel one another and settle slowly because they will not unite. *Coagulation* is one method of overcoming the negative potential. Since the particles are negatively charged, a positively charged *coagulant* helps. *See electrophoresis.*

surface detention The thin sheet of water that is held on the surface either of the ground or of vegetation during rain. It may eventually evaporate or infiltrate into the ground.

surface irrigation Watering the soil with water or sewage *effluent* flowing directly on to the ground, as opposed to *spray irrigation* or *subsurface irrigation.* Surface irrigation may flood the land (*flood irrigation*) or the furrows between the rows of crops (*furrow irrigation*). The latter method reduces *evaporation,* since only part of the soil is flooded. *See also drip irrigation.*

surface lining of wells *See well casing.*

surface loading rate The flow of liquid divided by the water surface area of a treatment tank. The figure, expressed in m^3 per m^2 per day or hour, etc., is commonly used in the design of *clarifiers, sand filters* and *primary sedimentation* or *secondary sedimentation tanks.*

surface overflow rate *Overflow rate.*

surface runoff *Overland flow.*

surface solids loading rate, solids l.r. In an *activated sludge* settling tank, the product of the *surface loading rate* and the *MLSS* (kg/m^3). It is a measure of the mass of solids to be handled for a given surface area of settling tank per day, and is the only design parameter of activated sludge settling tanks that includes the concentration of the *mixed liquor*. At maximum flow (usually three times dwf) it should not exceed 120 kg m^{-2} day^{-1}, which is equivalent to a MLSS of 3000 mg/litre (3 kg/m^3) at an *overflow rate* of 40 m^3 m^{-2} day^{-1}.

surface tension The tendency of liquids to shape themselves at their surface into a form with minimum surface area, to oppose spreading.

surface-tension depressant A *surfactant*.

surface water Water in a river, lake, stream, etc. In *sewers* it is also the rainwater from streets, roofs and paved areas.

surfactant, surface-active agent, dispersant, emulsifier, wetting agent Any *solute* that affects (ordinarily reducing) the *surface tension* in water. It is the most important part of soap, *synthetic detergents*, etc., helping laundering, mineral *flotation* and the making of *emulsions* or other processes that need a low *surface tension* in water.

surge (1) *See pressure surge in pipe flow.* (2) In the flow of bulk materials such as minerals or refuse on a belt, a temporary sharp increase of the mass of material, sometimes provided for by a *surge bunker* that takes in the extra material during surges and returns it to the circuit when the flow slackens.

surge bunker In a line of conveyors, an intermediate bunker with a volume calculated to accept the largest *surge* without overflowing. For unpelletised *WDF* and raw refuse, surge bunkers are inadvisable; they block.

surge prevention The prevention of any *pressure surge in pipeflow*, aimed at eliminating the damage to pipework caused by these sudden variations in pressure. Several methods exist which may be used in combination. An air vessel connected to the pump delivery pipe reduces *surges* both when the pump starts and when it stops. Surface feeder tanks (surge tanks) above the main, connected to it, fulfil a similar duty and are located every 10 m or so of rise and fall on the *hydraulic gradient.* Other methods include valves that close slowly and are operated by electric motor (motorised valves), relief by-passes or a flywheel on the pump drive, to slow down any change of pumping speed.

surplus sludge, waste s., excess s. *Sludge* that has to be disposed of—in particular, that from *activated sludge* treatment which is in excess of the *return activated sludge.*

suspended magnet separator An *overband separator.*

suspended matter in air According to BS 1747 (Methods of measuring air pollution), any solid or liquid particles in air that have no appreciable falling velocity and therefore persist in air for a long time. The largest size is therefore a few microns, but it is mostly smaller than 1 μm and can include *flyash, dust,* acid droplets or other *particulates.*

suspended matter in water   Undissolved material that includes *suspended solids*, greases, tars or oils. Most British river waters have below 200 mg/litre of suspended matter.

suspended sediment   *Sediment; see also sediment load.*

suspended solids   The solids suspended in a liquid include *settleable* and non-settleable solids. The total suspended solids are measured by filtering the water through a glassfibre filter paper and drying the paper at 105°C for 1 h. Oil, tar, etc., may be removed beforehand by washing the sample with a volatile *organic* solvent, but this operation can usually be omitted. The volatile suspended solids (VSS) are estimated by heating the filter paper with the suspended solids to 550°C. The VSS are thus oxidised to *carbon dioxide. See volatile solids.*

suspension   A moving liquid or gas containing particles floating in it. If the particles are large enough to have a free-fall velocity under *Stokes' law*, they will sink and may thus be removed unless the fluid is kept moving, as in *fluidised-bed combustion* or a *sludge-blanket clarifier. See also* above and below.

suspension current   A *density current.*

suspension firing   The burning of particles tiny enough to hang in the air. Burning is several times faster than on a *grate.* A rough criterion for the size of fuel that can be fired in suspension is: the burnout time in the furnace should be less than the time taken for the gases to pass through it, usually about 2 s. In the *vortex incinerator* the time is probably longer. With suspension firing of coal or refuse the fuel is ordinarily blown into the furnace through pipes from the crusher or treatment plant.

sustained yield of a water source   *See borehole yield, yield of a water source.*

SVI   The *sludge volume index.*

sweating   The separation of metals by the use of their different melting points—e.g. lead at 327°C or its alloys at lower temperatures.

swimmer's itch   *Schistosome dermatitis.*

swimming ciliates   *See ciliate protozoa.*

swimming water   Water for swimmers is judged on the basis of its *coliform count* as an indication of its general bacteriological quality. About half the ailments of swimmers are of the ear, nose, eye or throat, and about one-fifth are gastro-intestinal. In natural waters the coliform count should be not greater than 1000 per 100 ml. In swimming baths there should be no coliforms, and less than 200 general *bacteria* per ml incubated on *agar-agar* at 37°C. Swimming baths should have 0.4 to 1.0 mg/litre of *chlorine residual* and a pH of 6.5 to 8.3. The maximum of 8.3 reduces irritation of the eyes. *Bromine, chlorine dioxide, iodine* and *silver* are also used, though less often, for swimming pool *disinfection.*

Sylvicola fenestralis, Anisopus fenestralis A dipteran *filter fly* that is common on *trickling filters. See grazing fauna.*

symbiont Either of the two organisms that benefit from a *symbiosis.*

symbiosis, symbiotic relationship, mutualism Any relationship between two organisms (a symbiotic pair) from which both benefit and neither suffers. For example, *algae* in an oxidation pond consume *carbon dioxide* and release *oxygen*, while the

Figure S.17 Symbiotic relationship between algae and bacteria

bacteria convert the oxygen back into carbon dioxide (*see Figure S.17*). Algae suffer in an excess of oxygen and bacteria suffer from a lack of it, so both profit.

syndets *Synthetic detergents.*

synecology The *ecology* of a community of organisms, as opposed to *autecology.*

synergism (1) Any interaction of two or more substances such that their total effect is greater than their effects separately. (2) The ability of two or more organisms to produce effects that neither could produce alone.

Synergistic reactions may be *symbiotic.*

synthesis The building of complex substances from simple ones—e.g. *photosynthesis.*

synthetic data In *hydrology,* data obtained from computer models, often used in *unit hydrographs.*

synthetic detergents, syndets Water-soluble substances, used for laundering, which are often preferred to soap because they do not form a *scum* on hard water, their *calcium* salts being soluble. (Soap forms a scum because its calcium and *magnesium* salts are insoluble and float.) Syndets contain a *surfactant* plus a *builder* and possibly also a bleach or a brightener. Only 15 to 35% of commercial syndet is surfactant. Surfactants are long-chain organic compounds containing water-soluble and oil-soluble groups. The three types—anionic, cationic and non-ionic—are so called because of their electrical charge or lack of one when they dissolve in water. Anionic detergents are common in domestic washing powder and liquid detergents. Originally, as hard detergents they were made up of branched *alkyl benzene* *sulphonates*, which are only slowly *biodegradable,* and caused *foaming* at sewage *outfalls* and elsewhere. After 1964 *linear alkyl sulphonates* were used—soft detergents readily biodegradable in conventional *sewage treatment.* Non-ionic detergents—about 25% of syndets in 1978—usually contain numerous ethylene oxide polymers plus water-soluble groups. They are relatively biodegradable but are found only in liquid detergent, because it is difficult to form them into free-flowing powders. Cationic detergents, usually *quaternary ammonium compounds*, are expensive and used only where a bactericidal detergent is needed, as in hospitals, and are not easily biodegradable. The UK average concentration of syndets in sewage in 1970 was about 18 mg/litre. When the content rises above 25 to 30 mg/litre in sludge, *anaerobic sludge digestion* is inhibited and gas production is reduced. This has happened in the south of England, where the water is hard and domestic consumption of syndets is high. The presence of syndets in drinking water is now unusual, since they are mostly biodegradable, but foam, when it occurs, seriously reduces the *oxygen* intake of water in *biological treatment.* Syndets can be removed by *foam fractionation. See* Table of Allowable Contaminants in Drinking Water, page 359.

synthetic textile wastes   In general, synthetic textile wastes are less polluting than *cotton textile wastes* or *woollen industry wastes,* because the natural fibres have to be well washed before they can be processed. Wet processes include scouring (washing), dyeing or bleaching followed by rinsing. The *BOD₅* of the mixed waste waters is unlikely to exceed 300 mg/litre, with *suspended solids* of 100 mg/litre or less.

Synura   A green *alga* which may occur in large numbers (*blooms*), resulting in objectionable tastes and smells in a drinking water. In temperate climates it occurs in spring.

T

tactic movement   *Taxis*

Taenia   A genus of parasitic *tapeworm. Taenia saginata* larvae are found in cattle and *T. Solium* in pigs, but both are also *parasites* in man.

taeniasis   Infection by the tapeworm *Taenia.*

tagging, labelling   The use of *tracers.*

tailings   (1) In refuse, tailings are the material left after ashes and dust have been screened off, and iron, steel and any other material salvaged. This material then goes to the tip for disposal or to the *incinerator* for reduction in volume. (2) Waste material from *mineral preparation* plants, mainly fine solid *gangue* that becomes fluid if mixed with water.

tall stack policy The policy in most countries of building chimneys so high that in fine weather no pollution reaches the ground nearby.

tanker ballast water See *oil tanker washings.*

tank sewer A short length of *sewer* with a considerably larger diameter than most of its run, used for storing *stormwater.*

tannery wastes *Chromium* salts (e.g. chromium sulphate) are often now used, although some heavy leather products still require natural vegetable materials for tanning. The mixed waste waters of pre-tanning and tanning are high in *total solids* (5000 to 20 000 mg/litre), 5% being *suspended solids,* 95% dissolved. The BOD_5 varies between 400 and 4000 mg/litre. Vegetable tanning produces more BOD than chrome tanning. The mixed wastes contain about 30 to 70 mg/litre of trivalent chromium and 100 to 1000 mg/litre of sulphide. *See beamhouse wastes, fellmongering.*

Tanypus A midge found in rivers with mild *sewage* pollution. The *larvae* may survive even in waters with 3 to 6 mg/litre of lead.

tapered aeration (*Figure T.1*) Provision of more air at the entrance to an *aeration tank* and less near its outlet. The *oxygen demand* of the *mixed liquor* is proportional to the concentration

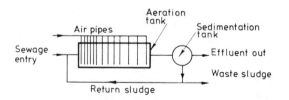

Figure T.1 Tapered aeration of activated sludge, flow diagram

of the unreacted *sewage.* Thus if all the sewage enters by the main inlet to the aeration tanks, the mixed liquor needs more air there than near the outlet. This can be achieved by having more *diffusers* or *cone aerators* of larger capacity at the beginning of the tank. *Compare step aeration.*

tape sampler, AISI *sampler.* A strip of white paper that catches dust from air drawn automatically through it at pre-set intervals. The result is a sequence of spots on the tape, made every hour or half hour. The relative blackness of the spots can be measured by a light meter (photometer) and they can also be chemically analysed.

315

tapeworms, Cestodes *Parasites,* worms that are found in a variety of animals or fish. Some are parasitic in man, especially *Taenia.* Tapeworm eggs are excreted by the infected human and so are found in *sewage.* Many of the eggs are removed by the *sludge* in *sedimentation,* but they remain alive in *raw* or *digested sludge* for many months. Tapeworm populations can be controlled by keeping pigs and cattle away from sewage *outfalls* or from ground recently treated by sewage sludge. *See Diphyllobothrium latum, Hymenolepsis nana.*

taste and palatability of water Human taste can distinguish only between four sensations on the tongue: bitter, sour, salt and sweet. However, the sense of smell combined with that of taste enables us to acquire an enormous range of apparent tastes that give palatability to what we eat and drink. Consequently, the palatability of a water depends almost entirely on its smell, since few waters have any real taste. Many harmless but unpalatable tastes and smells exist in water, caused by *actinomycetes, algae, trace organic* compounds, *chlorination,* etc. They can sometimes be removed by *filtration* with *activated carbon* or *ozonation* or even *breakpoint chlorination.*

taxis, tactic movement Movement of a *cell* or organism in response to a stimulus. Phototaxis is movement towards light. Chemotaxis is caused by a chemical such as a *pheromone.*

taxon (plural **taxa**) A class or type of organism. *See* below.

taxonomy The classification of living organisms by their body structure, in a hierarchy of seven major divisions (*taxa*), reducing in range from (1) *kingdom* to (2) *phylum,* (3) *class,* (4) *order,* (5) *family,* (6) *genus* and (7) *species.* Thus, *Escherichia coli* is a species of the genus *Escherichia* in the class of Eubacteria, in the phylum of *bacteria.* For bacteria the standard textbook is *Bergey's Manual of Determinative Bacteriology* (8th edn, 1974). An earlier classification was to divide all small beings into plants and animals but now all microbes are categorised as *protists.*

TDS *Total dissolved solids* (or salts).

telemetry Remote measurement, including the long-distance transmission of information from rain gauges, flow meters, level gauges, etc.

telescopic valve A type of *decanting valve* consisting of a vertical length of pipe outside a *sedimentation tank,* with a bellmouth at the top, which can be raised or lowered manually or by electric power. It is used for withdrawing sludge especially from *primary sedimentation tanks.* The lower end of the pipe is connected to the bottom of the *sludge hopper.* When the bellmouth is lowered below the liquid level in the tank, the sludge naturally rises up into the bellmouth; the operator can see it and judge when de-sludging should stop, by the appearance of the sludge.

telesmoke A simple optical instrument that enables *Ringelmann chart* readings of *dark smoke* to be taken fairly easily and by a single observer.

temporary hardness *Carbonate hardness.*

Tenten filter A *moving-bed sand filter.*

teratogenic Descriptive of a substance that deforms a foetus or otherwise causes offspring to be mis-shapen. Some *organo-chlorine compounds* are teratogenic.

terminal pond, t. lagoon A *waste stabilisation pond* or *lagoon* with no overflow. The *effluent* evaporates or percolates into the soil.

terminal velocity Freely falling masses increase in downward speed to a maximum, the terminal velocity, reached when the fluid resistance equals the weight of the falling mass. *See Stokes' law.*

tertiary treatment, polishing *Sewage treatment* that improves on, and follows, efficient *secondary treatment.* Conventional sewage treatment with *primary sedimentation,* plus secondary treatment in the form of biological oxidation with final *sedimentation,* can reliably produce a final *effluent* with *suspended solids* below 30 mg/litre and *BOD$_5$* below 20 mg/litre. If, e.g., a 10 : 10 effluent is needed, tertiary treatment will polish the 30 : 20 effluent by removing the fine suspended matter that remains in the effluent from the *humus* tanks or activated sludge treatment and, consequently further reduce the BOD. The commonest tertiary processes are *micro-straining,* use of various sand filters, also running the effluent on to *grass plots,* or *land treatment* or *maturation ponds.* Sometimes tertiary treatment enables water to be immediately re-used for an industrial or semi-industrial purpose—e.g. farming. Otherwise it may be necessary because the *receiving water* has a relatively small flow and consequent low *dilution. See advanced waste water treatment.*

tetra-ethyl lead and **tetra-methyl lead** *See lead in petrol.*

Tetrahymena pyriformis A *ciliate protozoon* that is common in freshwater and salt marshes. The numbers of this species may be counted in *bio-assays* of marshes—e.g. for assessing the effects of toxic chemicals.

textile industry wastes *See cotton textile wastes, synthetic textile wastes, woollen industry wastes.*

textile removal device An endless rope continously moving through a *trommel,* which catches pieces of rope, textiles or wire and dumps them outside the trommel where they cannot block it.

theoretical oxygen demand The *oxygen demand* that would be calculated from the chemical equation if all the parts of the sample were completely oxidised. *See ultimate oxygen demand.*

therm A unit of heat especially for gas; for other fuels it has been superseded by the megajoule (MJ). 1 therm = 100 000 British thermal units = 105.5 MJ.

thermal conditioning of sludge *See heat treatment of sludge.*

thermal oxidation *Incineration.*

thermal pollution Discharge of hot water to a *receiving water.* It may be undesirable, because warming the water reduces its *air saturation value,* its capacity to dissolve oxygen, which may harm fish. Few fish will survive temperatures above 30°C. One UK requirement is that water discharged to a river should not exceed 25°C. The main culprit is industrial *cooling water.* In the USA in 1980 the cooling water requirements of the power stations were expected to equal one-fifth of the country's *runoff.* Since many of these warm *outfalls* are concentrated in a few rivers, the fish would be killed unless another solution was found to the problem—e.g. utilisation of the surplus, low-grade heat. Nuclear power stations, being thermally very inefficient, require more cooling water than fossil fuel-fired power stations. Thermal pollution eliminates or greatly reduces the *periphyton* where it does not kill all life.

thermal precipitator A dust-sampling instrument containing an electrically heated wire, which determines the amount of dust in air.

thermal processing of refuse *Incineration* or *pyrolysis.*

thermal stratification Water is most dense at 4°C and less dense at all other temperatures, a difference which causes many *density currents.* Large *reservoirs,* lakes or seas may stratify in summer into three distinguishable layers, the *epilimnion* at the top, the *thermocline* further down and the *hypolimnion* at the bottom *(see Figure T.2).* During the spring the water is cool and mixed by the storms of the preceding winter but the surface water is warming up, creating the warm epilimnion above the

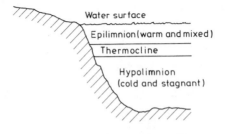

Figure T.2 Thermal stratification of a lake or sea

thermocline. The thermocline is a relatively narrow strip of water with a rapid change of temperature from the warmth above to the cold, stagnant hypolimnion below. In autumn the upper layers cool and tend to sink, until eventually the body of water becomes completely mixed in the autumn turnover. During the summer stratification the best water comes from the thermocline. The epilimnion may have many *algae* and the hypolimnion may be *anaerobic*. During the autumn turnover it is often difficult to obtain any good water. Lakes or reservoirs used for water supply are therefore often de-stratified as a matter of policy during the summer. Winter stratification may occur also, because the densest water, at 4°C, remains at the bottom. The water at 0°C floats on top and sometimes freezes. *See de-stratification of lakes and reservoirs.*

thermal wheel   A *heat-recovery wheel.*

thermocline, metalimnion   The thin horizontal layer of water that separates the warm *epilimnion* above from the cool *hypolimnion* below in summer in a lake, sea or *reservoir* deeper than about 20 m. Up or down through the thermocline the temperature varies much more rapidly than elsewhere. *See thermal stratification.*

thermo-osmosis   Other things being equal, water and water vapour underground move in the direction of flow of heat, towards cooler ground, because this also is the direction of increasing *surface tension.* Thermo-osmosis is important in icy regions such as northern Canada, where it causes accumulations of ice underground.

thermophile   Any *bacterium* that thrives at temperatures above 40°C.

thermophilic digestion   *Anaerobic sludge digestion* at about 50°C, a more difficult process to control than *mesophilic digestion* and more expensive because of the extra heat needed.

thermoplastics   *Plastics* that become soft when heated and harden again when cooled. They make up more than half of plastics output in western countries, and include polypropylene (6%), polythene (30%), polystyrene (18%) and vinyls, including *polyvinyl chloride* (19%). Waste, reclaimed thermoplastics have been re-shaped by hot pressure at 150 to 170°C with a pressure between 200 to 270 atm, using mixed polythene, polystyrene and PVC.

thickener, thickening   *See sludge thickening.*

thio-   Prefix indicating a connection with sulphur.

Thiobacillus   A genus of *sulphur-oxidising bacteria* that are *aerobic* and *lithotropic* autotrophs, although some species also may be *heterotrophs.* Different species may oxidise sulphur or sulphides or thiosulphates or sulphites. *T. ferro-oxidans* derives energy also from the oxidation of *ferrous salts* (Fe^{2+}) to ferric (Fe^{3+}). It lives in water at pH of 2 to 4 and is abundant at discharges of *acid mine drainage* water which contain much iron sulphide. *T. thio-oxidans, T. neapolitanus* and *T. concretovorus* are important to the *sulphide corrosion* of sewers. The first two can grow at pH 6.0 but their production of sulphuric acid lowers the *pH* to about 3.0, when *T. concretovorus* begins to flourish, lowering the pH further to 1.0 *T. concretovorus* cannot exist at pH above 4. Some authors state that *T. concretovorus* may be only a strain of *T. thio-oxidans.*

Thiorhodaceae *Chromatiaceae.*

Thiothrix This *filamentous organism,* a sulphur *bacterium* related to *Beggiatoa,* is a *chemosynthetic* autotroph and *lithotroph.* It oxidises *hydrogen sulphide* to sulphur and can be recognised under the microscope by the yellow globules of intracellular sulphur. It may occur in *bulking* activated sludge.

thiourea *Allyl thiourea.*

third pollution The pollution of land by litter, whether abandoned cars, newspapers, soft drink *cans* in flower beds or badly *controlled tips.*

thixotropic Description of a fluid that acts like a gel when stationary but like a viscous liquid when stirred. It is a useful property for drilling fluids in boreholes. Thick sewage *sludge* also may be thixotropic.

threshold limit value, TLV The highest concentration of an airborne pollutant to which a worker may be subjected. The UK Health and Safety Executive and their Factory or *Alkali Inspectorate* accept the values regularly published by the American Conference of Governmental Industrial Hygienists. These apply only to works, not the general public. Each TLV applies to one pollutant alone in air. If other pollutants are present, their TLVs may have to be reduced. Other pollutants may increase the harmful effect of the first, or reduce it.

threshold odour number A subjective measure of smell—the number of times a sample can be diluted with pure water (or clean air) before the smell ceases to be detectable in the water (or air).

tile field A *leaching field.*

tilted-plate separator A *lamella separator.*

time of concentration For a *catchment* during a storm, the time taken for the flow rate in a given channel to reach the maximum. It is the *overland flow* time plus the open channel or pipe flow time. For *stormwater sewers* there is also the time of entry or inlet time, during which the rainwater flows through gutters and *gulleys* into the *sewer* (some 3 to 4 min in towns), also the time taken to flow to a given point in the sewer. *See unit hydrograph.*

time lag of a catchment, basin lag The total delay between the centroid of a rainstorm and the peak *runoff,* seen in a *hydrograph* of the storm.

tipper US term for a person who empties *dust bins,* a dustman.

tipping at sea *See ocean dumping.*

tipping trough An automatic device with the same function as a *dosing siphon,* used at small works for applying *effluent* from *sedimentation tanks* to a *trickling filter.* The trough fills with *sewage* effluent, tips over when full and discharges as a plug to the trickling filter. The flow is then strong enough to swing the *distributor* round.

tip temperature In the outer, well-aerated part of a tip, 2 weeks after it is laid down, the tip temperature may exceed 65°C, while the centre of the material, laid at the same time, may be much cooler if it is well compacted and airless. Temperatures of 40 to 60°C in tips may continue for many months. The high temperature helps to kill *microbes* that might cause disease, and shows that *garbage* and perhaps paper are decomposing and that some immediate settlement may be expected. If the tip temperature rises excessively and fire is suspected, it can be damped down, if not extinguished entirely, by compaction with heavy vehicles. Sealing of the exposed surfaces with clay or similar impervious material also helps.

TKN Total nitrogen by the *Kjeldahl technique.*

TL$_m$ *Median tolerance limit.*

TLV *Threshold limit value.*

TNO The Dutch public health association.

TNO rotor, cage r. A *brush aerator* with a horizontal shaft, which superseded the *Kessener brush.* It is used in *oxidation ditches* with channel depths of 1 to 2 m. At maximum immersion the power required is about 1.5 kW per metre length of rotor. For plants serving a population above 10 000, it has been superseded by the *Mammoth rotor. See oxygenation efficiency.*

TOC *Total organic carbon.*

TOD *Total oxygen demand.*

tortuous flow *Turbulent flow.*

tortuosity The path of a fluid through a *sand filter* is long and twisting (tortuous), the length being a fair measure of its tortuosity. This, if the bed thickness if T and the length of the flow path is L, the tortuosity is L/T, a value greater than 1.

total alkalinity, methyl orange a. In water supply this is defined as the *alkalinity* measured above pH 4.5, the endpoint of titrations using methyl orange as an indicator. It represents the *pH* when bicarbonate ions (HCO_3^-) have reacted with strong acid to form H_2CO_3.

total chlorine residual *Chlorine residual.*

total coliform count A *bacterial count* of all the different coliform bacteria present in a sample, as opposed to that of one species—e.g. *Escherichia coli.*

total dissolved solids, t.d.salt, TDS TDS are measured by evaporation at 103 to 105°C of a filtered sample of water and weighing the residue, but they can calso be estimated quickly from the *electrical conductivity* of the water. Continuous recorders of conductivity immediately indicate any change in conductivity that might be a sign of new pollution. *See* Table of Allowable Contaminants in Drinking Water, page 359.

total head *See head of water.*

total incineration, slagging i. A US way of reducing refuse to *slag,* completely molten ash, by high-temperature burning, which results in the lowest volume of ash and metal residue. The temperature required is at least 1650°C, about 700°C more than conventional *incineration.* The slag can be poured into water to make slag granules or may be broken up by other means. The volume reduction is claimed to be 96%, leaving 2% slag and 2% flyash, a considerable improvement on other methods; but extra fuel is needed, oxides of nitrogen *(NOx)* are likely to pollute the air and the *refractory* lining may be damaged by the slag. More *dust* and *grit* are emitted than with conventional incineration, and extra equipment is needed to handle the extra fuel. Probably the same or more *gas-cleaning plant* is needed; consequently, it must be even more expensive than conventional incineration. The high temperature can be achieved by the use of oxygen, by highly pre-heated air or by extra fuel. Most systems use a *shaft furnace.*

total Kjeldahl nitrogen The sum of *ammoniacal nitrogen* and unoxidised *organic nitrogen,* measured by the *Kjeldahl technique.*

total organic carbon, TOC A fast method of measuring the *organic* content of a water. A small quantity of the liquid sample or a dilution of it is injected into a stream of hot air or *oxygen* in the instrument. The water is vaporised and the organic matter oxidised to *carbon dioxide,* CO_2. The concentration of CO_2 in the gas stream is measured by an infra-red device. Alternatively, the CO_2 may be reduced in a catalytic column to *methane,* CH_4. The methane concentration can then be measured. This latter technique is more complex but more accurate at low concentrations of organic matter.

total oxygen demand, TOD TOD can be measured, according to

ASTM 31, from the depletion of *oxygen* in a *nitrogen*–oxygen carrier gas in a platinum catalysed combustion chamber.

total solids   Total solids in water are measured, like *total dissolved solids,* by evaporation at 103 to 105°C. It is the sum of the *suspended solids* and dissolved solids.

totting   Investigating the contents of other people's dust bins or waste tips, an activity illegal in the UK since the Public Health Act 1936.

toxic   Poisonous.

toxic sludge   *See sump oil* and below.

toxic waste   Poisonous waste. In the UK it is illegal to transport or tip it without notifying the local authority in the area where it is, the local authority in the area where it will be tipped and the *water authorities* in both areas (*see Figure T.3*). Harmful wastes in the UK have been estimated at 0.5 to 2.5 million tonnes a year.

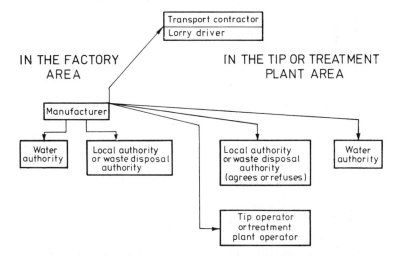

Figure T.3   Notifications of intention to deposit toxic waste, required under the UK Deposit of Poisonous Waste Act 1972, are shown by arrows. The tip operator must advise the manufacturer that the deposit has been made. The operator must also notify the two local and two water authorities. The local authority in the tip area must notify agreement to the other local authority, the tip operator, the transport contractor and the manufacturer

The corresponding figure in the USA has been put at 10 million tonnes by the *EPA,* excluding *nuclear reactor wastes.* In the UK about 90% of toxic waste, liquid or solid, is tipped on *controlled tips,* 5% at sea, 1% is incinerated and less than 1% is chemically treated. Solid toxic wastes are most dangerous when they dissolve and enter a watercourse or *groundwater.* Liquid wastes should therefore not be tipped near any toxic solids that are at all soluble. *See* below.

toxic waste disposal Since *toxic wastes* are not all allowed to be left on a *controlled tip,* there are six or so methods of treatment, some of which make them suitable for tipping, including: recovery (*regeneration*) of the chemical; chemical fixation or *encapsulation;* chemical or biochemical treatment; dumping at sea under licence; *incineration; pyrolysis.* Physical treatments include de-watering, *filtration, sedimentation* or *distillation,* all of which reduce the final volume for disposal. Recovery is best because it enables the substance to be re-used and eliminates the poison from the tip. Biochemical treatment has been used for pharmaceutical wastes and those from *spent liquor,* including phenols. *See Dumping at Sea Act 1974.*

toxin A poison produced by an organism.

trace element An element at a concentration, in general, below 1 mg/litre. Many of them are *micro-nutrients.*

trace organics Some *organic compounds* found in tiny quantities in drinking water can be smelly or taste unpleasant—e.g. *phenols.* They are difficult to remove by *chlorination* and they may also form poisonous *organo-chlorine compounds.* Most of them can be removed in *carbon adsorption beds,* before chlorination, and this has been done at *Lake Tahoe. See foaming, foam fractionation.*

tracer A substance injected into a tank, river, etc., to identify the water, and to show where it travels to and how quickly. Tracers can show *density currents* or other flow patterns in a tank. Tracer tests in *groundwater* are usually slow and may not help in *monitoring wells.* A groundwater tracer should move at the same rate as the water, and not be adsorbed on clays or other soils (like many *cations* and dyes) and so be lost. Tracers should be cheap, easily injected and harmless to drink. A chemical tracer should be harmless, and if radioactive, should be so much diluted that it cannot harm the drinker. The commonest radioactive tracers are phosphorus-32, carbon-14, *iodine*-31.*Tritium oxide* is also used. Chemical tracers include dyes such as *fluorescein* (sodium salt), *rhodamine B* or metallic ions rarely found in natural water, such as *lithium salts.*

trade effluent, industrial e. Used water, not necessarily polluting, discharged by a trade or industry. A wide variety of pollutants are, however, discharged and need to be considered if the river quality is to be maintained. In 1973 some 70% of trade effluents flowed into *sewers* in Britain, the remainder direct to rivers or the sea. No widely accepted standards exist for trade effluents discharged to sewers, but the limits stated by some *water* *authorities* are: not warmer than 42°C; *pH* between 6 and 11; sulphide, cyanide in *cyanide compounds* and available *sulphur dioxide* each less than 10 mg/litre as S, CN, or SO_2; free *ammonia* below 100 mg/litre as NH_3; *sulphate* below 1500 mg/litre as SO_4; available *chlorine* less than 100 mg/litre as Cl; formaldehyde less than 20 mg/litre as HCHO; grease or oil together less than 500 mg/litre; *settleable solids* less than 500 mg/litre. Absolutely prohibited are substances that are dangerous in sewers, such as petrol, calcium carbide and the organic solvents.

trade-waste incinerator A batch *incinerator* that burns furniture, animal carcases, bulky trade waste and other refuse unsuitable for *controlled tipping.* Even if *incineration* is not the authority's main method of refuse treatment, most waste disposal authorities need at least one such incinerator. They are also installed at hospitals.

train A sequence. *See sampling train, treatment train.*

tramp iron Broken bolts, loose nuts or other stray iron or steel, often easily extracted from material flowing on a conveyor belt by a *magnetic separator.*

tramp oil Oil that should not be present in a fluid. It may sometimes be removed by *flotation.*

transect A strip of land that is surveyed for its plant and animal populations, often by frequency estimations. Re-surveys will indicate whether pollution has affected the life. A beach, for example, has been surveyed in a transect 10 m wide from low water at spring tides to above high water.

transfer station, t. loading s. (or USA **transfer loading facility**) A site in or near a city, at which collecting lorries give up their refuse for removal to the disposal site by bulk transporter or barge or railway.

transpiration Loss as water vapour through the leaves of plants, of water brought up from the soil through the roots. It increases with increasing wind, sunshine and temperature and as the plant increases its area of foliage. Transpiration varies from plant to plant. Soil moisture does not greatly affect it until the *wilting point* of the soil is reached. *See evapo-transpiration, hydrological cycle.*

trash can US term for *dust bin.*

travelling-bridge scraper The commonest method in the UK of scraping the *sludge* along the floor of *horizontal-flow sedimentation tanks.* A bridge spanning the width of the tank has *scraper* blades hung from it. It scrapes from the outlet end of the tank to the *sludge hoppers* at the inlet end. The blades are then lifted off the bottom and the bridge returns. The mechanism is electrically driven through either a rack and pinion or a rope drive. (*Refer to Figure H. 4.*)

travelling grate   One of the many types of *grate* used on modern continuous *incinerators*. It may be a succession of chain grates or a rocking grate or a reciprocating grate.

Travis hydrolytic tank   A *hydrolytic tank.*

treatability   For liquid wastes, usually this means the ability to undergo *biological treatment.*

treatment train   A sequence of treatment processes.

Trematoda   *Flukes.*

TREMcard, TRansport EMergency card   A card accompanying a lorry carrying a dangerous load, which shows the type of load, its dangers and the action to be taken after an accident. TREMcards are prepared by the European Council of Chemical Manufacturers' Federations. *See also Hazchem.*

trenching   Disposal of liquid *sludge* by partly filling a trench with it and ploughing and cultivating the land after it has dried.

trench method, sloping t.m. of sanitary landfill   A standard US method of *controlling tipping,* used where *groundwater* is not near the surface and there is no danger of flooding the trench. A bulldozer or other digging tool digs a trench of (say) 2 m deep and 2.5 times as wide as the length of the dozer blade (about 5 m) with the ends of the trench sloping at 1 in 3. The trench is kept to a length of about 50 m. Earth is brought at the end of the day from the far end of the trench to cover the day's refuse. Thus the

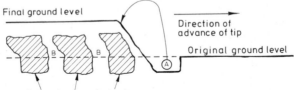

Figure T.4 Trench method of controlled tipping (vertical section)

method provides its own *cover* automatically from the trench dug for the next day's refuse, unlike the *area method,* but it is extravagant with land, about 2 m of unused ground is left between neighbouring trenches. (*See Figure T.4.*)

Trent biotic index   A way of biologically classifying rivers according to their degree of pollution, by the frequency and types of animals (invertebrate) found in them, such as *caddis flies,* beetles, *may flies, stone flies,* worms, blood worms, leeches, *freshwater shrimps.* The index ranges from 0 for very polluted water without animals through nine other classes to class 10, without pollution and with abundant life. The Department of the Environment and the Scottish Development Department in their *river classification* grouped the biotic indices in the following way so as to verify whether the biological classification tallied with their mainly chemical classification: indices 9 to 10, unpolluted; 6 to 7 slightly polluted; 3 to 5 polluted; 0 to 2 grossly polluted. Both methods use a *biological index of pollution.*

Trichuris trichiura Whipworm.

trichloramine, NCl₃ A *chloramine* formed during *breakpoint chlorination* only if the water is acid, below pH 5.0.

trichlorofluormethane, trichlorofluoromethane, fluorotrichloromethane, fluorocarbon-11, Freon 11 An *aerosol propellant* with the chemical formula CCl_3F, of which up to 100 000 tonnes are released yearly into the atmosphere. *See haloform.*

Trichoptera Caddis flies.

Trichosporon A fungus which is common in *activated sludge* plants and *trickling filters.*

trickling filter, percolating f., sprinkling f., biological f., continuous f., bacteria bed A development from *land filtration* which was first used at Salford, England, in 1893. It is, with the *activated sludge* process, one of the two basic *biological treatments* for *sewage.* In the typical standard-rate filter *(Figure T.5),* the *effluent* trickles down over coarse stones that fill a 2 m

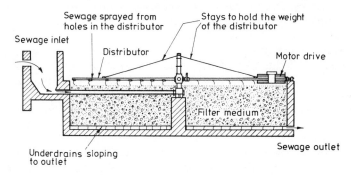

Figure T.5 Trickling filter, cross-section showing electrically driven distributor

deep tank. Many other types of trickling filter exist, some with plastics packing and up to 8 m high. The size of the stone packing is usually about 24 mm cube at the top, increasing to 100 to 140 mm for the lowest 0.3 m. *Biological films* develop on the surfaces which are all freely accessible to upward-flowing air. The slimes *adsorb* and absorb both soluble and *suspended matter* and biodegrade it. Trickling filters are thus not true *filters.* Unlike its intermittent predecessor the *contact bed,* the trickling filter is continuous. The effluent flows to a *humus tank.* The conventional way of working a standard-rate trickling filter is to pass *settled sewage* through it once, in single-stage filtration. Design is usually on the basis of *organic loading* of 0.1 kg BOD_5 per day per m³ of filter volume, or a flow of 0.4 m³ per day per m² of plan area. This should reduce the effluent to a BOD_5 below 20 mg/litre. In summer the effluent will probably be nitrified but in winter this is less likely. The attractions of the standard-rate trickling filter are its low power cost and simplicity. Its disadvantages are the capital cost for large works, the large land area compared with activated sludge and the absence of versatility in operation. The trickling filter is not wholly *aerobic,* except when it starts up. After the slime has formed, there is an *anaerobic* layer at the surface of the stones which little or no oxygen can reach. The main micro-organisms are *aerobes, anaeobes* and *facultative anaerobic bacteria,* but *fungi* are present also on aerobic surfaces, where they compete with bacteria for food. *Algae* occur at the top of the filter where there is sunlight. Trickling filters have been used to nitrify activated sludge effluent, and *anaerobic filters* can be used for *de-nitrification. See dosing siphon, filter flies, maturing, ponding, sloughing, sludge production, tipping trough,* and *compare alternating double filtration, enclosed aerated filtration, high-rate trickling filtration, nitrifying filtration, re-circulation.*

tri-media filter   A *multi-media filter.*

tripolite   *Diatomite.*

trippage   The number of trips made by a *returnable container.* In the UK glass milk bottles are refilled about 25 times, but plastics milk bottles in the USA may be refilled 100 times.

tritium oxide either **T₂O** or **HTO**   An unsual form of *heavy water* which is *radioactive* because of its tritium content. It can be used as a *tracer.*

trommel (German for 'drum'), **rotary screen**   A rotating steel plate cylinder with holes punched in it (*see Figure T.6*) which functions as a *screen.* Because of its rotation it is not easily blocked and is often used for separating municipal refuse into different sizes. One trommel may be inside another; if so, the bigger holes are in the smaller trommel. When two sizes of hole are used on one trommel, the larger holes are downhill from the smaller ones. Material is often screened out of municipal refuse at about 20 mm. This minus 20 mm material is mainly ash,

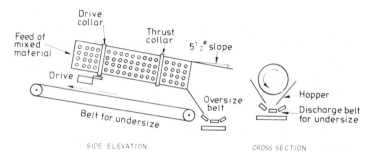

Figure T.6 Trommel for sorting mixed material into two sizes (casing ommitted).

A trommel for sorting into three sizes has two different opening sizes, with the larger openings downhill

broken glass and other dense material suitable for *daily cover*. *See rotary screen magnet, textile removal device.*

trophic Concerned with nutrition.

trophic chain A *food chain*.

trophic level The location of an organism in the *food chain*. In the *ecological* pyramid each level of the pyramid feeds off the level below it.

trunk main A supply *main* from a *water treatment* plant to a *service reservoir* or *water tower*, not normally used for supplying water to a consumer.

trunk sewer A large *sewer* into which many branch sewers flow.

Trypanosoma A genus of flagellate *protozoa* which includes *T. zambiense* (or *gambiense*) and *T. rhodesiense*. They cause *sleeping sickness*.

trypanosomiasis *Sleeping sickness.*

tse-tse fly *See sleeping sickness.*

tube diffuser *See diffuser* (1).

tube module An *inclined-tube settling tank*.

tube precipitator An *electrostatic precipitator* that collects liquid such as sulphuric acid mist, and is equipped with tubes instead of the plates of the precipitator for collecting solids.

tube press, multiple-tube press A *slurry*-dewatering device that has remarkable performance in de-watering difficult, fine, polluting industrial slurries such as steel furnace scrubber liquid (from 1.5 % solids concentrated to 15 % solids), etc. It is built up from pairs of concentric tubes, either 2 or 2.4 m long and of 0.3 and 0.2 m diameter. The inner tube is perforated and surrounded by stainless steel mesh supporting a replaceable filter *membrane*. The *filter cake* collects outside it and is periodically removed by lowering the inner tube, which allows the cake to drop off. The slurry has to be pumped into the annulus between the tubes. On the outer edge of the annulus, just inside the outer tube, is another tube, of strong rubber. Pressure is applied outside the rubber tube by hydraulic fluid, compressing the rubber, reducing its diameter and forcing the clear water through the filter into the inner tube, from which it flows away to treatment and the river. Pressures much higher than have been possible in any previous filter are used, about 150 atm, in spite of which the energy consumption is not high. The major disadvantage is high capital cost compared with other mechanical de-watering devices.

tuberculation Bacterial *corrosion* of metal pipe in the form of hard but brittle bubbles (tubercles or nodules) of rust up to 20 mm high above the surface. In iron or steel pipe the rust is hard ferric hydroxide sometimes strengthened by *calcium carbonate* and *manganese* dioxide.

tuberculosis Tuberculosis can be *waterborne,* caused by swimming in polluted water. Some of the *bacteria* that cause it are *Mycobacterium tuberculosis, M. balnei* and *M. bovis,* obligate *aerobes* that multiply in lesions within the lung.

tube settler An *inclined-tube settling tank.*

tube sheet, t. plate A sheet or plate of steel through which holes have been punched or drilled. Boiler tubes are inserted into these holes and a gastight joint made between the tube and the plate. Tube plates are also used in *fabric filters* to carry the thimbles that hold the filter bags.

Tubificidae A family of red worms (e.g. *Tubifex*) typical of severely sewage-polluted mud which may occur also on highly overloaded *trickling filters.*

tularaemia An illness of rodents caused by various bacteria, named *Bacterium tularense, Francisella tularensis* and *Pasteurella tularensis.* Animal populations can be decimated in a season by this disease, which they catch from ticks. Humans normally acquire it by contact with skins or carcases, occasionally by tick bite or possibly from water.

tumbling bay A *back drop.*

Turbellaria, planarians Free-living flatworms. Some species may be found in water that is clean or mildly polluted with sewage—e.g. *Dendrocoelum lacteum.* The main ones are *flukes* and *tapeworms.*

turbid Water that is not transparent is turbid. *Turbidity* is caused usually by particles of *colloid* size, including micro-organisms, which are removed from drinking water by *coagulation.* The Delhi epidemic of *infectious hepatitis* showed that this is essential.

330

turbidimeter, turbidity meter An instrument such as a *nephelometer* which measures the *turbidity* of water.

turbidity Dirt in water, suspended particles that reduce its transparency. It can be measured by the reduction of the strength of a light ray passing through the water. Several ways of measuring turbidity exist, including the *Jackson turbidity unit* and *nephelometer* methods. The *EPA* laid down in 1976 that the turbidity of drinking water should exceed neither 1 turbidity unit as a monthly average nor 5 units as the average of two consecutive days, determined nephelometrically.

turbidity breakthrough Excessive *turbidity* (dirt) in a *filtrate.*

turbidity current A *density current.*

turbine aerator A high-speed, vertical-shaft submerged *aerator,* which mixes *sewage* with air or *oxygen* bubbles introduced into the *aeration tank* through *spargers* or *diffusers.* Sometimes a *draft tube* (2) is fixed below it.

turbine pump A multi-stage *centrifugal pump* with *diffuser* passages, which is often used in boreholes. It may be fitted with mixed-flow *impellers* as opposed to purely *centrifugal* impellers in order to prevent overloading of the motor if the *head of water* drops below the design level.

turbulent-contact absorber A *moving-bed scrubber* with several beds about 30 cm deep when still.

turbulent flow, eddy f., sinuous f., tortuous f. Flow faster than *laminar flow,* in which the fluid eddies and does not move steadily.

turnover *See thermal stratification.*

turnover time The time required for water to lose a quantity of *phosphorus* (or other *nutrient*) equal to the total amount present.

tuyeres, twyers A horizontal ring of holes at hearth level through the wall of a *cupola* or blast furnace, which provides the hot or cold blast of air or oxygen for burning the fuel.

TWL Top water level.

two-drum magnetic separator A *magnetic separator* that recovers iron and steel in clean condition without additional treatment, by the use of two magnetic drums. The first drum at the end of the refuse conveyor picks up the iron and steel and drops them on to a short subsidiary belt that feeds the second drum. Much of the paper, etc., drops off, and of what remains, the bulk is removed at the second drum, rotating against the direction of the subsidiary belt. The second drum lifts up the magnetic material and drops it on to the main belt for iron and steel.

two-layer filter A *dual-media filter.*

two-stage digestion *See anaerobic sludge digestion.*

two-stage filtration The use of two *trickling filters* in series, usually with a *sedimentation tank* between them. *Compare alternating double filtration.*

two-stage incinerator, controlled-air i., starved-air i. When metals are reclaimed by *incineration,* this can take place in two simultaneous stages, the first at a relatively low (550° C maximum) temperature to avoid spoiling the metal, the second in an *afterburner* with added air to eliminate air pollution and to burn the *pyrolysis* products from the first stage. This is done when the insulation is burnt off copper wire, when brake linings are burnt off brake shoes or when 200-litre steel drums are reclaimed. Gas or oil firing may be needed for starting up or closing down without polluting the air.

two-stage sedimentation The use of two sets of *sedimentation tanks* in series. For the first-stage tanks *de-sludging* is more frequent than for the second stage, where there is a smaller quantity of more watery *sludge.* The treatment is little used in *sewage* works.

two-stage shredding, double s. Shredding the same refuse in two *mills* successively, sometimes with *magnetic separators* between. It is roughly twice as expensive as single shredding but a product small enough to make *waste-derived fuel* is difficult to obtain by single shredding. *See multi-stage shredding.*

two-storey tank A tank with one compartment over the other—e.g. the *hydrolytic tank,* the *Imhoff tank.*

twyers *See* tuyeres.

tyndallisation Steaming a culture *medium* for short periods at 100° C on three successive days, to avoid the damage to heat-sensitive substances in the medium which might be caused by the longer period in an *autoclave.* The process is intended also to kill *cysts.* Cysts are encouraged to grow by the first heating and develop out of the cyst form but they are killed by the second or third.

typhoid fever The most serious of the *enteric diseases.* Infection is by intake of food or water containing the bacterium *Salmonella typhi* which has been excreted by an infected person. Even after recovery from the disease, a small proportion of humans continue to pass *S. typhi* in their *faeces* for several months. The *bacteria* can live but not multiply for some months in water.

typhus *See rickettsiae.*

tyres *See rubber tyres.*

U

Ulothrix A wire-shaped green *alga* found in a variety of fresh waters, polluted or unpolluted, and on *trickling filters.*

ultimate oxygen demand, UOD Either the *theoretical oxygen demand* or the ultimate biochemical oxygen demand, the maximum value of the *BOD* test, ordinarily reached after 25 to 30 days' *incubation.* The UOD is the sum of the ultimate *carbonaceous BOD* and the ultimate *nitrogenous BOD.*

ultra-centrifuging A technique for centrifuging very small suspended particles, consequently used for *concentrating viruses.* Because of their small size, only 0.2 to 0.02 μm, *viruses* need to be subjected to a centrifugal force of about 60 000 times gravity for 1 h.

ultra-filtration, hyper-filtration The use of *membrane* filters capable of selectively sieving out particles smaller than 10 μm down to 0.002 μm. To prevent the membrane blocking too rapidly with built-up solids, the water flows past the upstream side of the membrane, sweeping most of the solids with it in the more concentrated waste stream. The membranes are made up of polysulphone or polyacrilonitrile or *cellulose acetate,* and need frequent, laborious cleaning. The operating pressure can be between 1 and 7 atm. *See membrane electret.*

Ulva, **sea lettuce** A green seaweed that grows in sewage-polluted, nutrient-rich estuaries and gives off *hydrogen sulphide* when it rots. It is a multi-cellular *green alga,* attached to sediments or rocks on the bottom.

uncased hole, open h. A borehole without *well casing.*

unconfined groundwater *Groundwater* that is overlain by a *zone of aeration,* unlike *confined groundwater.*

underdrains Pipes under a *trickling filter* or true *filter,* which remove the water. In the *rapid-gravity sand filter, dual-media filter* and *multi-media filter* they also provide access for the backwash water. Their design should be such as to provide adequate backwash water without upsetting the filter *media* above. In *sand filters* underdrains should be covered with gravel and possibly with *garnet* as well. *Sludge drying beds, grass plots,* and similar *sewage treatment* facilities also have underdrains but without backwash.

underfire air, primary a. The air passing through the grate of a furnace. It should be kept to a minimum so as to reduce the *carry-over* to the *dust arrestors.*

underflow In *sewage treatment,* the flow from the bottom of a tank—e.g. *sludge* from a *secondary sedimentation tank* or gravity *sludge thickener.* In *mineral dressing* it may be dense

333

slurry that flows out of the bottom of a gravity thickener or the undersize from a *wet classifier.*

undersize The material that passes through the holes of a *screen. Compare oversize.*

undulant fever *Brucellosis.*

uniform flow Flow in an open channel of constant cross-section or in a closed conduit with an established velocity distribution across the conduit, which is constant along it. *See non-uniform flow.*

uniformity coefficient, c. of u. A measure of the uniformity or variability of the grains of sand in a batch. It is the size of opening through which 60% of the sand by weight passes, divided by the size of opening through which 10% of it will pass. In other words, it is the D_{60} *effective particle size* divided by the D_{10} size.

United States Public Health Service, US PHS This US federal agency in 1962 listed the main requirements of drinking water. Some of them are stated in the Table of Allowable Contaminants in Drinking Water on page 359. The EPA has stated them more recently. *See also drinking water standards, irrigation, swimming water.*

unit hydrograph, u. graph A *hydrograph* that shows for a particular basin the varying *overland flow* that results from unit rainfall (e.g. 25 mm) over it in a given time. Its early shape is determined by the make-up of the *time of concentration.*

unloading *Sloughing off.*

Unox system An *oxygen-activated sludge* process.

UNSCEAR The United Nations Scientific Committee on the effects of Atomic Radiation.

UOD The *ultimate oxygen demand.*

upflow contactor An upward-flow *carbon adsorption bed.*

upland catchment water Water from land that is more than 100 or 200 m above sea level is in Britain likely to be unpolluted by *sewage,* because most large cities are near sea level. It is therefore impossible for much sewage to flow into upland streams. Surface upland water, especially from moorland, is likely to be bacteriologically pure enough for drinking, but it may contain *organic* acids if the ground is peaty and may also be soft and plumbo-solvent. All English *water authorities* have some upland areas except for the Anglian. *See plumbo-solvency.*

upward-flow clarifier (1) A *pebble-bed clarifier.* (2) A *sludge-blanket clarifier.* (3) An *upward-flow sedimentation tank.*

upward-flow floc blanket tank A *sludge-blanket clarifier.*

upward-flow sand filter A *sand filter* in which, as in other types, the coarser sand particles are below but, unlike them, the flow is upwards, providing *coarse-to-fine filtration.* As the water flows up, large *suspended solids* are held in the pea shingle but the finest dirt is held only at the top by the finest sand. *Backwashing* is done by increasing the upward flow rate, creating a tendency to lift the *filter* bed. A metal retaining grid may be added about 10 cm below the top of the bed to prevent the sand being lost, but it allows the sand to expand and be washed. Backwashing, which occurs once or twice daily, can be air-assisted. During *filtration,* difficulties occur when the head available at a given level in the filter is more than the weight of the bed above it. The bed then expands (fluidises) and some of the retained dirt may pass into the clean water. Consequently, upward-flow filters are little used for drinking water treatment except possibly in the USSR, where sufficient assurance against these breakthroughs has been obtained by using very deep filter beds—e.g. 2.5 m. They have been used for *tertiary treatment.*

upward-flow sedimentation tank *(Figure U.1)* A *sedimentation tank* in which the *sewage* enters above the *sludge* level but substantially below the top water level, as in the *Dortmund tank.*

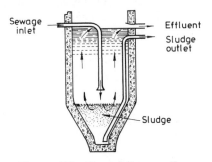

Figure U.1 Upward-flow sedimentation tank (cross-section) for raw water or sewage

Many *pyramidal sedimentation tanks* or *circular sedimentation tanks* have a mixture of radial and upward flow.

upward flow velocity *See overflow rate.*

uranine *See fluorescein.*

uranyl ion, UO_2^{2+} *See* Table of Allowable Contaminants in Drinking Water, page 359.

urea, $CO(NH_2)_2$ Urine contains about 2 to 5 % of urea, although it is rarely found except in *fresh sewage,* because it changes to *ammonia* in water.

urease An *enzyme* possessed by some *bacteria,* which converts *urea* to *ammonia.*

urinal flushing   *See flush cistern.*

US EPA   The United States *EPA (Environmental Protect Agency).*

US PHS   The *United States Public Health Service.*

US PHS 1962   Such a reference in this book implies the list absolute and desirable maximum concentrations of *heavy met* and other pollutants in water, published by the *United Sta Public Health Service* in 1962 for inter-state buses, aircra trains, etc. This has largely been superseded by later stateme of the *US EPA. See* Table of Allowable Contaminants Drinking Water, page 359.

V

vaccination of water   *See stabilisation* (1).

vacuum filter   A horizontal-shaft, continuously rotating dr with filter cloth or metal coil stretched around it, subjected vacuum inside the drum. The lower 25 to 40 % of the drum submerged in the fluid being filtered or de-watered. Rotation slow, at 2 to 9 rev/min allowing a cake of *sludge* to form outs the cloth. There are several methods of discharging the cal with, e.g., drum-type filters having the cloth attached all the w round the drum, discharging by pressure blowback just ahead the knife discharge. String discharge filters use strings abc 1.3 cm apart passing around most of the drum but departi from it at the top over discharge rolls to carry the cake aw from the drum. Belt-type filters use a continuous woven cloth metal belt which acts as the filter *medium* but it also is led aw

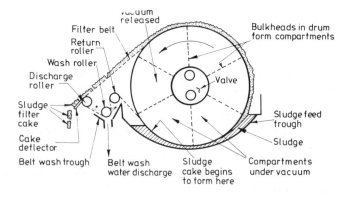

Figure V.1   Vacuum filter with a continuous filter belt, for concentrating sludge

336

from the drum to discharge and washing rolls. This facility for continuously washing the cloth reduces *blinding.* Coil type filters use two layers of stainless steel coil springs which leave the drum in the same way as on the belt filter. The two layers are then separated and the cake discharges from the upper layer. The coils are washed and re-applied to the drum by grooved aligning rolls. Vacuum filters are used in many industries besides *sewage treatment.* Sometimes a series of vertical rotating discs (disc filter) replaces the horizontal cylinder. The discs are faced with nylon cloth and a knife discharges the cake from them. They have a high surface area for a given cost but are little used for de-watering *sewage* or waterworks sludges. Coil or belt filters are commoner and they may reduce *activated sludge* to about 85 % water, or raw primary sludge to an easily handled cake with up to 70 % water. *Conditioning* with suitable chemicals is often essential. As with the *filter press,* the *effluent* is highly polluting and needs treatment.

vacuum flotation *See dissolved-air flotation.*

vacuum latrine A Swedish type of flush toilet with a negative pressure in the *sewer* to suck the *sewage* down. A flush of only 1 or 2 litres is thus possible instead of the 9 litres usual for WCs and drains do not have to slope at the *self-cleansing gradient.*

vacuum sewerage With the *vacuum latrine, sewage* flows in separate *sewers* from *sullage* and undergoes different *sewage treatment.*

vacuum transport *See pneumatic transport of solid waste.*

vadose water Water in the *zone of aeration,* suspended against the force of gravity.

vadose zone The *zone of aeration.*

vagrant waste US term for litter.

valve *See air-release valve, decanting valve, isolating valve, non-return valve, plug valve, sluice valve, telescopic valve, washout valve.*

van der Waal's forces Forces of attraction between molecules or particles of comparable size, which can, among other things, help in *filtration.*

vane-axial cyclone An *axial-inlet cyclone.*

vector control Elimination or reduction of *vectors of disease.* The removal of stagnant water eliminates *mosquitoes* and reduces or eliminates the mosquito-borne diseases. *Shredding* or *controlled tipping* of refuse discourages rats, cockroaches and flies. *DDT* has been used as a vector control but *persistent pesticides* are not desirable, and vectors of *malaria* are becoming resistant to DDT.

vector of disease A fly, *insect, mosquito,* mouse, rat, louse or other form of life that carries a *bacterium, fungus, virus* or other

337

cause of disease from one *host* to another.

vee-blade scraper (*Figure V.2*) A *scraper* for a flat-bottomed *radial-flow sedimentation tank* or *horizontal-flow sedimentation tank.* The scraper blades are connected to form a series of about four vee-shapes. The moving scraper pushes *sludge* which has settled on the bottom of the tank to the back of the vee, from which it is continuously removed by an *air-lift pump* or a *siphon.*

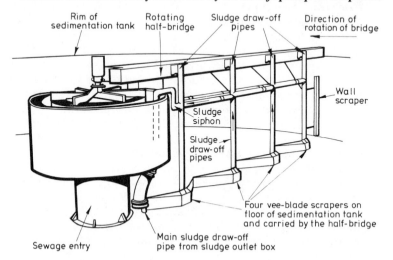

Figure V.2 Vee-blade scrapers for sewage sludge sedimentation tank, viewed in perspective from inside the tank (Ames Crosta Babock Ltd.)

Sludge can be removed very quickly after it has settled but its solids concentration may be low.

vee-notch weir (1) A vee-shaped notch for measuring the flow of fluid (*see Figure V.3*) (2) A type of *weir plate.*

velocity head The energy per unit mass of fluid, resulting from its speed of flow. It is related to kinetic energy as follows:

$$\frac{\text{kinetic energy}}{\text{mass}} = \text{velocity head} = \frac{v^2}{2g}$$

where v is the speed and *g* is the acceleration of gravity.

ventilation *See air changes, sewer ventilation.*

venting of landfill The gas emitted from *controlled tips* is about half *methane* and half *carbon dioxide* by volume, sometimes with a little *nitrogen.* To avoid all possibility of explosive concentrations of gas accumulating below ground, US designers vent the fill, using occasional thick layers of gravel to cover the *cells,* and pipes at the surface to pass through the thicker *final*

338

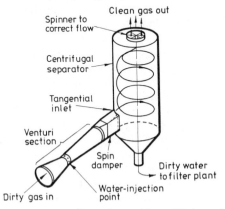

```
                              25 mm
                    125 —
                        -
                    100 —
                        -        21. 2
                    75 —         17. 6
                        -        11.8
                    50 —         7. 5
                        -        4. 2
                    25 —         1. 7
                        -        0.55
                     0 —
```

Head of   Flow litres
water mm   per min.

Figure V.3 Measurement of a water supply by a vee-notch shown (left) using a box with a V-notch cut out of the side for measuring the delivery from the pipe. (From 'Water Treatment and Sanitation; simple methods for rural areas', Intermediate Technology Publications Ltd., London.)

cover. To remain permeable the gravel must never be saturated.

Venturi flume   An open channel specially shaped so that the flow in it can be calculated from the heads (depths) upstream and at the throat. If the level of the water flowing away is low enough to prevent backing up into the flume, only the upstream head needs to be measured. The upstream part of the Venturi flume converges to a narrow throat and widens out again to the usual parallel width. *Compare standing-wave flume.*

Venturi scrubber   A *wet scrubber* in which *flue gases* pass through a Venturi throat at very high speed, 60 to 90 m/s. atomising

Figure V.4 Venturi scrubber (Parker A, 'Industrial Air Pollution Handbook')

339

water injected at the throat to a fine spray. The spray is carried with the gases to a separator chamber where some of the water settles out with the impurities it has picked up *(see Figure V.4)*. The remainder passes on to an **entrainment separator.** An expensive pressure drop of 1 m or more water gauge may be expected in the gases but Venturi scrubbers can extract tiny particles from 0.2 to 1 μm. *See ejector Venturi.*

vernal slough   The *sloughing off* in the spring, at *trickling filters.*

vertical cell digester   A *mechanical composting* unit consisting of a building of six or more storeys, on each of which the refuse remains for 24 h, starting at the top storey. At 24-h intervals the shredded refuse is swept off to the next floor by mechanical rakes. A plant in Jersey, Channel Isles, of this type with six storeys has an intake of 90 tonnes a day of refuse. Metals and glass are removed beforehand, and sewage *sludge* is added to increase the water and *nitrogen* content. The temperature in the digester reaches 74° C.

Vexillifera   An *amoeba* in seawater that consumes *Escherichia coli.*

viable bacteria   *Bacteria* that are alive and can multiply.

vibrating screen   A *screen* for mineral or *sewage treatment,* set at a slight incline to the horizontal. The water passes down through the screen with the fines, the coarse solids travelling down to the lower end, where they drop off.

Vibrio cholerae, V. comma   A *bacterium* that flourishes in the small intestine and causes *cholera.*

virion   One *virus,* a viral cell.

virological examination   Examination for the presence of *viruses.*

virology   The science of *viruses.*

virus (plural **viruses**)   Important disease-producing *microbes,* even smaller than *bacteria.* They can be from 10 to 250 *nanons* long (0.01 to 0.25 μm) and can be seen only under an *electron microscope.* They are totally parasitic, depending on living cells even for reproduction, and can infect all animals, plants and bacteria. They can be separated from other micro-organisms by passing through a *filter* with pore size of 0.45 μm through which only viruses pass. A *virion* is basically nucleic acid in the core and *protein* outside. Although they reproduce in a *host* cell, they are not truly living, and are not killed but inactivated. Very little is known about them so far as *waterborne diseases* are concerned, and their numbers are estimated by counting the effects of their activity, the *plaques* they form in a colony of bacteria. They are sparse compared with bacteria but are more dangerous to man, since the effective dose can be much smaller. Viruses are excreted only by infected people. In humans they cause the common cold, influenza, *poliomyelitis, infectious hepatitis,* some cancers possibly; and in animals, foot and mouth disease and rabies. In *sewage* viruses do not multiply and their concentration is never large, but they are harder to destroy than bacteria. *Coagulation* of sewage *effluent* or *raw water* with *alum* or *lime* or an iron salt reduces the viruses in the water and may raise the *pH* to a level that inactivates many of them although some can survive a pH of 10.5 to 11. Viruses may be removed by *adsorption* on to *activated carbon* or by the *biological film* of a *slow sand filter.* *Breakpoint chlorination* applied before and after carbon filtration, leaving 0.5 mg/litre of free *chlorine residual,* is sometimes relied on in Britain to inactivate viruses as the last process at the waterworks. In other countries *ozonation* is considered the complete solution. Ozone is 1000 times more efficient than *chlorination* for inactivating viruses. No *sewage treatment* completely eliminates them. If no plaque is formed in 1 litre of water it can be assumed that it is safe to drink, provided that a sample of at least 10 litres has been taken. Some viruses live many days in sewage or river water. *See also concentrating viruses, inactivation of viruses, pathogenic bacteria.*

virus detection   Coliform bacteria are less resistant than *viruses* to a hostile environement, including all *water treatment* processes, yet they are still used as the main indicator of *bacterial purity* of a water. Viruses are more enduring and methods are urgently needed to detect the few units that exist in a large volume of infectious water. Such a method should reliably determine whether a few viruses are present in 100 litres of water and concentrate all types with equal efficiency. No ideal method yet exists and there are many methods of *concentrating viruses.* One laboratory at least, in each country or region, should be able to make examinations for viruses in drinking water and to carry out research (*WHO*).

viscosity   The treacliness of a fluid. There are two SI ways of stating viscosity, either in poises for dynamic viscosity or in stokes for kinematic viscosity. (1 poise = 0.1 kg m^{-1} s^1; 1 stokes = 0.0001 m^2/s or 10^{-4} m^2/s). Most cold fluids are more viscous than when hot and water is no exception. Solid particles, settling in water, do so 2.3 times faster at 30° C than at 0° C simply because the water is much less viscous at 30° C.

viscous flow   *Laminar flow.*

void ratio, voids r.   The volume of the voids in a soil or *filter* or *trickling filter,* divided by the volume of the solids. *Compare porosity.*

volatile acids   *Organic* acids generally, sometimes those of low molecular weight—e.g. acetic, propionic and butyric acids.

volatile solids The solids, whether dissolved or *suspended solids,* which can be oxidised to *carbon dioxide* at a temperature of 550° C. Being *organic,* the ratio between them and the total solids is the organic fraction of the solids.

volume diameter The diameter of a sphere possessing the same volume as the particle under consideration. *Compare Stokes diameter.*

volumetric loading (1) In a *biological treatment* system, the *organic load* per unit volume of reactor, usually expressed as kg of BOD_5 applied per day per m^3 of tank, pond or *filter. Activated sludge* plants are more commonly designed on *sludge loading rate.* (2) *Hydraulic loading* (1).

volute In a *centrifugal pump* the volute is the spiral casing surrounding the *impeller.* The volute centrifugal pump has no *diffuser* passages and can therefore pass some solids. It is best suited to pumping large volumes against low heads.

vortex Swirling fluid.

vortex air cleaner A *straight-through cyclone.*

vortex finder In a cyclone *dust arrestor,* the tube projecting from the cylinder at the top as an outlet for the clean air; or in a *wet cyclone,* for the clean water; or in a *dense-medium cyclone,* for the 'floats'.

vortex grit separator A conical grit-separating tank with the apex of the cone downwards. The dirty grit is lifted from the apex at the bottom of the cone by an *air-lift pump* which delivers it to a separate *vortex grit washer.*

vortex grit washer A *wet cyclone* separator used for washing grit. *See above.*

vortex incinerator An *incinerator* without a *grate,* originally developed by the US Bureau of Mines for burning waste paper. It is a vertical cylinder into which the combustion air is blown tangentially above the burning material. The air spirals down to it and then rises up the middle of the cylinder as in a *cyclone,* carrying with it all the ash and some still-burning material.

Vorticella A stalked *ciliate protozoon.* Its presence usually indicates that an *activated sludge* is in good condition. It may also occur in *trickling filters* or in sewage-polluted streams.

VSS, volatile suspended solids *See MLVSS, volatile solids.*

W

Warburg respirometer A type of *respirometer.*

washing of sludge *Elutriation.*

wash load Suspended *sediment load,* consisting of particles smaller than those in the river bed, usually washed into the stream by rain. It travels downstream until a slow flowing area is reached, and is there dropped.

washout In air pollution, removal of particles or gases by collision with descending raindrops. *Compare rainout.*

washout valve, scour v. A valve provided at a low point in a *main* which allows debris to be flushed out at a point where the waste water can conviently drain away. *Compare air valve, scouring sluice.*

wash water (1) The water used for *backwashing* deep-bed filters, the mesh of *micro-strainers,* etc. *Rapid sand filters* may yield a backwash water with up to 1000 mg/litre of *suspended solids.* In *tertiary treatment* of sewage *effluents* backwashing water is usually returned to the works inlet or to the primary *sedimentation* tanks, which increases the quantity of *sludge.* In waterworks, backwash water is allowed to settle in *wash water recovery tanks.* (2) Any water used for cleaning an industrial process.

wash water recovery tank A tank where dirt in the *wash water* settles out and the top water is recovered to pass through the *water treatment* plant again, while the *sludge* is removed for de-watering.

wash water trough Troughs are provided above the *filter* sand or anthracite in modern *rapid* or *multi-media filters,* etc., to remove *wash water.* The troughs should be about 0.5 m above the level of the *medium* when it has settled. This allows for expansion of the bed during *backwashing* and ensures that sand is not lost with wash water.

waste compactor A *compactor* (2).

waste-derived fuel, WDF, refuse-derived fuel, RDF Shredded refuse from which unburnable material has been extracted. UK cement kilns and other industrial plants began to burn it (unpelletised) together with coal in about 1975. Pelletising makes the material more valuable, enabling it to be both carried in a lorry and stored. Otherwise it has to be blown after shredding direct from the *mill* by pipe to the furnace. *See enriched pulverised refuse.*

waste disposal authority, WDA The local government reorganisation of 1974 established a new legal framework for the licensing and planning of waste disposal sites. The WDA is the county council in England, but in Wales and Scotland it is the district or island council. The collection authority in England is rarely or. never the disposal authority.

waste disposal unit A *garbage grinder*

waste oil *See sump oil.*

343

waste reclamation   *See reclamation of solid waste, reclamation of water.*

waste sludge   *Surplus sludge.*

waste stabilisation pond, s. pond, lagoon, oxidation pond   The five types of lagoon for sewage treatment are discussed under *aerated lagoon, anaerobic lagoon, facultative lagoon, high-rate aerobic lagoon* and *maturation pond.* In general a waste stabilisation pond is a large shallow pond that receives either *raw water* or sewage *effluent,* where sewage is treated by the actions of *algae* and *bacteria.* The method is most profitable in sunny climates where land is cheap and plentiful, lowering the cost of the lagoon and simplifying its operation. *See also sludge lagoon.*

wastewater   Domestic *sewage* or industrial *effluent.*

wastewater engineering   *Sewerage* and *sewage treatment,* a US term coming to be accepted in Britain.

waste way, w.weir   A *spillway* (overflow channel) over a dam.

Water Act 1973   An Act which reorganised the water industry in England and Wales into *water authorities* responsible for drinking water, *sewage,* land drainage, flooding. The Act set up the *National Water Council* and other national bodies such as the *Water Space Amenity Commission,* but was not concerned with the *Scottish water industry.*

water authority, regional water authority   Ten bodies set up under the *Water Act 1973* to assume control of water services for their regions in England and Wales on 1 April 1974. They are concerned with river pollution, fisheries, land drainage and flooding, water supply and conservation, *sewerage* and *sewage* disposal, and the use of water for pleasure and nature conservation. Water authorities have to be consulted before a *waste disposal authority* decides on the location of a *controlled tip.* This should ensure that *leachate* will not contaminate any *aquifer.*

waterborne disease   Many infectious diseases are carried by water. In one Asian country waterborne disease caused 40% of the deaths in 1969 and 60% of the illnesses. About 90% of the country people or 72% of the whole population suffered from intestinal *parasites,* some of them brought by drinking water. Diseases carried by drinking water kill five million babies annually and make one-sixth of the world population ill. In London as recently as 1854, when the population was only 2.5 million, 10 000 died of *cholera* in one year because of a polluted well in Broad Street. There was then little or no piped water. Away from the warmth of the human body, most bacteria excreted by humans die more quickly than viruses. Any virus or bacterium excreted in human *faeces* or urine can probably pass

344

through water to other humans, but its survival is influenced by many factors, including *pH*, toxic metal ions, sunlight, *adsorption* on clay particles during *sedimentation*, and *predators*—other micro-organisms. Waterborne diseases include *ascariasis, amoebic meningo-encephalitis, cholera, dysentery, infectious hepatitis, leptospirosis, paratyphoid, salmonellosis, shigellosis, tuberculosis, tularaemia, typhoid.* The main defences against waterborne disease are *water treatment,* including *disinfection,* and proper sanitation. *See water-related diseases.*

water bugs, Hemiptera A family of winged insects. Different species have widely different tolerance to pollution.

water conservation The preservation, control and development of water resources above and below ground, the prevention of pollution, etc.

water consumption *See water demand.*

water cycle *Hydrological cycle.*

water demand, w. consumption UK domestic demand for water in the 1970s was increasing at 2 to 4% yearly, involving, theoretically, a doubling of demand by A.D.2000. Although drinking and cooking require no more than 5 litres daily per person, the 1970 domestic consumption was about 160 to 200 litres, distributed roughly as follows: body washing and closet flushing 50 litres each; waste in distribution 20 litres; dishwashing and laundry 15 litres each; garden watering, car washing, etc., 5 litres; drinking and cooking 5 litres. US consumption is higher, especially in summer, with the use of domestic *once-through cooling* equipment. US daily domestic demand averaged 303 litres per person in 1972, increasing from 265 litres in the east to 434 litres in the west, assuming households of three and deducting 25% for leakages. In low-income communities using *standpipes* the demand is about 60 litres per person per day or less. In general, water demand increases as the climate warms up and as the standard of living improves. In cities the daytime demand is about twice the night-time demand.

water gauge A water-filled ∪-tube, a *manometer* applied to gas or air flows.

water grid A network of several bulk water supplies, such that a supply to an area may come from lakes, *reservoirs* or rivers, depending on the availability and quality of each at the time.

water hammer *Pressure surge in pipe flow.*

water hog, w. louse *Asellus aquaticus.*

water hyacinth, *Eichhornia crassipes* A floating plant that can double in numbers every 10 days but will not grow much at any temperature below 10°C. Thus, in warm, rich *sewage* it produces nearly 18 tonnes per day per hectare of wet vegetation or about

345

212 tonnes per year dry. It also absorbs the salts of *heavy metals.* In the USA, experiments on the use of this plant for *sewage treatment* were in progress in the mid-1970s. It needs a warm climate and is a catastrophic weed in warm, slow-flowing rivers such as the Nile, the Congo or the Mississippi.

water pollution control   Apart from its general and obvious sense, this term also means *sewerage* and *sewage treatment* and disposal. *See pollution control.*

Water Pollution Control Act Amendments 1972, US Public Law 92-500   Far-reaching, complex and expensive legislation aimed at restoring and maintaining the physical, chemical and biological health of United States waters. A goal set by Congress was to minimise the discharge of pollutants by 1985. The agency to achieve it is the *EPA* helped by the state governments.

Water Pollution Control Federation   *See WPCF.*

water quality standards   *See drinking water standards, irrigation, swimming water.*

water reclamation   *See reclamation of water.*

water-related diseases   Infectious diseases which strictly, are not *waterborne diseases* but for which water is needed, such as *dengue, malaria, onchocerciasis* and *yellow fever.* The *vectors* of all these diseases flourish in rivers or lakes, and water management schemes may influence their existence.

Water Research Centre, WRC   A body set up under the 1974 reorganisation of the UK water industry to include the former Water Pollution Research Laboratory, Stevenage, and the Water Research Association, Medmenham. WRC is funded by its members, by the Department of the Environment and by external contracts of research for its members and others in the water industry.

watershed   (1) In the UK the line between two *catchments,* therefore a summit line, known in USA as a divide. (2) In the USA a *catchment* area.

water side   Description of the surface of a boiler tube or drum that is in contact with steam or water, not the *fire side.*

water softening   Removal of *hardness* from water. The *calcium* and *magnesium* ions can be removed by chemical *precipitation.* Alternatively, cation *ion-exchange resins* (e.g. *zeolites*) can be used to exchange the calcium and magnesium for sodium ions. Ion-exchange resins, which exchange other cations for hydrogen ions, produce an acid water, because the bicarbonates, *sulphates* and *chlorides* become carbonic acid, sulphuric acid and hydrochloric acid. However, by mixing this with hard water all the bicarbonates in the *raw water* can be removed and the acids in the softened water are neutralised—e.g.

$$Ca(HCO_3)_2 + H_2SO_4 = CaSO_4 + 2H_2O + 2CO_2$$

The *carbon dioxide* can be removed by *aeration*. This process reduces the *carbonate hardness* to zero, although some *non-carbonate hardness* remains. *See de-mineralisation, excess lime softening, lime-soda softening, lime softening, sodium restricted diet.*

Water Space Amenity Commission, WSAC A body with representatives from the *water authorities,* the Sports Council, the English Tourist Board, etc., which serves mainly as an information service on the uses of natural waters for pleasure.

water springtail *Springtail.*

water supply *See water treatment.*

water table, saturation line (mainly in USA **phreatic surface**) The surface, which may be undulating, of the *standing water* level of an *unconfined* groundwater. *See perched water table; compare piezometric surface. (Figures A.8 and W.1.)*

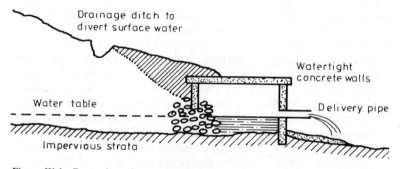

Figure W.1 Protection of water supply at the surface (For acknowledgement, see *Figure W.3)*

water tower A *service reservoir,* usually small, raised above ground.

water treatment *(figure W.2) Screening,* then *coagulation, flocculation* and *clarification,* followed by *filtration* and

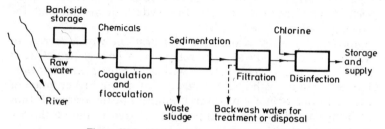

Figure W.2 Water treatment, flow diagram

disinfection, is the usual treatment which converts *raw water* to *drinking water.* It removes *suspended* and *colloidal* matter from reasonably clean water but cannot reduce their *salinity* without *de-salination,* a further treatment which is unusual. A *carbon adsorption bed* also may be needed. The *sedimentation* that occurs during *raw water storage* can also be regarded as part of the treatment.

water wall, watercooled wall Vertical water tubes, which form the walls of a steam boiler and improve its efficiency by reducing heat losses. Water walls are not confined to steam boilers but may also be used in *incinerators* to cool the *flue gases. Stainless steel* or *silicon carbide* shields may be placed in front of them to protect them from *corrosion.* The water temperature should be above 150°C to avoid pitting corrosion on the fire side of the tubes.

waterway A river, canal or stream. For the water engineer a waterway is a source of water, even if for the *sewage* works engineer it is a drain.

waterworks A site where *raw water* receives *water treatment* before it is sent to the *service reservoir.*

waterworks sludge The two main types of waterworks *sludge* are the *water softening* sludges and those from the *clarifier* that clean the water. *Wash water* from the *backwashing* or filters also produces sludge. Waterworks sludges are mainly inorganic. Those from *clarification* and backwashing are mainly metallic hydroxides. Softening sludges are largely *calcium carbonate* and *magnesium* hydroxide. They are as difficult to de-water as sewage sludges but do not gas or smell, in spite of which they are expensive to dispose of. *See sludge lagoon, sludge thickening.*

watt A unit of electrical power, a rate of consumption of energy equal to one joule per second. If provided as heat rather than electricity, it may be written W_{th} for 'thermal watt'.

waxes *See lipids.*

WDA *Waste disposal authority.*

WDF *Waste-derived fuel.*

wedge-wire screen A *screen* made of wire of wedge-shaped cross-section with the thin ends of the wedges down, so that the openings enlarge downwards.

weighbridge (in USA **scales**) A platform on which lorries or trailers, or on a railway, rail wagons can be weighed. One is provided at every *incinerator* as well as at most *controlled tips.*

weighting Many methods, in statistics, of giving higher value or weight to one or more of several numbers being averaged, thus ensuring that their greater importance is seen in the result. *See composite sample.*

Weil's disease, leptospiral jaundice A severe form of *leptospirosis,* a disease transmited from rats' urine and therefore a particular danger to sewermen. The *bacteria* that cause it may be present in polluted ponds or streams. Dogs, pigs and other vertebrates also suffer from the disease.

weir loading, w. overflow rate The flow of water or sewage *effluent* from a tank per unit length of outlet weir, thus:

$$\text{weir loading} = \frac{\text{maximum flow (m}^3\text{/day)}}{\text{total length of outlet weir (m)}}$$

If the weir loading is excessive, the approach velocity of the effluent may lift the *sludge* from the bottom of the tank. The usual design value is 220 m³ per m per day but *sedimentation tanks* with weir loadings above 1000 have worked efficiently. Weir loadings have much less effect than *overflow rates* on the efficiency of *sedimentation* tanks. *See* below.

weir plates Horizontal adjustable metal plates set on a *weir* to form a crest that can be accurately adjusted to level. At weir overflow rates below 100 m³ per metre length of weir per day the plain horizontal weir is affected by *surface tension* and is not self-cleaning. Consequently, for lower flows than this, *vee-notch* or castellated plates are usual. Such plates efficiently reduce the length of weir that is in use. A vee-notch reduces the weir length more as the flow rate decreases. Vee-notches should be used as outlet weirs for *sedimentation tanks* or *clarifiers.* Each notch should have a maximum depth of 35 mm. Enough notches should be provided to ensure a maximum flow per notch of 30 to 40 m³/day.

well casing, w. lining To prevent contamination of a water well or borehole, a steel tube, the casing, should be inserted to seal off

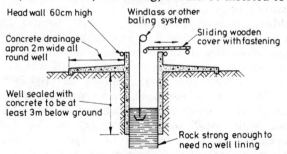

Figure W.3 Well head for hand-dug well providing drinking water, showing protection from pollution by surface drainage (From 'Hand-dug Wells and their Construction' by S. B. Watt and W. E. Wood, Intermediate Technology Publications Ltd., London, 1976)

349

any ground that might harm the water. Cement grout is forced into the gap between the casing and the ground to prevent vertical seepage of contaminants. The casing should extend above ground and the ground surface around the well head should be concreted. (*See Figure W.3.*)

well screen Tubular mesh or sheet steel with slots in it, to admit fluids and exclude soil, inserted into an oil or gas or water well at the producing depth (horizon). In a water well a coarse gravel filter may be placed outside the screen to catch sand particles and hold them.

wet air oxidation, wet oxidation, Zimpro process, conditioning with air A *heat treatment of sludge* due to F. J. Zimmermann of the USA in 1954. The *raw sludge* is pressurised with air and passed through heat exchangers to heat it to 125−140°C. The *sludge* then enters a pressurised reactor and steam is injected to raise it further to 150−250°C. The heat-treated sludge is easier to de-water and some of the *organics* in the raw sludge hydrolyse and dissolve. The liquor from the de-watering of heat-treated sludge can thus have a very high *BOD. See hydrolysis.*

wet arrestor, w. collector Any *dust arrestor* that uses water; not a *fabric filter,* nor an *electrostatic precipitator,* nor a dry *cyclone.*

wet classifier Any device in *mineral dressing* that separates solids in water into two or more size classes—e.g. a *screw classifier.*

wet cyclone (1) Or liquids cyclone, hydrocyclone. A *cyclone* (2) in which water or *dense medium* circulates rather than air. (2) A cyclone used as a *dust arrestor,* in which the walls are wetted to improve the dust collection and sometimes also to cool the gas. The water is sprayed in at the inlet to the cyclone.

wet pulveriser A *drum pulveriser.*

wet reduction Size reduction of materials in a *mill* with water, by *drum pulverisers,* etc.

wet scrubber, washer Many types of *dust arrestor* that also extract water-soluble gases such as *sulphur dioxide,* SO_2, from *flue gas,* especially if the wash water is *alkaline.* All wet scrubbers inevitably pollute a large amount of water, but they have the advantage that they eliminate fire risk when they recover flammable powders, explosive solvent vapours, etc. Mechanical aids to scrubbing may include fan blades that propel the gas, with *wash water* introduced at the hub, etc. Mechanically aided scrubbers need little space but they have a high power demand. They are also expensive to operate, more so even than *Venturi scrubbers,* which are equally efficient for catching dust. The temperature of flue gas leaving wet scrubbers is around 65°C and it is saturated with water vapour which can be seen as a fog, the *vapour plume. See baffle type scrubber,* *centrifugal scrubber, ejector Venturi, impingement scrubber, low-energy scrubber, moving-bed scrubber, packed tower scrubber, surface area scrubber; see also stack gas re-heating.*

wetted perimeter The length of the wetted cross-section of an open channel.

wetting rate Of a *trickling filter,* the flow of fluid per day per unit plan area of the filter. *Compare hydraulic loading.*

wet well The part of a pumphouse that contains the liquid to be pumped. In a wet-well pumping station, the pump is submerged in the wet well with a (dry) motor room above. There is no dry well. In a dry-well pumping station, the pumps are in a *dry well* and the adjoining wet well is used as a sump (storage well). The motor room is above the dry well. Wet wells should be small enough for the *sewage* not to become septic and for solids not to settle, yet large enough to eliminate frequent starting and stopping of pumps. Ten starts per hour is the maximum for a pump.

whipworm, Trichiuris trichiura A parasitic *nematode* worm which is common in warm humid, shady areas. Humans are infected by taking in the eggs in contaminated food or water. The eggs are passed in the *faeces.* It may be controlled by efficient *sewage disposal* and water supplies.

white goods Dumped refrigerators, gas cookers, etc., partly built of vitreous enamelled sheet steel, often white. The sheet steel is useless as *scrap* unless the enamel is removed by passing it through a *hammer mill,* breaking it into fist-sized pieces, and converting it to grade 1 scrap steel.

white products *Light oils.*

WHO *World Health Organisation.*

WHO International Standards 1971 A publication of the *World Health Organization* (3rd edn) which states minimum standards for drinking water.

wilting coefficient, w. point The water content of a soil, below which plants wilt and die even in damp air. It varies from about 2% of the dry weight of a coarse sand to about 15% for a clay. *See field capacity.*

Windhoek The capital of Namibia, in southern Africa. By 1954 it was realised that the town would have to invest in *reclamation of water.* From 1966 to 1976 3300 m^3/day of the *sewage* flow were reclaimed, and more recently 11 500 m^3/day. It is hoped that up to half the sewage flow will be reclaimed. The reclamation processes in sequence are: conventional primary and secondary *sewage treatment, maturation pond, recarbonation, algal harvesting, foam fractionation,* chemical *coagulation, breakpoint chlorination, sedimentation, sand filtration, carbon* *adsorption bed.* The water is then blended with the *raw water* for conventional *water treatment.*

windrowing, windrow composting Spreading prepared *garbage* or other moist *composting* refuse in long low heaps (windrows) on land, turning it once or twice a week for 5 weeks, then leaving it for 2 to 4 weeks more to complete the *aerobic digestion.* For a village of about 2000 people in the USA the area of windrowing land needed would be about 24 hectares, so it is unsuitable for countries where land is expensive. There may also be objections because of the unpleasant appearance of the windrows. In the Netherlands windrows may be 9 m high but elsewhere the usual height is 1.5 m, settling after a few days to 1.3 m because of microbial oxidation. *Compare mechanical composting.*

windrow turner A mobile machine resembling earthmoving equipment which can turn the windrows of composting material, when driven past them. Some windrow turners also shred the refuse.

windshield The outer part of a *multiple-flue stack.*

Winkler test The standard chemical method for determining how much *oxygen* is dissolved in a water or waste water. It involves adding *manganese sulphate, $MnSO_4$,* then a mixture of sodium hydroxide, NaOH, and potassium iodide, KI:

$$2NaOH + MnSO_4 = Mn(OH)_2 + Na_2SO_4 \quad \text{(at high pH)}$$

$$Mn(OH)_2 + O \text{ (dissolved in sample)} = MnO_2 + H_2O$$

Concentrated sulphuric acid is then added:

$$MnO_2 + 2KI + 2H_2SO_4 = I_2 + MnSO_4 + 2H_2O + K_2SO_4$$

Therefore one atom of oxygen liberates one molecule of *iodine,* I_2. I_2 is then measured by titrating against *sodium thiosulphate* with starch as an indicator. There are many modifications of the test to avoid interferences by other *ions,* particularly *nitrite.*

wood pulp *See paper and pulp manufacturing wastes.*

woollen industry wastes Raw wool contains many impurities and raw wool scouring therefore produces an *effluent* with a high content of inert and *organic* solids, greases and detergents. Pollution by finishing processes, including dyeing, varies with the process. Mixed wastes from raw wool scouring and finishing have a BOD_5 of 500 to 2000 mg/litre, *pH* above 9.0, and *suspended solids* up to 1000 mg/litre.

works liquors Liquors produced in a *sewage treatment* works include those from *lagoons,* storage tanks or *storm tanks,* liquor separated from *digested sludge* in *digestion tanks,* and other liquors separated from *sludges* in de-watering *(IWPC).*

World Health Organization, WHO WHO has published much information on *waterborne* and *water-related disease*. The WHO booklet International Standards for Drinking Water' (1971) states minimum standards to be achieved throughout the world, while higher standards are stated in the WHO's 'European Standards for Drinking Water'.

worm diseases *Helminthic diseases.*

WPCF, Water Pollution Control Federation The main US professional body for the treatment of waste waters and river water quality management.

Wuchereria bancrofti A parasitic *nematode* worm that causes *wuchereriasis.*

wuchereriasis A type of *filariasis* caused by *Wuchereria bancrofti.*

Y

yard waste US term for brushwood, shrub trimmings, grass cuttings, etc., from the house or garden.

yellow fever A very severe form of jaundice, caused by a *virus* carried by the *mosquito, Aëdes aegypti.*

yield/drawdown curve *See specific capacity of a well.*

yield of a water source There are numerous definitions of the term 'yield', A common one is: the steady supply that could safely be maintained through a drought of a given probability. A UK water undertaking should have a yield from its major sources that allows for a once in 50 years drought, but other countries may have other criteria. *See borehole yield.*

Z

zeolite Either a natural mineral or an artificial product, but always a hydrated sodium aluminium silicate which can give up its sodium in exchange for the *calcium* and *magnesium* in water, thus reducing *hardness* by *ion exchange.* When the zeolite ceases to soften the water it is 'spent', but can be regenerated by passing a strong solution of sodium chloride through it. Zeolites cannot be used for *turbid,* acid or iron-bearing waters, and *lime-soda softening* may be preferable for these. There is little difference in cost between them. Zeolites can produce water of zero hardness, but after the lime-soda process some hardness always remains. Zeolites were first used on an industrial scale by Gans in

Germany in 1905. *See clinoptilolite, ion-exchange resins, regeneration.*

zero-discharge layout Design of an industrial plant to release no *effluent,* whether *cooling water* (by *closed recirculation system)* or chemical effluent.

zero drift Movement of the zero point of an instrument in one direction, away from the correct zero, causing errors in all readings, although the difference between any two readings may be correct.

zig-zag separator An *air column separator* that is either zig-zag in shape or has baffles on alternate sides of the vertical shaft (*see Figure Z.1.*)

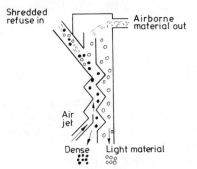

Figure Z.1 Zig-zag classifier for shredded refuse

Zimmermann process, Zimpro p. *Wet air oxidation.*

zinc, Zn Rare in natural waters, zinc quite often occurs in drinking water because of the common use of galvanised steel pipe and tanks — easily attacked by some waters which may thus contain as much as 3 mg/litre. This amount of zinc is bactericidal, although it is quite harmless to humans; consequently, such water should not be used for bacteriological testing. *See* Table of Allowable Contaminants in Drinking Water, page 359.

zinc equivalent A number related to the contents of *copper, zinc* and *nickel* in *sludge,* which can be used to estimate the safe amounts of these metals to be applied to the soil. The zinc equivalent should not exceed 250 mg/litre and is calculated by multiplying the copper content of the sludge by 2 and the nickel content by 8, and adding them to the zinc content. Various London sludges, some of which went to sea outfalls, varied from 2000 to over 15 000 mg/litre in 1973.

zone of aeration, unsaturated z., z. of suspended water, vadose z. The ground above the *water table.* It is not full of water nor is it completely dry, but, being partly filled with air in the pores, may retain solids from any water passing down through it and may oxygenate it in the same way as a *trickling filter*. Most water passes through the zone of aeration where there is one, before reaching the *zone of saturation*. In deserts the zone of aeration may be 500 m deep. The water in the zone of aeration is divided into *soil water, pellicular water* and the *capillary fringe*, in sequence downwards.

zone of saturation   The ground below the *water table* in *ground water* where the voids are filled with water under hydrostatic pressure.

zone settling   *See settling regimes.*

zones of water supply   *See pressure zones.*

zooflagellates   *Flagellate protozoa* that are not *photosynthetic.*

zoogloeal   A description of the gelatinous mass in a *biological film. See* below.

Zoogloea ramigera   A *bacterium* common in *activated sludge, trickling filters* etc. It is *heterotrophic* and produces a gelatinous slime. There is some confusion about whether this bacterium is a distinct species or merely a strain of other bacteria that also produce *zoogloeal* slime.

zooplankton   *Plankton* that are not *photosynthetic.* They graze on *phytoplankton* and can be one of the main limitations to their numbers.

zwitterion   A molecule which has positive and negative ions contained in its structure but has zero net charge—e.g. amino acids at the *iso-electric point.*

zymosis   (1) *Fermentation.* (2) Reactions induced by *enzymes.*

Appendix: Abbreviations and conversion factors of units

bhp	brake horsepower	kN	kilonewton
Btu	British thermal unit	kN/m²	kilonewtons per square metre
cc	cubic centimetre (millilitre)	lb	pound
cm	centimetre	m	metre
cuft, ft³	cubic foot	m³	cubic metre
cu.m, m³	cubic metre	mg	milligram
cu yd, yd³	cubic yard	mgd	million gallons per day
dia.	diameter	min	minute
fpm	feet per minute	ml	millilitre
ft	feet	mN/m²	meganewtons per square metre
g	gram (or gravity acceleration)	mPa	megapascal
gpm	gallon per minute	mm	millimetre
hp	horsepower	Pa	Pascal
imp. gal	imperial gallon	psi	pounds per square inch
in	inch	rev/min	revolutions per minute
J	joule	s	second
kg	kilogram	sq.	square
kJ	kilojoule	US gal	US gallon
km	kilometre	yd	yard

Length

25.4 mm = 1 inch
39.37 in = 3.281 ft = 1 metre
5280 ft = 1760 yd = 1.609 km = 1609 metres
1 international sea mile = 1852 m = 1.151 miles
1 fathom = 6 ft = 1.829 metre

Area

645 mm² = 6.45 cm² = 1 in²
10.76 ft² = 1 m² = 1.196 yd²
100 m² = 1 are = 119.6 yd²

$10,000 \text{ m}^2 = 100 \text{ hectares} = 1 \text{ km}^2$
$4840 \text{ yd}^2 = 1 \text{ acre} = 0.4047 \text{ hectare}$
$640 \text{ acres} = 1 \text{ mile}^2 = 2.59 \text{ km}^2$

Volume

1 imp. gal of water weighs 10 lb (4.54 kg)
62.4 lb of water occupy 1 cu ft = 6.24 imp. gal = 7.49 US gal
1 US gal = 0.833 imp. gal. = 3.785 litres
1 imp. gal = 4.546 litres = 1.201 US gal
1 litre of water weighs 1 kg and occupies 0.22 imp. gal = 0.264 US gal
1 acre-foot = $43560 \text{ ft}^3 = 1235 \text{ m}^3$
$1 \text{ m}^3 = 35.31 \text{ ft}^3 = 1.308 \text{ yd}^3 = 220$ imp. gal = 264 US gal

Weight

16 ounces = 1 lb = 454 g
112 lb = 1 hundredweight (cwt)
20 cwt = 2240 lb = 1 long ton = 1016 kg
1 short ton = 2000 lb = 2 kilopounds
1000 kg = 1 tonne = 2200 lb
grains per cubic foot A unit sometimes used for measuring the amount of solids suspended in *smoke* or air: 7000 grains = 1 lb, or 15 432 grains = 1 kg, or 15.43 grains = 1 g, and $1 \text{ m}^3 = 35.31 \text{ ft}^3$. Consequently,

$$1 \text{ g/m}^3 = \frac{15.43}{35.31} = 0.437 \text{ grain/ft}^3$$

or

$$1 \text{ grain/ft}^3 = 2.28 \text{ g/m}^3$$

Heat and Energy

1000 joules = 1 kJ = 0.948 Btu
$1 \text{ Btu/ft}^3 = 37.26 \text{ kJ/m}^3$
$1 \text{ MJ/m}^3 = 4.31$ Btu/imp. gal = 3.59 Btu/US gal
1 Btu/lb = 2.326 kJ/kg
1 joule = 1 newton-metre = 1 watt-second = 1 pascal-cubic metre
1 Btu = 0.252 kg-calorie = 1055 joules
1 kg-m = 7.22 ft-lb = 9.81 joules
1 hp (USA and British) = 746 watts = 76 kg-m/sec = 550 ft-lb/sec
1 metric hp = 75 kg-m/sec = 0.987 US and British hp.
°C = degrees Celsius, formerly degrees Centigrade
°F = degrees Fahrenheit = $(1.8 \times °C) + 32$
K (Kelvin) = °C + 273.15

Hydraulic Loadings and Flow

1 m³/m³ = 0.16 ft³/imp. gal = 0.13 ft³/US gal
     = 168 imp. gal/yd³ = 202 US gal/yd³
     = 6.22 imp. gal/ft³ = 7.47 US gal/ft³
1 m³/m² = 20.45 imp. gal/ft² = 24.54 US gal/ft²
     = 184 imp. gal/yd² = 221 US gal/yd²
     = 890,000 imp. gal/acre = 1.069 million US gal/acre
1 m³/m = 67.1 imp. gal/ft = 80.5 US gal/ft
1 m³/s (cumec) = 35.31 ft³/s (cusec) = 2119 ft³/min
     = 220 imp. gal/s = 19.0 million imp. gal/day
     = 264 US gal/s = 22.8 million US gal/day

Mass Loadings and Concentrations

1 kg/m = 0.67 lb/ft
1 kg/m² = 0.204 lb/ft²
1 kg/hectare = 0.89 lb/acre
1 mg/litre = 1 g/m³ = 8.33 lb/million US gal = 10 lb/million imp.
gal
1 kg/m³ = 1 g/litre = 1000 mg/litre = 0.0626 lb/ft³ = 1.69 lb/yd³

Pressure

1 pascal (Pa) = 1 newton per sq. metre (N/m²) = 0.000 145 psi
1 million Pa = 1 megapascal (MPa) = 1 N/mm² = 1 million
N/m²   = 1 MN/m² = 10 bar = 10000 millibar = 145
psi   = 0.065 ton/in² = 10.2 kg/cm²
1 'technical atmosphere' = 1 kg/cm² = 14.2 psi = 10 m head of
water
1 'standard atmosphere' = 760 mm mercury at 0°C = 101.3
kPa = 14.7 psi
1 psi = 6900 N/m² = 0.069 bar

MAXIMUM CONTAMINANTS IN DRINKING WATER ALLOWED BY VARIOUS AUTHORITIES (mg/litre)
(*In parentheses = desirable maximum*)

	EPA 1976	WHO 1971	US PHS 1962
arsenic	0.05	0.05	0.05 (0.01)
barium	1.0	—	1.0
boron	—	—	1.0
cadmium	0.01	0.01	0.01
calcium	—	200 (75)	—
carbon-chloroform extract			0.2
chloride	250	600 (200)	250
chromium (total)	0.05	—	—
copper	1.0	1.5 (0.05)	1.0
cyanide compounds	—	0.05	0.2 (0.01)
fluoride	from 1.4 to 2.4	from 0.6 to 1.7	—
	(varying with ambient temperature)		
hardness as $CaCO_3$	—	500 (100)	0.3
hexa-chrome	—	—	0.05
hydrogen sulphide	0.05	—	—
iron	0.3	1.0 (0.1)	0.3
lead	0.05	0.1	0.05
magnesium	—	150 (30)	—
manganese	0.05	0.5 (0.05)	0.05
mercury	0.002	0.001	—
nitrate (as N)	10	10	10
phenols	—	0.002 (0.001)	0.001
selenium	0.01	0.01	0.01
silver	0.05	—	0.05
sulphate	250	400 (200)	250
synthetic detergent (MBAS)	0.5	1.0 (0.2)	0.5
total dissolved solids	500	1500 (500)	500
uranyl ion (UO_2)	—	—	5
zinc	5	15 (5)	5

WITHDRAWAL